An Introduction to Metamaterials and Nanophotonics

Metamaterials have established themselves as one of the most important topics in physics and engineering and have found practical application across a wide variety of fields including photonics, condensed matter physics, materials science, and biological and medical physics. This modern and self-contained text delivers a pedagogical treatment of the topic, rooted within the fundamental principles of nanophotonics. A detailed and unified description of metamaterials and metasurfaces is developed, beginning with photonic crystals and their underlying electromagnetic properties, before introducing plasmonic effects and key metamaterial configurations. Recent developments in research are also presented, along with cutting-edge applications in the field. This advanced textbook will be invaluable to students and researchers working in the fields of optics and nanophotonics.

Constantin Simovski is Professor of Radio Science and Engineering at Aalto University, Finland. He has worked for 40 years in the field of radio science, plasmas and optics, recently with a strong focus on metamaterials – a topic on which he has published extensively. He has also taught a graduate-level course on this topic.

Sergei Tretyakov is Professor of Electrical Engineering at Aalto University, Finland. He has published more than 300 journal papers in the area of radiophysics, nanophotonics, and metamaterials and has authored or coauthored five research monographs. For several years, he has been teaching a graduate-level course on metamaterials at Aalto University.

An Introduction to Metamaterials and Nanophotonics

CONSTANTIN SIMOVSKI

Aalto University, Finland

SERGEI TRETYAKOV

Aalto University, Finland

CAMBRIDGE
UNIVERSITY PRESS

University Printing House, Cambridge CB2 8BS, United Kingdom

One Liberty Plaza, 20th Floor, New York, NY 10006, USA

477 Williamstown Road, Port Melbourne, VIC 3207, Australia

314–321, 3rd Floor, Plot 3, Splendor Forum, Jasola District Centre, New Delhi – 110025, India

79 Anson Road, #06–04/06, Singapore 079906

Cambridge University Press is part of the University of Cambridge.

It furthers the University's mission by disseminating knowledge in the pursuit of education, learning, and research at the highest international levels of excellence.

www.cambridge.org
Information on this title: www.cambridge.org/9781108492645
DOI: 10.1017/9781108610735

First published 2020

Printed in the United Kingdom by TJ Books Ltd. Padstow Cornwall

A catalogue record for this publication is available from the British Library.

Library of Congress Cataloging-in-Publication Data
Names: Simovski, Constantin, author. | Tretyakov, Sergei, author.
Title: An introduction to metamaterials and nanophotonics / Constantin Simovski and Sergei Tretyakov, Department of Electronics and Nanoengineering, School of Electrical Engineering, Aalto University, Finland.
Description: New York : Cambridge University Press, 2020. | "November 20, 2019." | Includes bibliographical references and index.
Identifiers: LCCN 2020014529 (print) | LCCN 2020014530 (ebook) | ISBN 9781108492645 (hardback) | ISBN 9781108610735 (epub)
Subjects: LCSH: Metamaterials–Textbooks. | Nanophotonics–Textbooks.
Classification: LCC TK7871.15.M48 S56 2020 (print) | LCC TK7871.15.M48 (ebook) | DDC 620.1/1–dc23
LC record available at https://lccn.loc.gov/2020014529
LC ebook record available at https://lccn.loc.gov/2020014530

ISBN 978-1-108-49264-5 Hardback

Contents

Preface

This textbook has been written as lecture notes for the master and post-graduate (licentiate) course, Metamaterials and Nanophotonics, which was taught at the School of Electrical Engineering of Aalto University (Finland) in 2016. The course material is suitable for students specializing in electrical engineering and applied physics – especially electromagnetics, nanoscience, and optics. The reader will develop an understanding of light-matter interaction at the nano (subwavelength) scale and acquire knowledge of recent developments of the theory of optical properties of nanostructures and nanostructured materials and surfaces, as well as their applications. The book presents a unified view covering both metamaterials and nanophotonics from the fundamental, basic-principles point of view, offering short, focused, basic explanations of fundamental ideas behind new developments in this field in its generality.

Chapters 5, 6, and 8–11 have been written mainly by C. Simovski, and Chapters 2–4 and 7 have been written by S. Tretyakov. The book's concept has been developed by both authors together, and the whole text has been edited by both authors.

Acknowledgments

We wish to express our gratitude for the input and help we received from many colleagues and students. Many members of our research groups contributed by creating problems and control questions, as well as by drawing illustrations. We received very helpful input from Viktar Asadchy, Ana Díaz-Rubio, Fu Liu, Masoud Sharifian, Francisco Cuesta Soto, Svetlana Tcvetkova, Pavel Voroshilov, and Xuchen Wang.

1 Introduction

1.1 Motivation

Modern electronics, radio, and microwave techniques show fantastic achievements in device integration, miniaturization, power efficiency, and information processing speed. A device containing millions of transistors can fit into a micrometer-size volume, which is extremely small – not only in practical terms but also as compared to the wavelength of waves that the device can control. In what concerns miniaturization, modern electronics is reaching the limits imposed by nature. In order to qualitatively improve existing telecommunication and computing systems, we need to develop other technologies, and optical (photonic) means appear to be the most promising. In modern telecommunication systems, the transmission of signals to long distances is performed by optical fibers, which have largely replaced telephone and data cables. The optical frequency range is preferred because the information capacity of a telecommunication channel based on conducting wires is fundamentally limited by the available frequency spectrum and the physical properties of metals (see a discussion later in this section) while modern optical fibers possess negligibly small scattering and dissipative losses in the range of wavelength $\lambda = 1260$–1675 nm, where six telecom frequency bands with the relative widths 3%–5% of each of them are located. The same physical limitations of electronics also call for the use of optical techniques for signal processing and information storage. However, optical devices are bulky, especially in terms of the wavelength scale, and the existing technologies do not allow the integration of many functions in a single device that would be as compact as its electronic analog. One of the main reasons for this bottleneck is the difficulty of confining light to a small volume compared to the wavelength. In radio and microwave technologies, we have, for example, coaxial cables whose cross-section sizes are very small compared to the wavelength. In optics, we have fibers, but their cross sections are noticeably larger than the maximal guided wavelength. If we only had "optical means" to transport electromagnetic energy, the diameter of the power supply cables would be of the order of many kilometers!

As another example, let us consider frequency filters. In radio and microwave engineering, we first squeeze the wave into a transmission line of a tiny diameter and then construct a filter using capacitors and inductors of tiny sizes. The overall size of a properly designed filter is extremely small compared to the wavelength, and it is also small in absolute terms. Alternatively, we can filter propagating waves using extremely thin reactive sheets, called frequency selective surfaces. What can we do in optics? First, modern optics does not have nanoguides – optical waveguides with the cross section smaller than the guided wavelength. Second, the industry does not deliver optical capacitors or inductors, yet. And we do not have optically thin frequency selective sheets available on the market of optical devices. We have to resort to stacks of carefully selected dielectric layers, each of which has the thickness comparable to the wavelength, and illuminate them by optical beams that are much wider than the wavelength. Again, if we used only "optical means" to design radios, every simple filter would be many meters in diameter! And we need many tens of such filters in every mobile phone.

We can understand from these two examples that the traditional optical approaches and the signal processing that is so elaborated in radio and microwave engineering are not compatible. Only transmission of signals to long distances is performed optically nowadays. The processing of signals and computations are still done electronically. The time has come when optical and radio engineers must meet and work together to create and develop *nanophotonics* devices for future optical information processors and fully optical telecom systems. Radio and optics converge in this area, and the breakthrough is possible only through interdisciplinary research.

The interdisciplinary science of metamaterials and nanophotonics deals with electromagnetic devices and components whose characteristic dimensions are smaller than or of the same order as (for example, in photonic crystals) the wavelength of radiation that they control. The current state of the art is illustrated in Figure 1.1. While, for the use at microwave and radio frequencies, we

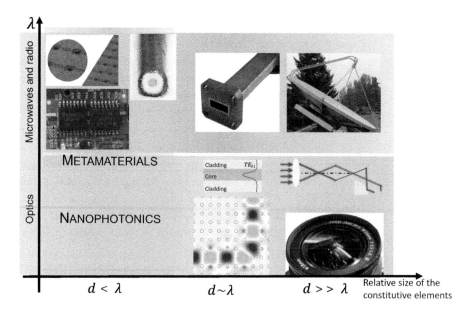

Figure 1.1 Metamaterials and nanophotonics devices are necessary for sub-micrometer integration, but this field is still largely unexplored.

have a basically complete set of components (waveguides and passive, active, and nonreciprocal components) as well as the means to integrate them into extremely small devices, at the optical frequencies, the corresponding box is only starting to be filled by optical nanoguides (such as plasmonic waveguides), metasurfaces, and metamaterial devices. Only the intermediate regime, where the characteristic sizes are comparable to the wavelength, is well studied and developed: the photonic crystal techniques are quite mature at this time. We see that any progress toward submicron integration of photonic devices is impossible without advances in metamaterials and nanophotonics.

One can ask why we need to develop integrated optical devices since we do have mature technology in electronics and microwaves. Why do we really need a breakthrough if there are optoelectronic converters transposing signals from the optical band to the band of radio frequencies and back? There are two main reasons why a breakthrough in this field is really demanded. First, optoelectronic converters, even advanced ones, form a bottleneck in the modern telecommunication systems (see Chapter 8). Second, there are two fundamental limitations of electronics and microwave technologies that block further qualitative progress in this field. One is the thermal noise. According to the Planck law, the spectral density of thermal radiation at a certain temperature is given by

$$P(f, T) = \frac{2hf^3}{c^2} \frac{1}{e^{\frac{hf}{k_B T}} - 1}, \tag{1.1}$$

where f is the frequency, T is the absolute temperature, h is the Planck constant, and k_B is the Boltzmann constant. In the optical domain, we have $hf \gg k_B T$ even at room temperature, and the thermal noise is practically negligible. In contrast, thermal noise is a serious problem at radio and microwave frequencies (here, $hf \ll k_B T$).

The other limitation is imposed by the dispersion and power loss in interconnects between electronic chip components. The signal will be transported without distortion and information loss if the pulse shape does not significantly change upon transmission and the signal power does not strongly decay. Ideally, we would like to have a transmission line without dispersion. In optics, free-space propagation is dispersion-free, and we have low-dispersion and low-loss optical fibers. Non-dispersive waveguides have purely real characteristic impedance (in free space, $\eta_0 = \sqrt{\mu_0/\epsilon_0}$). In electrical interconnects (metal transmission lines and cables), the impedance is determined by the per-unit-length parameters L, C, R:

$$Z_c = \sqrt{\frac{j\omega L + R}{j\omega C}} = \sqrt{\frac{L}{C}}\sqrt{1 + \frac{R}{j\omega L}}. \tag{1.2}$$

For propagation without pulse shape degradation, we need to ensure that

$$R \ll \omega L; \tag{1.3}$$

then, $Z_c \approx \sqrt{\frac{L}{C}} = \sqrt{\mu_0/\epsilon_0} \times$ a geometrical factor, and we have as an ideal situation as in free-space propagation of light. However, is this condition compatible with the submicron miniaturization of circuitry?

Let us make an estimate considering a short section of a transmission line (length l) formed by two metal conductors (the cross-section size $w \times w$). The resistance R can be estimated as $\frac{l}{\sigma w^2}$, where σ is the metal conductivity. The inductance $L \approx \mu_0 l$ (there is also a cross-section shape-dependent factor of the order of unity, which we neglect). Substituting the room-temperature conductivity of copper at microwaves ($\sigma \approx 5.8 \times 10^7$ S/m), we see that at microwave and millimeter-wave frequencies the diameter of the connecting wire w should be larger than about 0.1–0.5 mkm, which is a severe limitation on device integration. In addition to pulse shape distortion, inevitable losses in metal conductors lead to signal propagation loss and to device heating, which are other significant negative factors.

To summarize, one can expect that the next breakthrough in telecommunications, computing, imaging, etc., will come as a result of advances in nanophotonics and metamaterials, which will also enable the replacement of radio and microwave signals by optical signals in signal processing and computations. One may say that the knowledge of nanophotonics and metamaterials gives an engineer a unique chance to contribute to solutions of this global technological challenge.

1.2 Book Content

1.2.1 Metamaterials

Electromagnetic properties of natural or chemically synthesized materials are determined mainly by their chemical composition. For realizing optical devices, the range of accessible material properties is rather limited. We have dielectrics (with moderate values of the permittivity, between 1 and about 12), metals (rather lossy negative-permittivity materials), and weakly chiral materials, and that is basically all. There are no natural magnetics, not to speak about more exotic and interesting media (like mu-near-zero or extreme chirality materials) that an engineer would need in designing optical nanodevices. Within the metamaterial paradigm, it becomes possible to widen the material design opportunities by engineering "meta-atoms" as constitutive elements of artificial materials using available ordinary (natural) materials. As long as the meta-atom sizes remain sufficiently small on the wavelength scale of interest, the same approaches to the macroscopic description of electromagnetic properties of matter can be applied to metamaterials made of meta-atoms, just like to ordinary materials formed by atoms or molecules. However, within the metamaterial paradigm, the electromagnetic properties of materials can be controlled not only by varying chemical composition but also by engineering meta-atoms shapes and their internal structures, as well as mutual positions and orientations of meta-atoms in the composite material. In addition, the properties of meta-atoms can be modulated in time by some external force. The main appeal of the metamaterial concept derives from new possibilities to realize artificial materials with

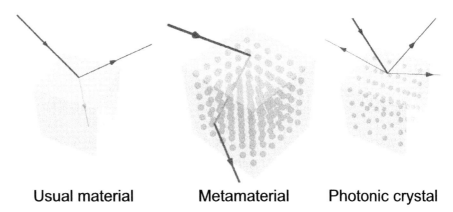

Figure 1.2   Reflection and refraction regimes for samples of bulk materials and metamaterials.

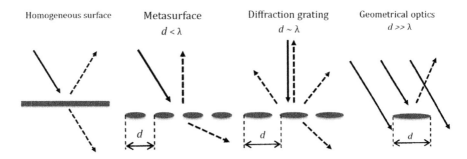

Figure 1.3   Reflection and refraction regimes: from homogeneous sheets to optically large screens.

electromagnetic properties that are not available in any natural material and to engineer properties that are optimal for particular applications. Basic possibilities to control reflection and refraction using "usual" materials, metamaterials, and photonic crystals are illustrated in Figure 1.2.

Research on artificial electromagnetic materials (now called *metamaterials* and *metasurfaces*) started at the end of the nineteenth century, very soon after formulation of Maxwell's equations, but developments were slow. At the present time, this research field is extremely active and developing quickly. Perhaps the main current challenges are to learn how to realize materials with precisely the optimal properties for particular applications and to explore the whole physically allowed spectrum of material parameter values [1] – in particular, including materials with extremely small or large values of the material parameters [2, 3].

At this time, it appears that the main research focus is shifting from volumetric electromagnetic metamaterials to metasurfaces [4, 5]; see a conceptual illustration in Figure 1.3. For realizations in the visible part of the spectrum, all-dielectric metamaterials and especially all-dielectric metasurfaces are actively studied [6]. Studies of nonlinear [7], bianisotropic [8], nonreciprocal, time-modulated [9], tunable, and programmable metamaterials and metasurfaces [10, 11] gain momentum. Fundamental research on metamaterials leads to developments of "metadevices" and systems; see, e.g., a review paper [12]. Last but not least, metamaterial technology comes into commercial products (e.g., products of Kymeta, Sensormetrix, Ecodyne). In the near future, one can hope to see

- Optimal (application-driven) designs of engineered materials and surfaces
- Reconfigurable, self-adapting, and software-defined metamaterials and meta-surfaces
- Extreme properties and extreme-performance metamaterials
- Active, nonlinear, and parametric (time-varying) structures, including the use of non-Foster elements
- Full exploration of spatial dispersion (especially mesoscopic regimes between metamaterials and photonic crystals)
- Going into quantum regime
- And more

1.2.2 Nanophotonics

Nanophotonics is the study of the behavior of light on the nanoscale and the interaction of submicron objects with light [13]. It is a multidisciplinary scientific and technical area that comprises the most advanced parts of modern optics – including quantum optics, radio science (especially applied electromagnetics), and nanotechnology. It also concerns electrical engineering, solid-state physics, physical chemistry, biophysics, and biochemistry. In this book, we mainly concentrate on the electromagnetic part of nanophotonics where classical optics intersects with radio science. However, we will also study nanostructures for light-energy harvesting and conversion into electricity. In Chapters 10 and 11, studies of light-matter interactions imply knowledge of some basic elements of solid-state and quantum physics.

In its optical part, one of the most important targets of nanophotonics is miniaturization of optical components (see also Section 1.1). For this purpose, one should learn how to squeeze macroscopic light beams – e.g., coming from an optical fiber, into a volume comparable to the wavelength or even subwavelength – and guide this concentrated light. In optoelectronic converters guided signals should be transmitted to photodetectors: advanced photovoltaic diodes, CMOS, or charge-coupling devices that have been available with submicron sizes since 1980s [14–16]. In prospective all-optical signal processors and in optical quantum computers, the squeezed signal needs to be transmitted to optical memory cells that may have the minimal size as small as 10 nm with the maximal size of the order of 500 nm [17]. In order to record and erase information, such a cell needs subwavelength concentration of the field in a nanoguide with essentially a submicron cross section [17]. So sharp light concentration is not achieved by usual lenses or curved mirrors. The most elaborated approach to subwavelength light concentration in optics is the use of so-called *plasmonic* structures (see Chapter 7). These structures can be in the form of bulk plasmonic metamaterials or plasmonic metasurfaces. Thus, there is no boundary between nanophotonics and nanostructured metamaterials. This is why the present book unifies both nanophotonics and metamaterial topics.

An important target of nanophotonics is related to medical and biological applications – especially optical nanosensing and nanoimaging – sensing and imaging of submicron objects. These tools are of extreme importance for genetics, micro- and molecular biology, and also for medical diagnostics. The challenge of a very weak interaction of light with so small objects as living cells, leucocytes, bacteria, and even single molecules is resolved using nanophotonics structures. Besides flat plasmonic surfaces used in a branch of optical nanoimaging called surface-plasmon microscopy, all other nanophotonic components used for optical nanosensing and nanoimaging have submicron structures. They are photonic crystals, metamaterials, and even isolated plasmonic nanoparticles (in two branches of optical nanosensing called plasmon-enhanced fluorescence and plasmon-enhanced luminescence). However, probably the hottest topic of nanophotonics nowadays is metasurfaces for molding light at nanoscale. Here, nanophotonics strongly intersects with a branch of optical nanosensing that had appeared long before the concept of nanophotonics and metamaterials was elaborated. This type of optical nanosensing is called *surface-enhanced Raman scattering* (SERS). SERS employs nanopatterned or textured plasmonic surfaces that perfectly match the basic definition of metasurface. Nowadays, SERS is a part of nanophotonics (see Chapter 9) as well as all other methods of optical nanosensing and nanoimaging.

Returning to data processing, one should recall that squeezing the optical signal to the nanoguide is not enough. The signal has to be filtered, separated from other signals, preventing their cross talks, amplified, etc. All these problems refer to nanophotonics. Photonic crystals enable very good nanoguiding and separation of signals in devices based on these waveguides. The cross section of photonic-crystal waveguides is still comparable to the wavelength: practically, of the order of one micron. However, traditionally, they are subjects of nanophotonics because the absolute majority of photonic crystals operating in the visible and telecom ranges are composed of submicron inclusions.

Filtering and amplification of squeezed signals is the subject or the so-called *metatronics*, which also covers subwavelength waveguides and some computations functionalities (see the corresponding chapter). Nanophotonics considers prospective submicron generators of optical signals called *nanolasers*. Next, nanostructures that enhance the light conversion into electricity form an important field of nanophotonics. It is known that the solar light can be converted to electricity using photovoltaic devices called *solar cells*. The invisible (infrared) light produced by very hot bodies can be converted as well. These devices are called *thermophotovoltaic generators*. In both photovoltaic and thermophotovoltaic generating systems, nanostructures open the door to new mechanisms of light harvesting and will enable future technical breakthroughs. Therefore, corresponding chapters of this book cover these two topics.

Of course, our choice of topics reflects the scientific interests of the authors, and it is simply not possible to include all important subjects of this broad area of science and technology. In particular, we do not describe optical memory cells, nanolasers, optical transistors, switches, modulators, and logical gates, which are important nanophotonics components, mainly because these topics belong to the domain of solid-state physics and quantum physics.

Problems and Control Questions

1. Explain the meaning of the terms *metamaterial* and *nanophotonics*.
2. Why optical phenomena in nanostructures are fundamentally different from phenomena in micrometer or millimeter-scale structures?
3. What is the physical origin of thermal noise?
4. How does the power density of thermal noise depend on the frequency and temperature? Use MATLAB, Mathematica, Excel, or another numerical tool to represent the spectral density of thermal radiation.
5. Let us consider a communication system at room temperature, in terms of the level of thermal noise. What is the optimal frequency range to operate it? What technology do we need to fabricate devices in this frequency range?
6. Why is frequency dispersion in waveguides a problem for applications? What is the physical meaning of wave propagation dispersion in waveguides?
7. How can we reduce dispersion in waveguides and transmission lines? Consider quasi-TEM waves, for example, in microstrip lines or coaxial cables. Express the complex propagation constant in terms of reactances and resistances per unit length and identify the conditions for reducing dispersion. How this condition affects the characteristic impedance?

Bibliography

[1] A. Sihvola, S. Tretyakov, and A. de Baas, "Metamaterials with extreme material parameters," *Journal of Communications Technology and Electronics* **52**, 986–990 (2007).
[2] R. W. Ziolkowski, "Propagation in and scattering from a matched metamaterial having a zero index of refraction," *Physical Review E* **70**, 046608 (2004).
[3] A. Alù, M. G. Silveirinha, A. Salandrino, and N. Engheta, "Epsilon-near-zero metamaterials and electromagnetic sources: Tailoring the radiation phase pattern," *Physical Review B* **75**, 155410 (2007).
[4] S. B. Glybovski, S. A. Tretyakov, P. A. Belov, Y. S. Kivshar, and C. Simovski, "Metasurfaces: From microwaves to visible," *Physics Reports* **634**, 1–72 (2016).
[5] S. Tretyakov, V. Asadchy, and A. Díaz-Rubio, "Metasurfaces for general control of reflection and transmission," in *World Scientific Handbook of Metamaterials and Plasmonics, Vol. 1: Electromagnetic Metamaterials*, ed. E. Shamonina, World Scientific Publishing Co., pp. 249–293, 2018.

[6] S. Jahani and Z. Jacob, "All-dielectric metamaterials," *Nature Nanotechnology* **11**, 23–36 (2016).

[7] N. M. Litchinitser and J. Sun, "Optical meta-atoms: Going nonlinear," *Science* **350**, 1033–1034 (2015).

[8] V. S. Asadchy, A. Díaz-Rubio, and S. A. Tretyakov, "Bianisotropic metasurfaces: Physics and applications," *Nanophotonics* **7**, 1069–1094 (2018).

[9] D. L. Sounas and A. Alù, "Non-reciprocal photonics based on time modulation," *Nature Photonics* **11**, 774–783 (2017).

[10] Q. He, S. Sun, and L. Zhou, "Tunable/reconfigurable metasurfaces: Physics and applications," *Research* Article ID 1849272 (2019).

[11] F. Liu, A. Pitilakis, M. Mirmoosa, et al., "Programmable metasurfaces: State of the art and prospects," *2018 IEEE International Symposium on Circuits and Systems (ISCAS),* Florence, Italy, May 27–30, 2018.

[12] N. I. Zheludev and Y. S. Kivshar, "From metamaterials to metadevices," *Nature Materials* **11**, 917–924 (2012).

[13] L. Novotny and B. Hecht, *Principles of Nano-Optics,* 2nd edition, Cambridge University Press, 2012.

[14] B. C. Burkey, W. C. Chang, J. Littlehale, et al., "The pinned photodiode for an interline-transfer CCD image sensor," *Technical Digest of International Electron Device Meeting IEDM 1984*, San Francisco, CA, December 9–12, 1984, pp. 28–31.

[15] A. J. P. Theuwissen, *Solid-State Imaging with Charge-Coupled Devices*, Kluwer, pp. 92–100, 1995.

[16] T. Lule, S. Benthien, H. Keller, F. Mütze, et al., "Sensitivity of CMOS based imagers and scaling perspectives," *IEEE Transactions on Electron Devices* **47**, 2110–2122 (2000).

[17] C. Rios, M. Stegmaier, P. Hosseini, et al., "Integrated all-photonic non-volatile multi-level memory," *Nature Photonics* **9**, 725–732 (2015).

2 Electromagnetic (Optical) Properties of Materials

2.1 Constitutive Relations and Material Parameters

In macroscopic electromagnetics, all the field vectors are averaged over electrically (optically) small volumes, each containing many atoms or molecules forming the material. Actually, only if such averaging is possible can we speak about electromagnetic properties of *materials*, natural or artificial. If the macroscopic fields change in space so quickly that the distances between the atoms or meta-atoms are comparable to the wavelength, we deal with photonic crystals or diffraction gratings, and it is not possible to introduce the notion of macroscopic material parameters, connecting macroscopic, volume-averaged or surface-averaged field vectors [1]. In this chapter, we will discuss only homogenizable natural and artificial materials. The properties of photonic crystals, holograms, and other similar structures are mainly determined by interference and diffraction phenomena because the characteristic spatial scale of inhomogeneity is comparable to the wavelength.

Electromagnetic properties of natural materials are described in terms of the constitutive relations, which define linear relations between the four field vectors: **E**, **H**, **D**, and **B**. In the most general case of anisotropic linear materials, the relations are usually written either expressing the induction vectors in terms of the fields,

$$\mathbf{D} = \bar{\bar{\epsilon}}(\omega) \cdot \mathbf{E} + \bar{\bar{a}}(\omega) \cdot \mathbf{H}, \qquad \mathbf{B} = \bar{\bar{\mu}}(\omega) \cdot \mathbf{H} + \bar{\bar{b}}(\omega) \cdot \mathbf{E}, \tag{2.1}$$

or, equivalently, as

$$\mathbf{D} = \bar{\bar{\epsilon}}'(\omega) \cdot \mathbf{E} + \bar{\bar{C}}(\omega) \cdot \mathbf{B}, \qquad \mathbf{H} = \bar{\bar{\mu}}'^{-1}(\omega) \cdot \mathbf{B} + \bar{\bar{D}}(\omega) \cdot \mathbf{E} \tag{2.2}$$

(the prime sign indicates that the permittivity $\bar{\bar{\epsilon}}$ and permeability $\bar{\bar{\mu}}$ are, in general, different in the two formalisms) [2, 3]. All the material parameters depend on the frequency, to account for frequency dispersion. Weak spatial dispersion appears due to the final size of the structural elements and distances between them, and it is modeled by the magneto-electric coupling dyadics, $\bar{\bar{a}}, \bar{\bar{b}}$, and the permeability dyadic $\bar{\bar{\mu}}$ (e.g., [4, 5]). Effective magnetic response may arise in composite media without any natural magnetism, as a reciprocal spatial-dispersion effect. The material parameters of one set can be expressed in terms of the parameters of the other set through simple algebraic relations. Form (2.1) is sometimes more convenient in solving engineering problems, while form (2.2) is more appropriate if the covariant description is preferred.

The general bianisotropic relations (2.1) are obviously well suited to describe linear materials that can be modeled macroscopically. Most often, much simpler descriptions are sufficient: dielectric materials and conductors are described in terms of the permittivity tensor (or scalar for isotropic media) only. The field coupling coefficients $\overline{\overline{a}}$ and $\overline{\overline{b}}$, as well as $\overline{\overline{C}}$ and $\overline{\overline{D}}$, describe such important effects as optical activity of chiral materials and nonreciprocal bianisotropic effects in some natural ferrimagnetic crystals and corresponding metamaterials.

2.2 Frequency Dispersion

2.2.1 Linear Response and Causality

Let us consider constitutive relations for linear media in the time domain; that is, dependence of induced polarizations at a particular moment of time t on the fields. In vacuum, we simply define

$$\mathbf{D}(t) = \epsilon_0 \mathbf{E}(t), \tag{2.3}$$

since, in vacuum, there is no induced polarization at all. If electric fields exist in some electrically polarizable medium, we add

$$\mathbf{D}(t) = \epsilon_0 \mathbf{E}(t) + \mathbf{P}(t), \tag{2.4}$$

where $\mathbf{P}(t)$ is the polarization vector (electric dipole moment of a unit volume). If our linear system is stationary (that is, its properties do not change with time), we can write an integral relation between the polarization vector at time t and the values of the macroscopic electric field at all *past* moments of time as a convolution integral:

$$\mathbf{P}(t) = \epsilon_0 \int\limits_{-\infty}^{t} \chi(t - t') \mathbf{E}(t') \, dt'. \tag{2.5}$$

Excluding a possibility of actions "from the future" by restricting the integration to the region $t' \in [-\infty, t]$, we ensure that the response is always causal.

Adding the free-space term $\epsilon_0 \mathbf{E}(t)$, we have, for the displacement vector, a similar linear integral relation:

$$\mathbf{D}(t) = \int\limits_{-\infty}^{t} \epsilon(t - t') \mathbf{E}(t') \, dt'$$

$$= -\int\limits_{\infty}^{0} \epsilon(\tau) \mathbf{E}(t - \tau) \, d\tau = \int\limits_{0}^{\infty} \epsilon(\tau) \mathbf{E}(t - \tau) \, d\tau. \tag{2.6}$$

Here, $\tau = t - t'$, and $\epsilon(\tau) = \epsilon_0[\delta(\tau) + \chi(\tau)]$, where $\delta(\tau)$ is the Dirac delta function.

The frequency-domain relations follow from Fourier transform

$$f(\omega) = \int\limits_{-\infty}^{\infty} f(t) \exp(-j\omega t)\, dt, \qquad f(t) = \frac{1}{2\pi} \int\limits_{-\infty}^{\infty} f(\omega) \exp(j\omega t)\, d\omega \qquad (2.7)$$

of these time-domain relations. The Fourier transform of a convolution integral gives the product of two transforms:

$$\mathbf{D}(\omega) = \epsilon(\omega)\mathbf{E}(\omega), \qquad (2.8)$$

where

$$\epsilon(\omega) = \int\limits_{0}^{\infty} \epsilon(\tau) \exp(-j\omega\tau)\, d\tau \qquad (2.9)$$

is the complex permittivity. Let us stress again that the integration is only over the positive values of time delay τ, as dictated by the causality requirement. As we will see in Section 2.2.2, this principle restricts physically allowed frequency dispersion of permittivity and other material parameters. Indeed, similar convolution integrals to those in (2.6) can also be written for the magnetic response and the magneto-electric coefficients because all of these effects are linear and causal.

Frequency-dependent polarizability describes material response to monochromatic fields. However, since the medium response is assumed to be linear (weak fields), the theory is applicable to any time dependence of fields and polarizations. To analyze arbitrary transient phenomena, it is possible to use Fourier expansions, relating the Fourier components of fields and polarizations by frequency-dependent permittivity.

2.2.2 The Main Dispersion Mechanisms and Models

Here, we concentrate on dielectrics and metals and study how the permittivity depends on the frequency (frequency dispersion). It can be understood from a classical model of movements of electrons under the applied electric field. We assume that the material response is linear – that is, the applied field is small enough that it induces polarization but does not change the material properties.

Let us first consider polar dielectrics – that is, the case when molecules have permanent electric moments even in the absence of applied electric field (for example, water). If the material is a gas or liquid, then without any polarizing field, the average dipole moment is zero because of random thermal motion of molecules. External fields orient molecules so that some averaged electric dipole along the external electric field is created; see an illustration in Figure 2.1. If at some moment of time (let us assume it is $t = 0$) we switch off the external field, the polarization vector will exponentially decay due to thermalization:

$$\mathbf{P}(t) = \mathbf{P}_0\, e^{-t/T}, \qquad (2.10)$$

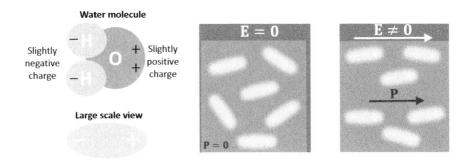

Water molecule

Slightly negative charge

Slightly positive charge

Large scale view

E = 0

P = 0

E ≠ 0

P

Illustration of the polarization processes in polar dielectrics. Courtesy of A. Díaz-Rubio.

where T is the relaxation time, which depends on the material properties and the temperature. It means that the kernel function in (2.6) and (2.9) is an exponential function:

$$\epsilon(\tau) = \epsilon_0 \left(\delta(\tau) + A e^{-\tau/T} \right),\tag{2.11}$$

where A is a constant depending on the particular material. The Fourier transform (2.9) gives

$$\epsilon(\omega) = \epsilon_0 + \frac{AT}{1 + j\omega T}.\tag{2.12}$$

More commonly, this result is written as

$$\epsilon = \epsilon_0 + \frac{\epsilon_s - \epsilon_0}{1 + j\omega T},\tag{2.13}$$

where ϵ_s is the permittivity at zero frequency (static permittivity). This expression is called the *Debye model*. An illustration of the typical dependence on the frequency is given in Figure 2.2.

Let us next turn our attention to the case where molecules get polarized only when an external electric field is applied. Consider oscillations of an electron near its equilibrium point $\mathbf{r} = 0$ under the influence of an external electric field $\mathbf{E}$; see Figure 2.3. The following forces act on the electron: the Lorentz force $\mathbf{F} = e\mathbf{E}$ (e is the elementary charge), a "return force" $\mathbf{Q} = -q\mathbf{r}$, and "friction force" $\mathbf{S} = -\Gamma \frac{\partial \mathbf{r}}{\partial t}$. Return force is present if the electron is bound (as in dielectrics). The classical equation of motion reads

$$m\frac{\partial^2 \mathbf{r}}{\partial t^2} + \Gamma \frac{\partial \mathbf{r}}{\partial t} + q\mathbf{r} = e\mathbf{E}\tag{2.14}$$

(m is the electron mass). Its time-harmonic (assuming $\exp(j\omega t)$ time dependence) solution is

$$\mathbf{r} = \frac{\frac{e}{m}}{\frac{q}{m} - \omega^2 + j\omega\frac{\Gamma}{m}}\mathbf{E} = \frac{\frac{e}{m}}{\omega_0^2 - \omega^2 + j\omega\nu}\mathbf{E}.\tag{2.15}$$

The dipole moment, by definition, equals $\mathbf{p} = e\mathbf{r}$. Assuming that there are N electrons per unit volume (N is called *concentration*), we can find the dipole

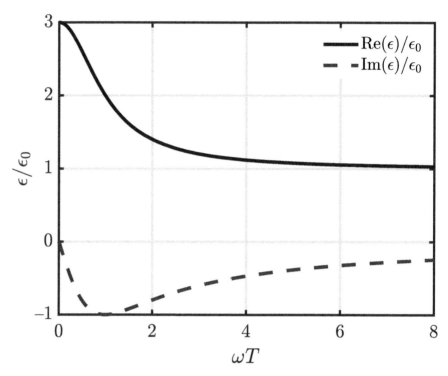

Figure 2.2   Generic frequency dependence of permittivity on the frequency, Debye model. In this example, $\epsilon_s = 3\epsilon_0$.

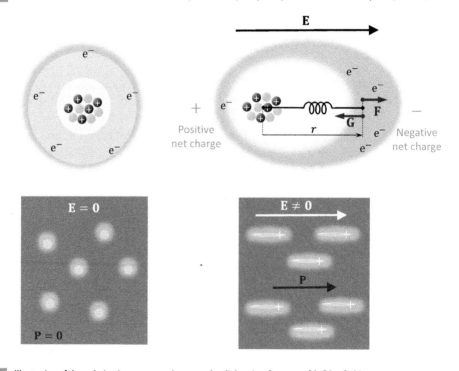

Figure 2.3   Illustration of the polarization processes in non-polar dielectrics. Courtesy of A. Díaz-Rubio.

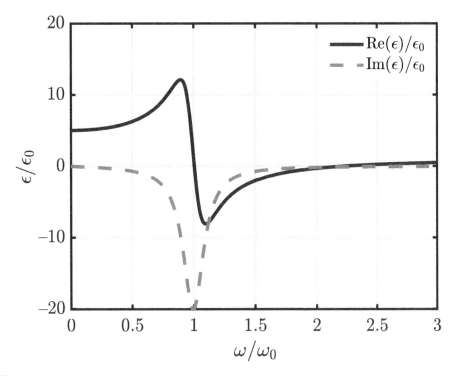

Generic frequency dependence of permittivity on the frequency, Lorentz model. In this example, $\omega_p/\omega_0 = 2$ and $\nu/\omega_0 = 0.2$.

moment per unit volume (called *polarization*): $\mathbf{P} \approx N\mathbf{p}$. Now, by definition of the displacement vector,

$$\mathbf{D} = \epsilon_0\mathbf{E} + \mathbf{P} = \epsilon\mathbf{E}. \tag{2.16}$$

Thus, the permittivity reads

$$\epsilon = \epsilon_0\left(1 + \frac{\omega_p^2}{\omega_0^2 - \omega^2 + j\omega\nu}\right), \tag{2.17}$$

where $\omega_p^2 = \frac{Ne^2}{me_0}$. The parameter ω_p is called *plasma frequency*.

This expression is called the *Lorentz dispersion model*. A typical frequency behavior is illustrated in Figure 2.4. It is a typical dispersion pattern for dielectric materials. More generally, dispersion for media with bound electrons can be modeled by a sum of several terms of this type, accounting for resonances at different frequencies.

Let us next consider metals, where conduction electrons are free to move. This means that there is no return force, and we should set parameter q in the preceding equations to zero. This is tantamount to setting the resonance frequency for freely moving electrons to zero, because $\omega_0^2 = q/m$. The permittivity function becomes

$$\epsilon = \epsilon_0\left(1 - \frac{\omega_p^2}{\omega^2 - j\omega\nu}\right). \tag{2.18}$$

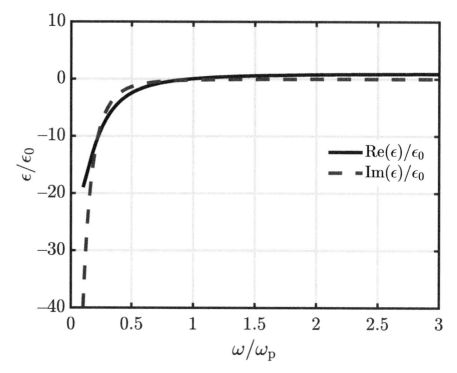

Figure 2.5   Generic frequency dependence of permittivity on the frequency, Drude model. In this example, $\nu/\omega_p = 0.2$.

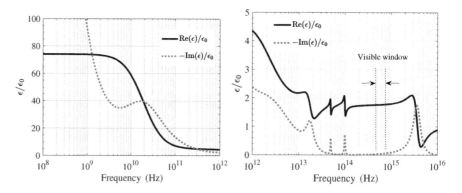

Figure 2.6   Permittivity of seawater. Study and graphics by X. Wang, based on data from [6, 7].

This expression is called the *Drude model*. An illustration of the typical frequency dispersion can be seen in Figure 2.5.

Most commonly, different physical mechanisms of dispersion dominate in different frequency regions. A good example is water. Figure 2.6 shows the permittivity of seawater in a broad frequency range, from statics to ultraviolet. At low frequencies, the response is dominated by conductivity (Drude model).

At higher frequencies, we see the relaxation resonance in the microwave range (Debye model) and a number of Lorentz resonances in the infrared and ultraviolet ranges.[1] At extremely high frequencies, the permittivity tends to unity.

2.2.3 Kramers-Kronig Relations

Let us consider the frequency-domain permittivity $\epsilon(\omega)$ (2.9) as a function of the complex variable ω, denoting $\omega = \omega' + j\omega''$ and $\epsilon(\omega) = \epsilon'(\omega) + j\epsilon''(\omega)$. From the definition (2.9), it is clear that $\epsilon(-\omega) = \epsilon^*(\omega)$, $\epsilon'(-\omega) = \epsilon'(\omega)$, $\epsilon''(-\omega) = -\epsilon''(\omega)$. In passive media, $\epsilon''(\omega) < 0$ at $\omega > 0$, which ensures that energy can be only dissipated but not generated by the medium.

The function $\epsilon(\tau)$ describes the effects of "material memory": the polarization at a given moment of time depends not only on the field at this same moment ($\tau = 0$) but also at past values $\tau > 0$. When τ increases, this function usually decreases, describing "fading memory." In any case, in linear materials, we can assume that this function is limited for all $\tau > 0$. Under this assumption, we see from the definition (2.9) that $\epsilon(\omega)$ is an analytical function in the lower half-plane of ω because the integrand has no singularities and exponentially decays at infinity, provided that $\omega'' < 0$.

It is most important to realize that the physical reason for this property is *causality* of the medium response. It means that the polarization at a given moment t can depend only on the fields in the past, meaning that $\tau \geq 0$. Obviously, if we also extend the integration area in (2.9) over the negative values of τ, the Fourier product would not possess this analyticity property.

If the medium is lossy, the function is analytical also on the real axis (for ideally lossless media, it can have poles at the real axis). Note that the properties of $1/\epsilon$ are complementary: this function has no nulls in the lower half-space.

Let us consider lossy dielectrics except conductors. For the susceptibility $\chi(\omega) = \frac{\epsilon(\omega)}{\epsilon_0} - 1 = \epsilon_r(\omega) - 1$, we consider the function

$$\frac{\chi(\xi)}{\xi - \omega} \tag{2.19}$$

(real $\omega > 0$). This function is analytical in the whole lower half-plane ξ (including the real axis) except for one simple pole at $\xi = \omega$. We can integrate this function along a closed contour going along the real axis and closing over the half-circle at infinity. The integral over the half-circle is 0, because the function exponentially decays at infinity. Thus,

$$\mathrm{PV} \int_{-\infty}^{\infty} \frac{\chi(\xi)}{\xi - \omega} \, d\xi = -j\pi\chi(\omega). \tag{2.20}$$

Here, ξ is a real variable and PV denotes the principal value of the integral. Splitting $\chi = \chi' + j\chi''$ into the real and imaginary parts, we get

$$\chi'(\omega) = -\frac{1}{\pi} \mathrm{PV} \int\limits_{-\infty}^{\infty} \frac{\chi''(\xi)}{\xi - \omega}\, d\xi, \tag{2.21}$$

$$\chi''(\omega) = \frac{1}{\pi} \mathrm{PV} \int\limits_{-\infty}^{\infty} \frac{\chi'(\xi)}{\xi - \omega}\, d\xi. \tag{2.22}$$

In case of conductors, the permittivity has a pole at $\omega = 0$, which brings an additional term in (2.22), equal to the conventional expression for the conductivity contribution $[-j\sigma/(\omega\epsilon_0)]$. These two relations are attributed to R. de L. Kronig (1926) and H. A. Kramers (1927); see a historical review in [8].

Using the property $\chi''(-\omega) = -\chi''(\omega)$, we can also write

$$\chi'(\omega) = -\frac{1}{\pi} \mathrm{PV} \int\limits_{-\infty}^{\infty} \frac{\chi''(\xi)}{\xi - \omega}\, d\xi = -\frac{2}{\pi} \mathrm{PV} \int\limits_{0}^{\infty} \frac{\xi\chi''(\xi)}{\xi^2 - \omega^2}\, d\xi. \tag{2.23}$$

For nonconductors (assuming that $\chi''(\xi)$ is regular at $\xi = 0$), we can let $\omega = 0$ and come to the relation

$$\chi'(0) = -\frac{2}{\pi} \int\limits_{0}^{\infty} \frac{\chi''(\xi)}{\xi}\, d\xi > 0. \tag{2.24}$$

This relation is called *sum rule* (there are several sum rules that can be derived in a similar way). This result shows, in particular, that in all passive (lossy) media, the static value of the susceptibility is positive and the permittivity is larger than unity: $\chi'(0) > 0$ and $\epsilon'_r(0) > 1$.

Let us study transparent frequency regions (negligible losses in the vicinity of ω). In this case, the pole contribution in (2.23) is zero (since $\chi''(\xi) = 0$ at $\xi = \omega$). The relation for the real part becomes

$$\chi'(\omega) = \epsilon'_r - 1 = -\frac{2}{\pi} \int\limits_{0}^{\infty} \frac{\xi\chi''(\xi)}{\xi^2 - \omega^2}\, d\xi \tag{2.25}$$

($\chi'' = \epsilon''_r$), which is a usual, well-behaving integral. Thus, we can differentiate under the integral sign. The result reads

$$\frac{d\epsilon'_r(\omega)}{d\omega} = -\frac{4\omega}{\pi} \int\limits_{0}^{\infty} \frac{\xi\epsilon''_r(\xi)}{(\xi^2 - \omega^2)^2}\, d\xi > 0. \tag{2.26}$$

It tells that in low-loss regions, the dispersion is always "normal": the real part of the susceptibility or permittivity grows with the frequency. This result is called the *Foster theorem* (because it was first established by R. M. Foster for reactances of electrical circuits [9]),

2.3 Electromagnetic Waves in Materials

2.3.1 Dispersion Equation

Let us consider an anisotropic dielectric material characterized by a matrix (dyadic) permittivity $\overline{\overline{\epsilon}}$. The permeability is assumed to be the same as in free space (μ_0). Let us look for solutions to Maxwell's equations of a uniform space in the absence of sources in the form of plane waves $e^{-j\mathbf{k}\cdot\mathbf{r}}$. For the complex amplitudes of plane-wave fields, the equations read (replacing ∇ with $-j\mathbf{k}$)

$$\mathbf{k} \times \mathbf{E} = \omega\mu_0\mathbf{H}, \qquad \mathbf{k} \times \mathbf{H} = -\omega\overline{\overline{\epsilon}} \cdot \mathbf{E}. \tag{2.27}$$

The divergence equation tells that

$$\mathbf{k} \cdot \mathbf{D} = \mathbf{k} \cdot \overline{\overline{\epsilon}} \cdot \mathbf{E} = 0. \tag{2.28}$$

Eliminating vector $\mathbf{H} = \frac{1}{\omega\mu_0}\mathbf{k} \times \mathbf{E}$, we get

$$\mathbf{k} \times (\mathbf{k} \times \mathbf{E}) = -\omega^2\mu_0\overline{\overline{\epsilon}} \cdot \mathbf{E}. \tag{2.29}$$

Expanding the double cross product using the relation $\mathbf{k} \times (\mathbf{k} \times \mathbf{E}) = \mathbf{k}(\mathbf{k} \cdot \mathbf{E}) - \mathbf{E}(\mathbf{k} \cdot \mathbf{k})$, the same equation also can be written as

$$\mathbf{k}(\mathbf{k} \cdot \mathbf{E}) - k^2\mathbf{E} = -\omega^2\mu_0\overline{\overline{\epsilon}} \cdot \mathbf{E}. \tag{2.30}$$

These equations can be satisfied for non-zero vectors $\mathbf{E}$ only if the determinant of the matrix (dyadic) acting on the field vector is zero:

$$\det[\mathbf{k} \times (\mathbf{k} \times \overline{\overline{I}}) + \omega^2\mu_0\overline{\overline{\epsilon}}] = 0, \tag{2.31}$$

where $\overline{\overline{I}}$ is the unit matrix (dyadic). This equation is called the *dispersion equation*, and it defines the propagation vector as a function of the frequency. Another form follows from (2.30):

$$\det[\mathbf{k}\mathbf{k} - k^2\overline{\overline{I}} + \omega^2\mu_0\overline{\overline{\epsilon}}] = 0. \tag{2.32}$$

For isotropic dielectrics, where $\overline{\overline{\epsilon}} = \epsilon\overline{\overline{I}}$, the double cross product simplifies to $\mathbf{k} \times (\mathbf{k} \times \mathbf{E}) = \mathbf{k}(\mathbf{k} \cdot \mathbf{E}) - \mathbf{E}(\mathbf{k} \cdot \mathbf{k}) = -k^2\mathbf{E}$ (since, in this case, from (2.28), it follows that $\mathbf{k} \cdot \mathbf{E} = 0$), and the wave equation (2.29) reads

$$(-k^2 + \omega^2\mu_0\epsilon) \cdot \mathbf{E} = 0. \tag{2.33}$$

For the dispersion equation, we have simply

$$k^2 = \omega^2\mu_0\epsilon \tag{2.34}$$

and

$$k = \pm\omega\sqrt{\mu_0\epsilon} = \pm\frac{\omega}{v}. \tag{2.35}$$

Here, $v = \frac{1}{\sqrt{\mu_0\epsilon}} = c\sqrt{\epsilon_0/\epsilon}$ is the phase velocity (see Section 2.3.3) of plane waves in medium with the permittivity ϵ. The two solutions, differing by sign, tell that waves can travel (with the same speed) in opposite directions along any straight line in isotropic space.

In reciprocal materials, the permittivity matrix is symmetric and can be diagonalized and fully characterized by its three eigenvectors and three eigenvalues: the permittivities along the eigenvector directions, called *optical axes*. If all three eigenvalues are different, the medium is called biaxial (biaxial crystal). It is usually convenient to work in the coordinate basis formed by the eigenvectors of $\overline{\overline{\epsilon}}$ (although, in lossy media, the eigenvectors can be different for the real and imaginary parts of $\overline{\overline{\epsilon}}$).

Let us next consider probably the most interesting and often used case in applications, uniaxial media, where two of the eigenvalues are the same. It physically means that there is one preferred direction (one optical axis), while the structure is isotropic in the plane orthogonal to that axis. Such materials are characterized by uniaxial permittivity dyadics

$$\overline{\overline{\epsilon}} = \epsilon_t \overline{\overline{I}}_t + \epsilon_z \mathbf{z}_0 \mathbf{z}_0, \tag{2.36}$$

where $\mathbf{z}_0$ is the unit vector along the optical axis (we assume that it is a real-valued vector) and $\overline{\overline{I}}_t = \overline{\overline{I}} - \mathbf{z}_0 \mathbf{z}_0 = \mathbf{x}_0 \mathbf{x}_0 + \mathbf{y}_0 \mathbf{y}_0$ is the unit (identity) dyadic acting in the transverse plane. In the matrix form, relation (2.36) is written as follows:

$$\overline{\overline{\epsilon}} = \begin{pmatrix} \epsilon_t & 0 & 0 \\ 0 & \epsilon_t & 0 \\ 0 & 0 & \epsilon_z \end{pmatrix}. \tag{2.37}$$

The dyadic form (2.36) is usually more convenient because it defines the permittivity in terms of vectors that have clear physical meaning (in this case, the unit vector along the optical axis $\mathbf{z}_0$), independently of a particular coordinate system.

To study eigenwaves and the dispersion equation, it is most convenient to work in the coordinate system defined by the optical axis direction (one of the Cartesian axes is $\mathbf{z}_0$) and the wavevector $\mathbf{k}$ of the wave that we study. Thus, we select the other two axes along the projection of the wavevector $\mathbf{k}$ to the plane orthogonal to $\mathbf{z}_0$ (vector $\mathbf{k}_t = \mathbf{k} - (\mathbf{z}_0 \cdot \mathbf{k}) \mathbf{z}_0$) and along the vector $\mathbf{z}_0 \times \mathbf{k}_t$, which is orthogonal to both $\mathbf{z}_0$ and $\mathbf{k}_t$.

In this coordinate system, the electric field vector is written as $\mathbf{E} = \mathbf{E}_t + \mathbf{E}_\times + E_n \mathbf{z}_0$, where we denote the projections of the electric field to the directions of $\mathbf{k}_t$ and $\mathbf{z}_0 \times \mathbf{k}_t$ by $\mathbf{E}_t$ and $\mathbf{E}_\times$, respectively. Now we can write the wave equation (2.30) in this coordinate system, substituting the electric field vector and the wavevector $\mathbf{k}$ written in these coordinates as $\mathbf{k} = \mathbf{k}_t + k_z \mathbf{z}_0$. We obviously have $k^2 = k_t^2 + k_z^2$ and $\mathbf{k} \cdot \mathbf{E} = k_t E_t + k_z E_n$. The vector equation (2.30) splits into three scalar equations: projections on the three axes. Writing down its determinant (the three rows in (2.38) correspond to the projections of (2.32) to the directions of $\mathbf{k}_t$, $\mathbf{z}_0 \times \mathbf{k}_t$, and $\mathbf{z}_0$, respectively) and equating it to zero, we transform (2.32) to

$$\begin{vmatrix} -k_z^2 + \omega^2 \mu_0 \epsilon_t & 0 & k_z k_t \\ 0 & -k^2 + \omega^2 \mu_0 \epsilon_t & 0 \\ k_z k_t & 0 & -k_t^2 + \omega^2 \mu_0 \epsilon_z \end{vmatrix}$$

$$= (k^2 - \omega^2 \mu_0 \epsilon_t) \left[(-k_z^2 + \omega^2 \mu_0 \epsilon_t)(-k_t^2 + \omega^2 \mu_0 \epsilon_z) - k_z^2 k_t^2 \right] = 0. \tag{2.38}$$

This dispersion equation apparently splits into two equations:

$$k_z^2 + k_t^2 = k^2 = \omega^2 \mu_0 \epsilon_t \qquad (2.39)$$

and

$$k_z^2 \epsilon_z + k_t^2 \epsilon_t = \omega^2 \mu_0 \epsilon_t \epsilon_z. \qquad (2.40)$$

It means that two different modes can propagate in the medium. For one of them, which obeys (2.39), the wavenumber k does not depend on ϵ_z at all, and the dispersion equation is actually the same as in simple isotropic media (2.34). Physically, this means that the wave has no electric field component along $\mathbf{z}_0$, and for this reason, it does not feel that that component of the permittivity matrix is different. This mode is called the *ordinary wave*. The second mode, which obeys (2.40), is called the *extraordinary wave*. Its dispersion equation is often written in this form (obtained by dividing (2.40) by $\epsilon_t \epsilon_z$):

$$\frac{k_z^2}{\epsilon_t} + \frac{k_t^2}{\epsilon_z} = \omega^2 \mu_0. \qquad (2.41)$$

2.3.2 Dispersion Diagrams and Constant-Frequency Contours

Let us start from the solution (2.35) of the dispersion equation for an isotropic dielectric (or free space) $\omega = vk$, where k is the wavenumber and $v = 1/\sqrt{\epsilon \mu}$ is the speed of light in it. If we restrict the consideration by light propagating along one direction – e.g., the axis $x - \omega$ as a function of the only wavevector component, k_x, reads $\omega = v|k_x|$. Allowing the light to propagate in the plane $(x-y)$, we obtain the function $\omega = v\sqrt{k_x^2 + k_y^2}$ whose plot is called the light cone; see Figure 2.7(a). This cone describes all plane waves with $k_z = 0$. For any point inside this cone, $k_x^2 + k_y^2 < k^2$. Therefore, $k_z = \sqrt{k^2 - k_x^2 - k_y^2} > 0$; i.e., the wavevector is real, and the interior of the light cone covers propagating plane waves possible in the medium at any frequency. For any point beyond this cone, $k_x^2 + k_y^2 > k^2$. Therefore,

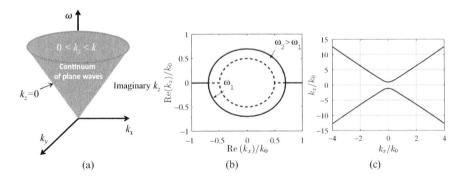

(a)   (b)   (c)

Figure 2.7 (a) Light cone: Dispersion diagram of isotropic dielectric space (here, $k = \omega\sqrt{\epsilon \mu}$). Continuum of plane waves (real k_z) is inside the light cone; continuum of evanescent waves (imaginary k_z) is outside it. (b) Constant-frequency contours for an isotropic dielectric space. (c) Constant-frequency plots for a hyperbolic medium. In this example, $\epsilon_z = -0.1\epsilon_0$ and $\epsilon_t = \epsilon_x = \epsilon_y = \epsilon_0$.

$k_z = \sqrt{k^2 - k_x^2 - k_y^2}$ is imaginary, and the exterior of the light cone corresponds to evanescent waves in the medium. A section of the light cone by a plane $\omega = $ const is a circle, and the family of such circles is called the *constant-frequency contours* of the medium; see Figure 2.7(b). Constant-frequency contours are plots in the plane $(k_x - k_y)$. For isotropic media, constant-frequency surfaces are spheres $\omega = v\sqrt{k_x^2 + k_y^2 + k_z^2}$ in the space $(k_x - k_y - k_z)$. This space – that of the wavevectors of waves propagating in the medium or crystal lattice – is called the *reciprocal space*. Notice that both constant-frequency contours and constant-frequency surfaces are often also called *isofrequencies*. It is important to know that isofrequencies make sense only for waves propagating in the medium or the lattice (eigenwaves).

For uniaxial media having positive eigenvalues of permittivity dyadic ϵ_t and ϵ_z, the constant-frequency surfaces are ellipsoids (more exactly, spheroids), as is evident from (2.41). If one of the eigenvalues is negative and the other one is positive, the shapes described by (2.41) are hyperboloids; see Figure 2.7(c). Such materials are called *hyperbolic media*, and they possess very interesting and practically important properties (see, e.g., [10]). Note that propagation constants of propagating modes can take large values, much larger (theoretically infinite) than the free-space propagation constant. This property opens up possibilities for far-field imaging with subwavelength resolution.

Some natural crystals possess permittivity of the form (in the coordinate system where the symmetric matrix $\overline{\overline{\epsilon}}$ is diagonal)

$$\overline{\overline{\epsilon}} = \begin{pmatrix} \epsilon_1 & 0 & 0 \\ 0 & \epsilon_2 & 0 \\ 0 & 0 & \epsilon_3 \end{pmatrix}. \tag{2.42}$$

In natural low-loss materials, all the three eigenvalues $\epsilon_{1,2,3}$ are positive, and the isofrequency surfaces are non-spheroidal ellipsoids. Such media are called *biaxial crystals*.

2.3.3 Phase and Group Velocities

The dependence of electromagnetic fields of a plane wave with complex amplitude $Ae^{-j\mathbf{k}\cdot\mathbf{r}}$ on the spatial coordinates (position vector $\mathbf{r}$) and time t is given by $\mathrm{Re}\left[Ae^{j(\omega t - \mathbf{k}\cdot\mathbf{r})}\right]$. That is, it is defined by the wave *phase* $\Phi(\mathbf{r}, t) = \omega t - \mathbf{k}\cdot\mathbf{r}$. At any moment of time, surfaces of constant phase are planes, defined by the equation

$$\mathbf{k}\cdot\mathbf{r} - \omega t = \text{const.} \tag{2.43}$$

With increasing time, the points at which the phase take a certain value propagate in space in the direction of vector $\mathbf{k}$. From (2.43), we see that the velocity of phase propagation is

$$\frac{dr}{dt} = v = \frac{\omega}{k}, \tag{2.44}$$

where k is the length of vector $\mathbf{k}$. This velocity is called *phase velocity*. On plots of constant-frequency contours, the vectors of phase velocity are directed simply along vectors $\mathbf{k}$.

The notion of phase velocity is defined for plane waves at fixed frequencies. If the medium is dispersive (in dielectrics, its permittivity depends on the frequency), phase velocity is different for waves at different frequencies. Thus, when we deal with propagation of packets of plane waves (say, finite-duration pulses expanded into Fourier spectra of plane waves), we need the notion of *group velocity*, which measures the speed of the wave package (the pulse). The maximum amplitude of a pulse appears at the point where the plane-wave components of the pulse interfere in phase, which corresponds to the extremum value of the phase $\omega t - \mathbf{k} \cdot \mathbf{r}$ as a function of the wavevector $\mathbf{k}$. We can find the extremum condition (such that, in a small vicinity of this particular point, the phase is stationary with respect to the wavevector) by equating the derivative of the phase to zero:

$$\frac{\partial}{\partial \mathbf{k}}(\omega t - \mathbf{k} \cdot \mathbf{r}) = \frac{\partial \omega}{\partial \mathbf{k}} t - \mathbf{r} = 0. \tag{2.45}$$

This derivative is the gradient of the phase along vector $\mathbf{k}$; that is,

$$\frac{\partial}{\partial \mathbf{k}} = \frac{\partial}{\partial k_x}\mathbf{x}_0 + \frac{\partial}{\partial k_y}\mathbf{y}_0 + \frac{\partial}{\partial k_z}\mathbf{z}_0. \tag{2.46}$$

We see that the point in space where, at a given moment of time, the plane-wave components all interfere constructively (the pulse maximum) propagates in space with the velocity

$$v_g = \frac{\partial \omega}{\partial \mathbf{k}}. \tag{2.47}$$

In actual calculations, it can be simpler to calculate the inverse quantity, differentiating components of vector $\mathbf{k}$ with respect to the frequency. On the constant-frequency contours, the vectors of group velocity are directed normally to these contours. Obviously, in anisotropic and dispersive media, the group velocity and phase velocity usually have different directions.

Problems and Control Questions

1. What is the physical meaning of the Kramers-Kronig relations? For what materials are they not applicable?
2. In what kinds of materials dispersion can be modeled by the Lorentz or Debye laws?
3. Vacuum is non-dispersive and has the constant relative permittivity 1 across all the frequencies. As a result, based on the Kramers-Kronig relations, vacuum is lossless – i.e., its permittivity has zero imaginary part. Can we design a non-dispersive material with the relative permittivity equal to 2?

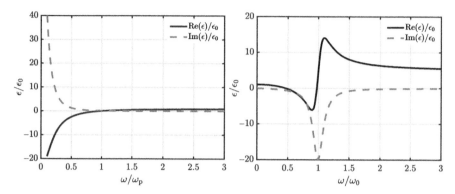

Examples of nonphysical behavior of material parameters.

4. Consider the permittivity of good conductors (metals) at microwave frequencies using the Drude model

$$\epsilon = \epsilon_0 - \frac{\omega_p^2}{\omega^2 - j\omega\nu}. \tag{2.48}$$

In this case, $\omega^2 \ll \omega\nu$ (justify this assumption for typical metals like silver or copper from literature data on their parameters). Under this assumption, derive approximate formulas for the imaginary and real parts of the permittivity (using the Taylor expansion). Discuss the results.

In the literature, metals at microwaves are most usually modeled by their conductivity σ, and the permittivity is assumed to be approximately

$$\epsilon = \epsilon_0 - j\frac{\sigma}{\omega}. \tag{2.49}$$

Does this approximation agree with your results following from the Drude model? The real parts of the permittivity are clearly (very) different.

Discuss which of the two models is closer to reality. Try to find experimental data on the real part of the permittivity of metals at microwaves (in the literature). Note that sometimes it is modeled by assuming that the conductivity σ is a complex number (while the permittivity equals ϵ_0).

Make overall conclusions from your study.

5. Figure 2.8 shows dependencies of permittivity as functions of frequency which look similar to the Drude and Lorentz dispersion plots. However, such dependencies cannot be realized in any passive material. Explain what basic physical principles forbid such behavior.[2]

Bibliography

[1] C. Simovski, "On electromagnetic characterization and homogenization of nanostructured metamaterials (review article)," *Journal of Optics* **13**, 013001 (2011).

[2] D. K. Cheng and J. A. Kong, "Covariant description of bianisotropic media," *Proceedings of the IEEE* **56**, 248–251 (1968).

[3] J. A. Kong, "Theorems of bianisotropic media," *Proceedings of the IEEE* **60**, 1036–1046 (1972).

[4] L. D. Landau and E. M. Lifshitz, *Electrodynamics of Continuous Media,* 2nd ed., Pergamon Press, 1984.

[5] A. N. Serdyukov, I. V. Semchenko, S. A. Tretyakov, and A. Sihvola, *Electromagnetics of Bi-Anisotropic Materials: Theory and Applications,* Gordon and Breach Science Publishers, 2001.

[6] J. E. K. Laurens and K. E. Oughstun, "Electromagnetic impulse response of triply-distilled water," in *Ultra-Wideband Short-Pulse Electromagnetics 4* (IEEE Cat. No.98EX112), ed. E. Heyman, B. Mandelbaum, and J. Shiloh, Kluwer Academic/Plenum Publishers, pp. 243–264, 1998.

[7] M. E. Thomas, "The electrical properties of seawater (including conductivity relaxation)," final report STD-R-1071, Submarine Technology Department, Johns Hopkins University, Applied Physics Laboratory, July 1984.

[8] C. F. Bohren, "What did Kramers and Kronig do and how did they do it?," *European Journal of Physics* **31**, 573–577 (2010).

[9] R. M. Foster, "A reactance theorem," *Bell System Technical Journal* **3**, 259–267 (1924).

[10] A. Poddubny, I. Iorsh, P. Belov, and Y. Kivshar, "Hyperbolic metamaterials," *Nature Photonics* **7**, 948–957 (2013).

[11] C. R. Simovski and S. A. Tretyakov, "Local constitutive parameters of metamaterials from an effective-medium perspective," *Physical Review* **75**, 195111 (2007).

Notes

1 The conductivity of seawater significantly affects the permittivity at low frequencies only: Above the microwave range, the permittivity of distilled water and seawater are nearly the same.

2 The curve for the real part of the permittivity on the right plot shows a typical example of the so-called *antiresonant* behavior, which can result from the use of inappropriate homogenization models for metamaterials with resonant unit cells; see discussions, e.g., in [11].

3 Metamaterials

3.1 Metamaterials Concept

Electromagnetic properties of natural or chemically synthesized materials are determined mainly by their chemical composition – that is, by properties of assembled atoms and molecules. Within the metamaterial paradigm, the role of atoms or molecules is played by "meta-atoms" as constitutive elements of artificial materials. Meta-atoms are macroscopic objects made from usual materials (metals, dielectrics, etc.) As long as the meta-atom sizes and distances between them remain sufficiently small on the wavelength scale of interest, the composites formed by meta-atoms can be considered as "materials," described by effective material parameters (such as permittivity or permeability), and the same approaches to the macroscopic description of electromagnetic properties of matter in terms of material relations (2.1) can be applied to metamaterials made of meta-atoms, just like to ordinary materials formed by atoms or molecules.

The electromagnetic response of arrangements of meta-atoms can be engineered by properly choosing their shapes, internal structures, sizes, mutual orientations, etc. Moreover, the response of individual meta-atoms can be controlled and tuned by external signals or internal, programmable microprocessors, opening up possibilities to engineer tunable, time-modulated, sensing, adaptive, and software-defined artificial materials and surfaces.

The main appeal of the metamaterial concept derives from new possibilities to realize artificial materials with electromagnetic properties that are not available in any natural material and to engineer properties that are optimal for particular applications.

Metamaterial can be defined as artificial, effectively homogeneous material formed by "an arrangement of artificial structural elements, designed to achieve advantageous and unusual electromagnetic properties" (this definition is adopted by the Virtual Institute for Artificial Electromagnetic Materials and Metamaterials; see www.metamorphose-vi.org). Optically thin layers (effectively sheets) formed by engineered meta-atoms are usually called *metasurfaces*, sometimes defined as two-dimensional versions of metamaterials [1, 2]. Arrangements of meta-atoms along a line in space can be called *metawires* [3]. The generality of the metamaterial concept is illustrated in Figure 3.1: The meta-atoms can have any shape and chemical composition, exhibit engineered nonlinear or active properties or even contain controllable and programmable microcontrollers (software-defined materials). The arrangements of meta-atoms in metamaterials

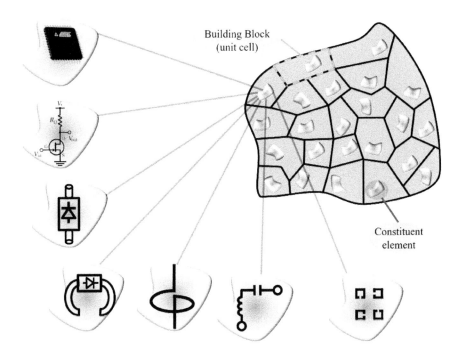

Figure 3.1   Illustration of the metamaterial concept. Courtesy of M. Lapine and M. Albooyeh.

and metasurfaces can be regular or random. The only limitations are on the sizes of the meta-atoms and on the (average) distances between them: they should be sufficiently small with respect to the wavelength of interest, to allow the homogenized, *material* description. Of course, we assume that meta-atoms are macroscopic objects, which are sufficiently large compared to the sizes of the natural atoms and molecules composing them. It is an important restriction for nanostructured metamaterials, whose meta-atoms can be as small as tens of nanometers. Particles smaller than a few nanometers are quantum rather than macroscopic objects.

If meta-atoms are arranged in a lattice whose period is comparable to the wavelength, the lattice response strongly depends on interference of waves reflected by unit cells. Dispersion curves and surfaces show stop bands in the frequency regions where the reflected waves interfere destructively. Such structures are called *photonic crystals* or *electromagnetic bandgap structures (EBG);* see Chapter 5. However, there is no sharp boundary between the notions of the metamaterial and photonic crystal. For example, in designing metasurfaces, it is sometimes required to engineer a dense distribution of small meta-atoms within a unit cell of a periodical structure whose period is comparable to the wavelength. One can perhaps define such structures as photonic crystals made of metamaterials.

For natural materials, the wavelength is very large as compared to the distances between atoms and molecules even for the visible light and ultra-violet waves. However, working with metamaterials, it is not always easy or even possible to create desired structures of small enough dimensions (small compared to the

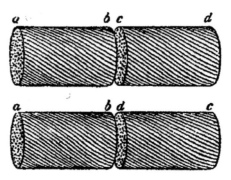

Figure 3.2   Chiral bundles of jute by Bose. Reprinted from [4] with permission.

wavelength), which is an important limitation. Fabrication of optical metamaterials always means nanofabrication.

Although the name *metamaterial* was coined only in the year 2000, when the notion of artificial materials with negative refractive index was introduced [5, 6], the idea of creating artificial electromagnetic materials from macroscopic elements made from usual materials is quite old, dating back to the nineteenth century.

One of the earliest interesting examples of artificial electromagnetic materials is related to the understanding and studies of optical activity and circular dichroism in chiral media. Optical activity is observed in materials whose microstructure lacks mirror symmetry. In natural materials, the effects of chirality are very weak because the molecules are very small as compared to the wavelength of light, and effects are not resonant. Moreover, in the nineteenth century, there were no means to experimentally study the relations between the geometrical symmetry of atoms and molecules and the optical properties of materials. However, after the works of H. Hertz, it became possible to work with radio and microwave frequency waves, and it was realized that it became possible to create artificial materials whose microwave properties could emulate the optical properties of natural materials. J. Bose did very interesting pioneer experimental studies of artificial chiral molecules and artificial chiral materials made of twisted jute bundles [4]; see the illustrations in Figure 3.2. In the beginning of the twentieth century, K. Lindman, in Finland, experimented with artificial chiral materials whose meta-atoms were made from metal springs embedded in cotton balls [7]. More information on the history of metamaterials research can be found, e.g., in [8]. Next, we discuss artificial materials with negative permittivity and permeability, the first metamaterial actively studied in the twenty-first century.

3.2 Double-Negative Materials

Double-negative materials are, by definition, isotropic materials obeying the constitutive relations

$$\mathbf{D} = \epsilon\mathbf{E}, \qquad \mathbf{B} = \mu\mathbf{H}, \tag{3.1}$$

Figure 3.3   First DNG/Veselago material. Reprinted from [9] with permission of AAAS.

where the material parameters ϵ and μ are real and negative. Since all materials are lossy, it is more accurate to say that both $\text{Re}\{\epsilon\} < 0$ and $\text{Re}\{\mu\} < 0$. These materials have many alternative names in addition to *double-negative (DNG) media*: materials with negative parameters, backward-wave media, materials with negative refraction index (NRI), left-handed materials, and Veselago media. The plurality of names perhaps reflects broad interest and interesting physical properties. The first experimental realization of double-negative materials (at microwaves, for a single polarization and in-plane propagation directions) was built using split rings and metal wires; see Figure 3.3.

As can be seen from Maxwell's equations for plane waves in form $e^{-j\mathbf{k}\cdot\mathbf{r}}$

$$\mathbf{k} \times \mathbf{E} = \omega\mu\mathbf{H}, \qquad \mathbf{k} \times \mathbf{H} = -\omega\epsilon\mathbf{E}, \tag{3.2}$$

plane waves in double-negative materials are *backward waves*; that is, the direction of the Poynting vector $\mathbf{S}$ is opposite to the wavevector $\mathbf{k}$:

$$\mathbf{S} = \frac{1}{2}\text{Re}(\mathbf{E} \times \mathbf{H}^*) = \frac{|E|^2}{2\omega\mu}\mathbf{k} = \frac{|H|^2}{2\omega\epsilon}\mathbf{k}. \tag{3.3}$$

This concept is illustrated in Figure 3.4.

Plane waves at an interface between vacuum or a usual (double-positive) isotropic medium and an isotropic double-negative half-space exhibit negative refraction, see Figure 3.5. This phenomenon can be understood from the continuity of the tangential component of the wavevector at the interface (which is necessary to make sure that the tangential field components are continuous across the interface). In the upper half-space filled with vacuum, both the Poynting vector and the wavevector are directed from the source (at infinity) to the interface. In the bottom half-space, the Poynting vector must be also directed down from the source. But the wavevector is now directed oppositely to

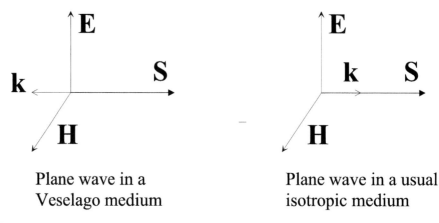

Plane wave in a
Veselago medium

Plane wave in a usual
isotropic medium

Figure 3.4 Forward and backward waves. Reproduced from [10] with permission. © 2003 Artech House, Inc.

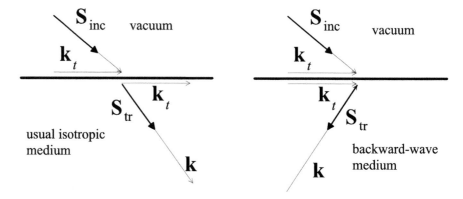

Figure 3.5 Negative refraction of plane waves. Reproduced from [10] with permission. © 2003 Artech House, Inc.

the Poynting vector; thus, the only possible configuration that ensures continuity of the tangential component of the wavevector is the configuration of negative refraction, as shown in Figure 3.5.

While this consideration appears to be trivial, the phenomenon of negative refraction is not so simple, and the first publications on experiments on negative refraction caused an intensive discussion. To understand this effect better, it is necessary to consider the propagation of pulses, restricted both in time and space. The result is illustrated in Figure 3.6, where is it shown that the beam gets distorted. Although the Poynting vector indeed exhibits negative refraction, the phase front is bent in the usual way.

Next, we will discuss surface plasmon modes at interfaces with double-negative materials. Here, we can make use of the theory of surface plasmon polaritons of Chapter 7. Let us recall the notions of propagating and evanescent waves. From (7.2), it is obvious that if $|k_t| \leq |k|$ (here, we consider, for simplicity, waves in media with negligible losses, where k is a real number, positive or negative), k_n is a real number, and waves propagate without amplitude decay. However, if $|k_t| > |k|$, k_n is an imaginary number, and waves exponentially decay in the direction of **n**, away from the surface (we consider a single wave in homogeneous

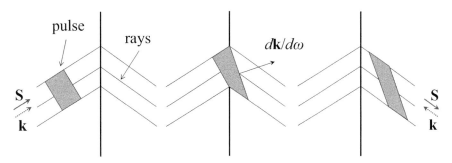

Figure 3.6 Propagation of a space-time modulated pulse: increasing moments of time, from left to right. Concept by S. Maslovski, courtesy of S. Maslovski and S. Tcvetkova.

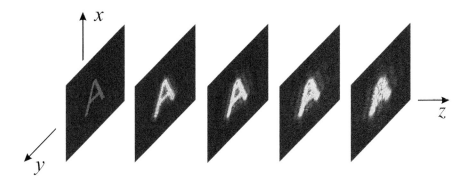

Figure 3.7 Image transformation in free-space propagation. Calculations and graphics by S. Maslovski.

space). Such waves are called *evanescent* waves. In problems of image transfer, we can envisage that a certain distribution of electric field, defined in a plane orthogonal to **n**, is expanded into the spatial Fourier integral, and $\mathbf{k}_t$ is the two-dimensional Fourier variable. Each of these components will propagate as a plane wave in space, carrying information about the image. However, only propagating components with small enough $|k_t| \leq |k|$ will propagate far away from the image plane. All evanescent waves will decay at distances considerably larger than the wavelength. This conclusion means that information about tiny details of the object will be lost. To resolve small details that have the size of the order a, we need to include in the Fourier spectrum harmonics of the order $k_t \approx 2\pi/a$. The smallest detail that will be retained at optically long distance from the source plane corresponds to $k_t \sim k$, meaning that the smallest $a \approx 2\pi/k = \lambda$. This limitation is called the *diffraction limit*.

Washing out image detail in free-space propagation can be considered analytically by calculating the corresponding Fourier integral

$$\mathbf{E}(x, y, z) = \frac{1}{4\pi^2} \int\limits_{-\infty}^{+\infty} \mathbf{E}(k_x, k_y) e^{-j(k_x x + k_y y \pm \sqrt{k^2 - k_x^2 - k_y^2} z)} \, dk_x \, dk_y, \qquad (3.4)$$

where x, y are the coordinates of the source and image planes, and wave propagation is along z (along **n**). An illustration is given in Figure 3.7.

Now we are ready to consider an interface between free space and a double-negative medium. We will write formulas for TM polarization. The other polarization is dual, and the results will be the same.

The reflection (R) and transmission (T) coefficients can be found in the usual way in terms of the wave impedances for tangential field components (requiring continuity). The result reads

$$R = \frac{\eta - \eta_0}{\eta + \eta_0}, \qquad T = \frac{2\eta}{\eta + \eta_0}, \tag{3.5}$$

where η_0 is the impedance in free space (see (7.3)) and η is the impedance in the double-negative medium. To find it, we should substitute negative values of the material parameters in (7.3):

$$\eta = \frac{k_n}{\omega\epsilon}, \qquad k_n = \sqrt{k^2 - k_t^2}. \tag{3.6}$$

The impedance in free space we write for the free-space parameters:

$$\eta_0 = \frac{k_n}{\omega\epsilon_0}, \qquad k_n = \sqrt{k_0^2 - k_t^2}. \tag{3.7}$$

Let us first consider propagating waves (real values of k_n). In a DNG medium, $\epsilon < 0$ and $\mu < 0$, and $k_n < 0$ (the backward-wave medium). It is important to note that if $\epsilon = -\epsilon_0$ and $\mu = -\mu_0$, we have $k = -k_0$, $k^2 = k_0^2$, $\eta = \eta_0$, and

$$R = 0, \qquad T = 1. \tag{3.8}$$

This result means that for these values of material parameters of the double-negative medium, we have total transmission and no reflection *for all angles of incidence*. Indeed, this result is independent from the value of k_t, as long as it is smaller than k (that is, for all propagating incident waves).

But if we excite the same interface with an evanescent wave (with $|k_t| > |k|$), the result is dramatically different. For evanescent waves,

$$k_n = \sqrt{k^2 - k_t^2} = -j\alpha, \quad k_n = \sqrt{k_0^2 - k_t^2} = -j\alpha_0, \quad \alpha > 0, \quad \alpha_0 > 0. \tag{3.9}$$

The square-root branch is chosen so that the fields in both media decay away from the source, which is located in free space. Furthermore, the impedances are also imaginary:

$$\eta = \frac{k_n}{\omega\epsilon} = \frac{-j\alpha}{\omega\epsilon}, \qquad \eta_0 = \frac{k_n}{\omega\epsilon_0} = \frac{-j\alpha_0}{\omega\epsilon_0}. \tag{3.10}$$

Now we can substitute these values in the formulas for the reflection and transmission coefficients (3.5) and analyze the results. Clearly, when $\epsilon = -\epsilon_0$ and $\mu = -\mu_0$, we have $\alpha = \alpha_0$ and purely imaginary wave impedances such that $\eta = -\eta_0$ for all k_t. A resonance occurs:

$$T = \frac{2\eta}{\eta + \eta_0} \to \infty, \qquad R = \frac{\eta - \eta_0}{\eta + \eta_0} \to \infty. \tag{3.11}$$

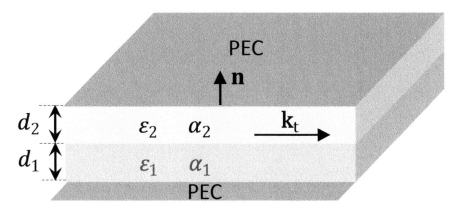

Figure 3.8 A pair of isotropic slabs between two perfectly conducting screens. Courtesy of A. Díaz-Rubio.

This resonance corresponds to existence of surface modes (surface plasmon-polaritons). In "usual structures" – for instance, for interfaces between metals and dielectrics, surface modes at a given frequency correspond to one or several particular values of the propagation constant k_t. In this case, however, we have a unique situation when the resonance condition is satisfied *identically for all* $k_t > k$. In other words, at the frequency where $\epsilon = -\epsilon_0$ and $\mu = -\mu_0$, the dispersion equation is satisfied for all wavenumbers of surface states. This unique property has important implications, and it is the basis of the so-called perfect lens concept; see Section 3.3.

As an example, let us consider a planar double-layer bounded on both sides by perfect mirrors, Figure 3.8. The dispersion equation for waves between the two mirrors can be derived by requiring the continuity of the tangential fields at the interface between the two slabs. The result reads

$$\frac{k_{n1}}{\epsilon_1} \tan k_{n1} d_1 + \frac{k_{n2}}{\epsilon_2} \tan k_{n2} d_2 = 0, \qquad \text{TM modes}, \qquad (3.12)$$

$$\frac{\mu_1}{k_{n1}} \tan k_{n1} d_1 + \frac{\mu_2}{k_{n2}} \tan k_{n2} d_2 = 0, \qquad \text{TE modes}, \qquad (3.13)$$

where $k_{n1,2} = \sqrt{k_{1,2}^2 - k_t^2}$. The indices $1, 2$ refer to the parameters of the two slabs.

Let us first study waves traveling along $\mathbf{n}$ (standing waves between the two metal boundaries) and let $\mathbf{k}_t = 0$ in (3.12) and (3.13). The eigenvalue equation simplifies to

$$\frac{\mu_1}{k_1} \tan k_1 d_1 + \frac{\mu_2}{k_2} \tan k_2 d_2 = 0. \qquad (3.14)$$

In particular, for thin layers, retaining the first terms of the Taylor expansions, we get $\mu_1 d_1 + \mu_2 d_2 = 0$. This is the resonance condition for standing waves bouncing between the two mirrors, but note that if one of the slabs is a double-negative one, the distance between the two mirrors can now be as small as we like [11], very much in contrast to usual resonators, whose minimum size equals half the

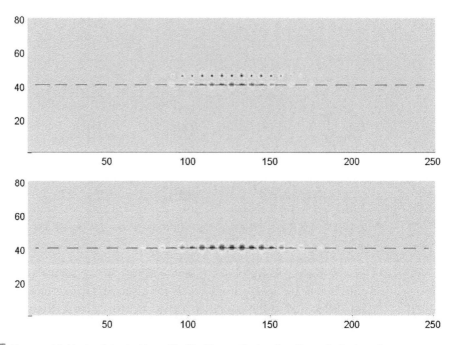

Figure 3.9　Evanescent fields stored at a double-positive/double-negative interface. Top: excitation by a given current source. Bottom: stored field distribution. Republished from [12] with permission of John Wiley and Sons, Inc.

wavelength. Physically, it happens because in one slab, waves experience phase advance, but in the other one, there is phase delay, and they can compensate each other.

Let us now study the waves with a non-zero tangential wavenumber $\mathbf{k}_t$ and consider the case when the material parameters of one of the slabs are negative of those of the other: $\epsilon_2 = -\epsilon_1$, $\mu_2 = -\mu_1$, and the first slab (1) is double-positive. Since the second slab is double-negative, we have $k_{n2} = -k_{n1}$. Analyzing Eqs. (3.12) and (3.13), we see that, in this case, both of them are satisfied identically for *all arbitrary* values of $\mathbf{k}_t$. This property has the same physical reason as the existence of surface modes with arbitrary tangential wavenumbers at a single interface.

An interesting property of "freezing" near-field distributions [12] is illustrated in Figure 3.9. Here, surface modes are excited by a current distribution with small, subwavelength detail in the transverse plane. At some moment of time, the excitation is switched off. However, the surface-mode oscillations induced at the resonant interface stay there for a long time, until dissipation consumes the stored reactive energy. It is important that the spatial profile of these oscillations is determined by the source, whereas in a conventional resonator, the field pattern of free oscillations of an excited eigenmode is determined by the resonator shape. Thus, one can say that the device "remembers" the information encoded in the spatial distribution of exciting currents.

3.3 Perfect Lens

The concept of perfect lens as a slab with the refraction index negative to that of the surrounding space was introduced by J. Pendry [13]. Actually, the focusing of propagating waves due to negative refraction on both interfaces was discussed already in 1968 in the review paper by V. Veselago [14], but the perfect focusing of evanescent modes was not noticed at that time.

The geometrical configuration of a planar perfect lens excited by a point source is shown in Figure 3.10. The slab is of infinite extent in the xy-plane. To understand the lens properties, we expand the incident field, created by the source, into the Fourier spectrum of plane waves. The focusing of propagating waves can be understood from the ray picture shown in this figure. The arrows indicate the direction of the wavevector. The wavevector inside the negative-index slab points toward the source because it is opposite to the Poynting vector direction, but the Poynting vector must always be directed *away* from the source.

Let us analyze the response to incident evanescent modes. Considering, for simplicity, excitation by a current line (two-dimensional problem) and directing the axis y along the current source, each evanescent plane-wave component has the fields of the form

$$\mathbf{E} = E\mathbf{y}_0 e^{-jk_x x - \alpha_0 z}, \qquad H_x = -\frac{\alpha_0}{j\omega\mu_0} E_y, \tag{3.15}$$

where $\alpha_0 = \sqrt{k_x^2 - k_0^2} > 0$.

In the usual way, looking for a solution inside the slab as a sum of two waves, decaying in the opposite directions, we can find their amplitudes satisfying the

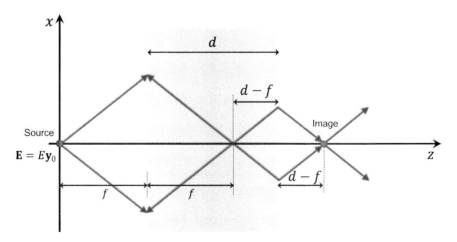

Figure 3.10 Perfect lens as a slab of a lossless negative-index material. Courtesy of A. Díaz-Rubio.

boundary conditions on the two interfaces and determine the reflection and transmission coefficients for each plane-wave component of the incident field:

$$R = \frac{\frac{1}{2}\left(\frac{\alpha_0\mu}{\alpha\mu_0} - \frac{\alpha\mu_0}{\alpha_0\mu}\right)\sinh\alpha d}{\cosh\alpha d + \frac{1}{2}\left(\frac{\alpha_0\mu}{\alpha\mu_0} + \frac{\alpha\mu_0}{\alpha_0\mu}\right)\sinh\alpha d},$$
(3.16)

$$T = \frac{1}{\cosh\alpha d + \frac{1}{2}\left(\frac{\alpha_0\mu}{\alpha\mu_0} + \frac{\alpha\mu_0}{\alpha_0\mu}\right)\sinh\alpha d}.$$
(3.17)

For $\epsilon = -\epsilon_0$ and $\mu = -\mu_0$, we get $R = 0$ and $T = e^{\alpha d}$. The total distance between the source and the image point is $2d$. One half of this distance is in free space, where the evanescent fields exponentially decay as $e^{-\alpha d}$. But as we have just found, evanescent fields inside the lens exponentially *grow* in exactly the same proportion as $e^{+\alpha d}$. This result holds for arbitrary values of α. Thus, the amplitude and phase of every plane-wave component of the incident field at the source plane is exactly the same as in the image plane. This conclusion means that the field distribution in the image plane is exactly the same as in the source plane: the lens is theoretically perfect.

The two phenomena that enable theoretically perfect focusing are illustrated in Figure 3.11. The special values of the material parameters $\epsilon_r = -1$ $\mu_r = -1$ correspond to a medium whose refraction index $n = \sqrt{\epsilon_r\mu_r} = -1$ while the wave impedance is the same as in free space. Propagating plane waves exhibit negative refraction without any reflection while plasmon-polaritons of an interface with free space resonate for any $k_t > k_0$, so the reflection coefficient from an interface with such negative-index half-space $R_{half} \to \infty$.

This result, however, is an idealization based on a number of unrealistic assumptions. The material filling the lens is assumed to be perfectly lossless while its relative permittivity and permeability are equal to exactly -1. Furthermore, the slab is assumed to be infinite in the transverse plane, and, what is more important, the effective material model is supposed to be valid for arbitrarily fast variations of fields in space. Actually, the DNG-material lens resolution is fundamentally limited by the granularity – finite sizes of the unit cells of the metamaterial structure, even for ideally lossless and infinite slabs.

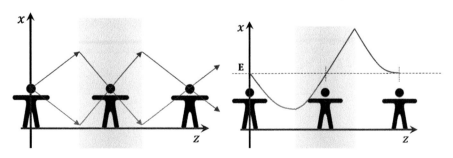

Figure 3.11 Two enabling phenomena for perfect-lens operation: negative refraction and surface plasmon-polariton resonance. Courtesy of A. Díaz-Rubio.

In the ideal scenario that we have just studied, even an infinitely thin line is imaged perfectly in the image plane. The distribution of fields in the source plane is singular, as the fields tend to infinity in the vicinity of a *point* or a *line* source. Thus, the field distribution is singular also in the image plane, but there is no source there. It means that the image is not a full two-dimensional or three-dimensional image of the source. Actually, while the fields *behind* the image plane are the same as that of the imaged source, the field between the image plane and the back side of the lens are very different. Actually, the evanescent fields decay from the back side of the lens toward the image plane (that is, they *grow* away from the image point), while in the vicinity of the actual source, evanescent modes decay along all directions away from the source. Now imagine that we have a perfect image of a line current. The evanescent field is infinite at the image point (since it is the same as at the source point). And this infinite field is exponentially *growing* toward the back side of the lens. Thus, the field is infinite everywhere between the lens and the image plane. And it is "infinitely more infinite" at the back side of the lens as compared to the field in the image plane or at the point-source position. The reactive energy stored in the lens is, of course, also infinite. In reality, some limited subwavelength resolution is achievable only for rather small focal distances, usually smaller than the wavelength, and in a very narrow frequency range, because negative values of material parameters in passive media imply strong frequency dispersion. The most critical factor is the level of dissipative losses in the lens material, since the key effect of the perfect lens – the effect of evanescent wave restoration – is due to a high-quality resonance of the system. Unfortunately, practical realizations of deeply subwavelength resolution with passive bulk double-negative metamaterials appear not to be realistic mainly because the needed resonance quality is not achievable.

3.4 Engineered and Extreme Material Parameters

The main appeal of the metamaterial concept is a possibility to create new materials with desired electromagnetic properties. Most interesting, naturally, are artificial materials with unusual and exotic properties: such media that we cannot find in nature or synthesize chemically. Basically, one works with usual materials, like metals or dielectrics, and forms electrically (optically) small particles made of usual materials, but showing desired polarizability properties due to their special shape, mutual arrangement, or external control.

3.4.1 Artificial Dielectrics

Probably the simplest example of artificial materials is the artificial dielectric. This is also arguably the earliest example of a practical use of metamaterials in electrical engineering (introduced by W. E. Kock in 1948 [15]). These metamaterials are used in the microwave range to replace bulky and heavy natural or chemically synthesized dielectric materials with dense arrays of small metal spheres or disks. The main idea behind this design is the use of electrically small but strongly polarizable metal particles (meta-atoms) to replace weakly polarizable molecules of natural dielectrics. An electric field at the position of a particular (reference) particle in the absence of this reference particle $\mathbf{E}_{\text{loc}}$ induces electric dipole moment

$$\mathbf{p} = \alpha_{\text{ee}} \mathbf{E}_{\text{loc}} \qquad (3.18)$$

when we bring the reference particle to its place. In other words, the induced dipole moment is linearly proportional to the exciting field. Here, for simplicity, we consider isotropic spherical particles. The proportionality coefficient is called the particle *polarizability*. If there are N spheres per unit volume, the polarization vector (the dipole moment of a unit volume) reads

$$\mathbf{P} = N\mathbf{p} = N\alpha_{\text{ee}} \mathbf{E}_{\text{loc}}. \qquad (3.19)$$

Using the Lorenz–Lorentz relation (e.g., [10]) between the local ($\mathbf{E}_{\text{loc}}$) and the volume-averaged macroscopic field ($\mathbf{E}$):

$$\mathbf{E}_{\text{loc}} = \mathbf{E} - \frac{\mathbf{P}}{3\epsilon_0}, \qquad (3.20)$$

where ϵ_0 is the permittivity of the host medium (not necessarily free space), we can find the effective permittivity of the metamaterial formed by a dense arrangement of small spheres. The volume averaging is made over an electrically small volume still containing many meta-atoms. The depolarization factor in the Lorenz–Lorentz formula connecting the macroscopic and local fields (1/3 in this case) comes from averaging the field of one inclusion over its own unit cell [10]. This effective permittivity is defined in the usual way via the relation

$$\mathbf{D} = \epsilon_0 \mathbf{E} + \mathbf{P} = \epsilon_{\text{eff}} \mathbf{E}. \qquad (3.21)$$

The result, obtained upon substitution of (3.19) and (3.20), reads

$$\epsilon_{\text{eff}} = \epsilon_0 + \frac{N\alpha_{\text{ee}}}{1 - \frac{N\alpha_{\text{ee}}}{3\epsilon_0}}. \qquad (3.22)$$

Note that this result is an approximation, where in addition to the quasi-static assumption the calculation of the interaction field between all the inclusions is based on the volume-averaging of the electric field of one inclusion over its unit cell (a detailed discussion of these approximations can be found in [10, chapter 5]). Other approximate expressions for the effective permittivity of mixtures formed by small particles exist, each with its own domain of applicability; see, e.g., [16]. More accurate, dynamic models are discussed, e.g., in [10] and [17, chapter 12].

Formula (3.22) tells that, tuning the value of the inclusion polarizability α_{ee} and/or the concentration of inclusions N, it is possible to vary the effective permittivity of the material theoretically without any limit: in the vicinity of the point where $N\alpha_{ee}$ is close to $3\epsilon_0$, the effective permittivity exhibits resonant behavior with respect to changing $N\alpha_{ee}$, with a swing from $-\infty$ to $+\infty$.

For small spheres, we can use the well-known quasi-static formula for the sphere polarizability

$$\alpha_{ee} = V(\epsilon - \epsilon_0)\frac{3\epsilon_0}{\epsilon + 2\epsilon_0}, \tag{3.23}$$

where V is the sphere volume and ϵ is the permittivity of the sphere material. To find the polarizability of a small perfectly conducting sphere, we can take the limit of infinite permittivity of the sphere material, which gives

$$\alpha_{ee}|_{\text{PEC}} = 3V\epsilon_0. \tag{3.24}$$

In this case, the product $N\alpha_{ee}|_{\text{PEC}} = 3\epsilon_0 f$, where $f = NV$ is the volume fraction (the part of the volume occupied by metal particles), and the effective permittivity is adjusted simply by changing the volume fraction. For very low volume fractions, the effective permittivity is close to that of the host medium, and it grows when the volume fraction increases. Obviously, in the considered low-frequency (radio and microwaves) regime and electrically small and non-resonant particles, the effective permittivity does not depend on the frequency.

Much more general and interesting possibilities open up if some electrically small but resonant inclusions are used. In nanophotonics, one can use plasmonic nanoparticles (see Chapter 7) or high-permittivity dielectric particles. At microwaves, the natural and most common approach is to use small particles made of metal strips or wires that are shaped so that they resonate at frequencies where their overall size is still electrically small. A classical example is a half-wave resonant dipole that is shaped as a meander line. While the total ("stretched") length is close to half of the wavelength – ensuring strong, resonant response – the length and width of the meandered strip or wire is much smaller. In this book, we use this example in Chapter 4 to demonstrate extraordinary reflection and transmission phenomena. Another example is a cap-loaded dipole.

Figure 3.12 shows several examples of metal-strip shapes suitable for realization of small but resonant particles. These examples are for interactions with vertically polarized electric fields, as shown in the diagram. To create isotropic composites, it is possible to modify the shapes or use mixtures of equal numbers of particles oriented along orthogonal directions.

Meta-atom properties can be engineered and made tunable or time-modulated by introducing bulk components (inductors, varactors, or other electronic or photonic components, as shown in the right of Figure 3.12).

Figure 3.12 Examples of small but resonant metal particles for microwave metamaterials. Courtesy of F. Cuesta.

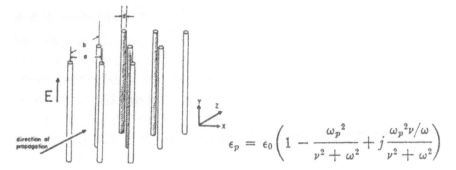

$$\epsilon_p = \epsilon_0 \left(1 - \frac{\omega_p^2}{\nu^2 + \omega^2} + j \frac{\omega_p^2 \nu/\omega}{\nu^2 + \omega^2} \right)$$

Figure 3.13 Wire medium and its effective permittivity. Reprinted from [18] with permission. © 1962 IEEE.

3.4.2 Wire Media

A possible route toward realization of artificial materials having negative or near-zero values of permittivity (artificial plasma) is the use of wire media. While artificial dielectrics formed by resonant inclusions can provide effective negative permittivity, but only close to the resonant frequency. Wire media have an advantage of weaker frequency dispersion. In the visible and infrared ranges, the real part of permittivity of noble metals is negative (the property utilized in plasmonics), but for applications at lower frequencies, metals are "too metallic" – meaning that the imaginary part of the permittivity, which is due to conductance, strongly dominates over the real part. The wire medium is an artificial material that behaves as a negative-permittivity material in desired frequency ranges, with controllable values of the permittivity.

Wire media were introduced in 1950s and 1960s as artificial dielectrics with the permittivity smaller than unity [19] or negative [18]; see Figure 3.13. The primary motivation was the need of such materials in the design of microwave lenses. Let us explain why the effective permittivity can be negative in wire media using a simple quasi-static model [20].

Wire Media as Artificial Plasma

The geometry of the problem is shown in Figure 3.14. The electric field inside the structure is polarized vertically, along the wires, and we assume that no waves travel along the wires; propagation is in the transverse plane. The voltage between the two imaginary planes orthogonal to the wires is connected to the wire current via the effective inductance: $U = Ij\omega L d$. Since $U = E_z d$, we have $E_z = j\omega L I$. Now we can find the effective permittivity, writing the definition $\mathbf{D} = \epsilon_0 \mathbf{E} + \mathbf{P}$, where the polarization density $\mathbf{P} = \mathbf{J}/j\omega = \mathbf{z}_0 I/j\omega a^2 = -\mathbf{z}_0 E_z/\omega^2 a^2 L$. Thus, the material relation for z-directed electric field and polarization takes the form

$$D_z = \left(\epsilon_0 - \frac{1}{\omega^2 a^2 L} \right) E_z. \tag{3.25}$$

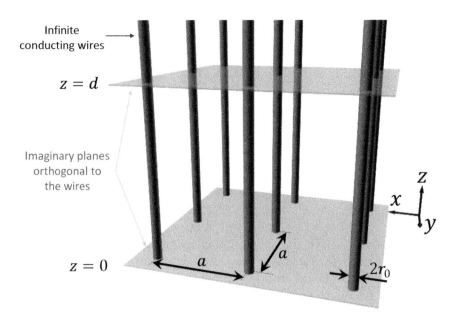

Infinite
conducting wires

$z = d$

Imaginary planes
orthogonal to
the wires

$z = 0$

Figure 3.14 Geometry of a wire medium sample: an array of parallel thin round conducting wires. Courtesy of A. Díaz-Rubio.

That is, the permittivity has the Drude dispersion form:

$$\epsilon_{zz} = \epsilon_0 \left(1 - \frac{\omega_p^2}{\omega^2} \right), \qquad \omega_p^2 = \frac{1}{a^2 \epsilon_0 L}. \tag{3.26}$$

To find the plasma frequency, we need to estimate the inductance per unit length L. Let us approximate the magnetic field distribution between any pair of wires as

$$H_y = \frac{I}{2\pi} \left(\frac{1}{x} - \frac{1}{a - x} \right). \tag{3.27}$$

Magnetic flux per unit length reads

$$\Psi = \mu_0 \int_{r_0}^{a/2} H_y(x)\, dx = \frac{\mu_0 I}{2\pi} \log \frac{a^2}{4 r_0 (a - r_0)}. \tag{3.28}$$

Finally, the inductance per unit length equals

$$L = \frac{\mu_0}{2\pi} \log \frac{a^2}{4 r_0 (a - r_0)}, \tag{3.29}$$

which defines the plasma frequency in terms of the geometrical parameters:

$$\omega_p^2 = \frac{2\pi}{a^2 \epsilon_0 \mu_0 \log \frac{a^2}{4 r_0 (a - r_0)}}. \tag{3.30}$$

Strong Spatial Dispersion

The derivation of formula (3.26)

$$\epsilon_{zz} = \epsilon_0 \left(1 - \frac{\omega_p^2}{\omega^2} \right) = \epsilon_0 \left(1 - \frac{k_p^2}{k^2} \right) \tag{3.31}$$

(here, $k = \omega\sqrt{\epsilon_0\mu_0}$ and $k_p = \omega_p\sqrt{\epsilon_0\mu_0}$) assumes that the fields do not change along the wire axis. However, it is obvious that waves can propagate along this array of thin conducting wires. The structure can be viewed as a multi-conductor transmission line (each pair of wires can function as a two-wire transmission line). Thus, if currents in the wires are excited at some points, these excitations will propagate along the wires at long distances *without decay* (neglecting dissipation in conductors). In terms of the effective medium response, this property is equivalent to strong, long-distance *spatial dispersion*, because the polarization vector at a certain point depends not only on the electric field at the same point but also on the fields at other points along the axis of the wire array. With this effect taken into account, the effective permittivity takes the form [21]

$$\epsilon_{zz}(\omega, k_z) = \epsilon_0 \left(1 - \frac{k_p^2}{k^2 - k_z^2} \right). \tag{3.32}$$

This formula is valid for field solutions in the form of plane waves. Here, k_z is the z-component of the propagation constant. In contrast to natural media with spatial dispersion, the terms that depend on the propagation constant are not small corrections. On the contrary, this function even diverges if the wave propagates along the wires. In the time domain, the corresponding relation (now valid not only for plane waves), reads [21]

$$\mathbf{D}(x,y,z,t) = \epsilon_0 \mathbf{E}(x,y,z,t) + \frac{\epsilon_0 k_p^2 c}{4} \mathbf{z}_0 \int_{-\infty}^{t} \int_{z-c(t-t')}^{z+c(t-t')} E_z(x,y,z',t')\, dz'\, dt', \tag{3.33}$$

which again shows that the induced polarization responds to the electric field in a strongly nonlocal fashion.

This property of strong spatial dispersion has important applications. Indeed, waves propagating along dense arrays of conductors allow transport of images at electrically long distances with a subwavelength resolution (the resolution is defined not by the wavelength but by the period of the wire array, which can be made much smaller than the wavelength). Experimental results on image transport can be found, for example, in [22]; see an illustration in Figure 3.15.

As we see, wire media offer a unique platform to control electric properties of artificial materials – including strong, long-distance spatial dispersion. A broad review of electromagnetic properties of wire media and their applications can be found in paper [23].

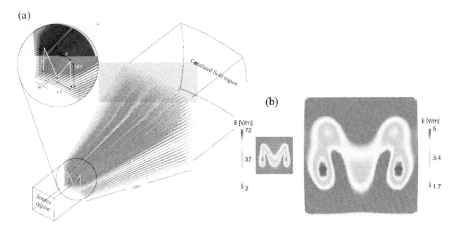

Figure 3.15 Enlarging superlens formed by an array of diverging metal wires. Reprinted from [22] with the permission of AIP Publishing.

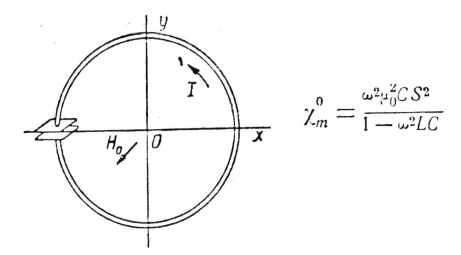

$$\chi_m^0 = \frac{\omega^2 \mu_0^2 C S^2}{1 - \omega^2 LC}$$

Figure 3.16 Split-ring resonator for microwave frequencies. Republished from [24] with permission of John Wiley and Sons, Inc.; permission conveyed through Copyright Clearance Center, Inc.

3.4.3 Split Rings

To control electric polarization, we use meta-atoms where induced currents flow predominantly along straight lines, forming electric dipoles. Obviously, in order to engineer artificial magnetic response, we should force the induced electric currents to form loops so that magnetic moments would be induced by applied time-varying magnetic fields. The most natural choice of the inclusion shape is a split ring.

Artificial magnetic meta-atoms were proposed in the 1950s for microwave applications [24]; see Figure 3.16. Later, strongly coupled pairs of rings were proposed [25–27]; see Figure 3.17. In double rings, the sum of the currents induced in the two rings forms a nearly uniform current loop, reducing usually undesirable

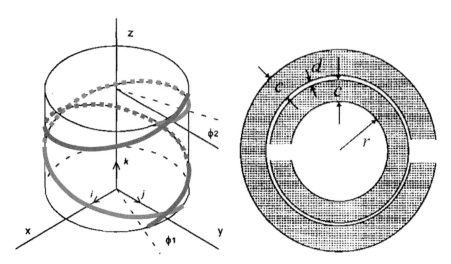

Figure 3.17 Typical topologies of double split rings. Left: adapted from [26] with permission by Taylor & Francis Ltd. Right: reprinted from [27] with permission. © 1999 IEEE.

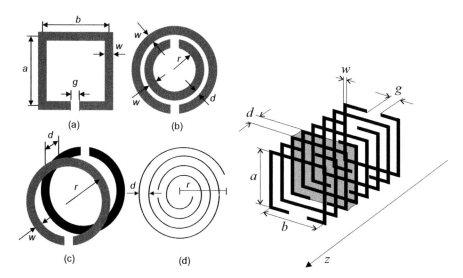

Figure 3.18 Various topologies of magnetic meta-atoms. (a) Single split ring [24]; (b) Double split ring [27]; (c) Two rings of identical sizes (suppressed bianisotropy); (d) Swiss roll. Right: Metasolenoid [28, 29]. Reproduced courtesy of The Electromagnetics Academy.

bianisotropic coupling (in bianisotropic meta-atoms applied magnetic field at frequencies close to the fundamental resonance induces not only a magnetic dipole but also an electric dipole).

A great variety of magnetically polarizable metal and dielectric particles have been proposed and studied. Some of them are shown in Figure 3.18. The magnetization effects in all of them can be understood using a simple circuit model, which we will consider next.

In order to ensure a non-zero magnetic moment of induced current distribution, we need to create loop currents. The current induced in a loop-shaped conductor is driven by the applied electromotive force, which is equal to the time-derivative of the external magnetic flux $-j\omega BS$. Here, B is the normal to the loop plane component of the external magnetic field, and S is the loop area. We assume that the loop is either closed or has a small gap or several small gaps so that the loop area is well defined. If the loop is closed and losses are small, the loop impedance is inductive, and the induced current amplitude reads

$$I = \frac{-j\omega SB}{j\omega L} = -\frac{SB}{L},$$

(3.34)

where L is the loop inductance. The induced magnetic moment is proportional to the induced current: $m = \mu_0 IS$. As we see, for closed loops the induced moment is out of phase with the applied magnetic field. Thus, using closed loops, we can realize only artificial diamagnetics, with the effective permeability $0 < \mu < \mu_0$. In order to control the induced current phase, the loop is made open. The small gap is modeled by an equivalent capacitance C, in series with the loop inductance. The induced current becomes

$$I = \frac{-j\omega SB}{j\omega L + \frac{1}{j\omega C}} = \frac{\omega^2 CSB}{1 - \omega^2 LC}.$$

(3.35)

The magnetic moment resonantly depends on the frequency and can take large values, both positive and negative, which opens a way to creation of metamaterials with desired magnetic response.

Let us look at the limiting values of the induced magnetic moment at very low and very high frequencies. The low-frequency limit at $\omega \ll \omega_0$ gives $I \approx \omega^2 CSB$, $m \approx \omega^2 \mu_0 CS^2 B$. This tells us that for all artificial magnetics, the magnetic response disappears in the static limit (proportionally to ω^2). In the high-frequency limit, at $\omega \rightarrow \infty$, we get $m \rightarrow -\mu_0 S^2 B/L$. This is an obviously nonphysical result because at extremely high frequencies, materials are not polarizable at all, due to inertia of electrons. This result indicates that the circuit model becomes invalid at high frequencies where the quasi-static assumptions are not satisfied.

Generalizing this approach, we can write, for any of the various particles shown in Figures 3.17 and 3.18, the same expressions for the magnetic polarizability. The induced current

$$I = \frac{\mathcal{E}^{\text{ext}}}{Z_{\text{tot}}} = \frac{\mathcal{E}^{\text{ext}}}{j\omega L_{\text{eff}} + \dfrac{1}{j\omega C_{\text{eff}}} + R_{\text{eff}}}$$

(3.36)

is determined by the applied electromotive force $\mathcal{E}^{\text{ext}}$ and the effective circuit parameters that depend on the particle shape, size, and material. The induced magnetic moment reads

$$m = \mu_0 nSI = \frac{\omega^2 \mu_0^2 n^2 S^2 C_{\text{eff}} H^{\text{ext}}}{1 - \omega^2 L_{\text{eff}} C_{\text{eff}} + j\omega R_{\text{eff}} C_{\text{eff}}},$$

(3.37)

where n is the number of loops in one meta-atom. Finally, the magnetic polarizability reads

$$\alpha_{mm} = \frac{m}{H^{\text{ext}}} = \frac{\omega^2 \mu_0^2 n^2 S^2 C_{\text{eff}}}{1 - \omega^2 L_{\text{eff}} C_{\text{eff}} + j\omega R_{\text{eff}} C_{\text{eff}}}. \tag{3.38}$$

Knowing the magnetic polarizability of a single particle, it is possible to use effective media models to estimate the effective permeability of metamaterials formed by many particles randomly or regularly distributed in a nonmagnetic matrix. The electromagnetic duality allows us to rewrite the formula for the effective permittivity of electrically polarizable particles (3.22) as a formula for the effective permeability of magnetically polarizable particles:

$$\mu_{\text{eff}} = \mu_0 + \frac{N\alpha_{mm}}{1 - \frac{N\alpha_{mm}}{3\mu_0}}. \tag{3.39}$$

Similarly to (3.22), here, N is the number of particles per unit volume of the composite.

The resonant frequency of the particles can be controlled by the particle size and shape. Considering an example of a single split ring (Figure 3.18(a)), we can estimate the inductance and capacitance as functions of the particle sizes. Assuming that the loop material is perfectly conducting, the inductance is proportional to the loop radius a, while the capacitance is proportional to the gap area w^2 (w is the diameter to the wire or strip) and inversely proportional to the gap width δ (see also Figure 3.19). Neglecting slow-varying logarithmic terms, we get an estimate for the resonant frequency

$$\omega_0 = \frac{1}{\sqrt{LC}} \approx \frac{1}{\sqrt{\epsilon_0 \mu_0 \frac{aw^2}{\delta}}}. \tag{3.40}$$

We see that if we keep the particle shape the same, assuming that the ratio w/δ is constant, and reduce the particle size, letting a and w tend to zero at the same rate ($a/w =$ const), then the resonant frequency is expected to increase as ($1/a$) when $a \to 0$.

Thus, it appears that only technological limitations of fabrication of extremely small particles limit the operational frequency of artificial magnetics. However, the assumption of perfect conductivity of metals does not hold at high frequencies (THz and beyond), which leads to other important limitations of these designs. If metals are not perfect conductors, fields penetrate inside metal,

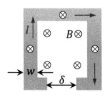 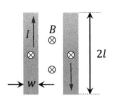 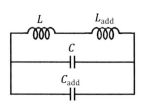

Figure 3.19 Split ring (left) and dual bars (center) at infrared and visible frequencies. Right: the equivalent circuit. Courtesy of A. Díaz-Rubio.

and the particle inductance is determined not only by the magnetic flux outside of the metal loop (the flux through the loop area S) but also by the flux inside metal wires or strips. The same is true for the total capacitance, because electric fields also are not zero inside non-ideal metals. One can expect that this effect will *reduce* the resonant frequency, since the total flux is larger for the same current in the loop, possibly preventing the creation of resonant magnetic particles for very high operational frequencies [30–32].

Let us estimate the effects of finite metal conductivity on the resonant frequency of split rings [32]. Assuming that the ring is made from a metal with Drude permittivity, we can write the induced current density inside metal as

$$\mathbf{J} = j\omega\mathbf{D} = j\omega\epsilon_0\left(\epsilon_r - \frac{\omega_p^2}{\omega^2}\right)\mathbf{E}. \tag{3.41}$$

The total current through the wire cross section is $I = Jwh$, and the voltage across the effective wire length $l_{\rm eff}$ is $V = El_{\rm eff}$. Thus, we have a relation between the current and voltage in form

$$I = \left(j\omega\frac{\epsilon_0\epsilon_r wh}{l_{\rm eff}} + \frac{\epsilon_0 wh\omega_p^2}{j\omega l_{\rm eff}}\right)V = \left(j\omega C_{\rm add} + \frac{1}{j\omega L_{\rm add}}\right)V, \tag{3.42}$$

which determines the *additional* inductance and capacitance of the particle due to the fields inside metal:

$$C_{\rm add} = \frac{\epsilon_0\epsilon_r wh}{l_{\rm eff}}, \qquad L_{\rm add} = \frac{l_{\rm eff}}{\epsilon_0 wh\omega_p^2}. \tag{3.43}$$

Here, h is the thickness of the metal strips (wh is the cross-section area of the metal conductor). Next, we estimate the "conventional," geometrical inductance and capacitance due to the fields in space, outside of the metal parts. We use the well-known formulas for the loop inductance (the loop radius a and the wire radius r_0)

$$L = \mu_0 a\ln\frac{8a}{r_0}, \qquad L \approx \mu_0\frac{l}{4}\ln\frac{8l}{w+h}. \tag{3.44}$$

Here, we have related the round loop radius to the effective length of the square loop $a \approx l_{\rm eff}/(2\pi) = l/4$ and substituted the equivalent wire radius $r_0 \approx (h+w)/4$. The gap capacitance we estimate as the parallel-plate capacitance

$$C \approx \epsilon_0\frac{wh}{\delta}. \tag{3.45}$$

Collecting these results, we find an estimate of the resonance frequency

$$\omega_0 = \frac{1}{\sqrt{(L+L_{\rm add})(C+C_{\rm add})}} \approx \frac{1}{\sqrt{\left(1 + \frac{\pi}{2}\frac{l}{\delta}\right)\left(\mu_0\epsilon_0\frac{wh}{2\pi}\ln\frac{8l}{w+h} + \frac{1}{\omega_p^2}\right)}}. \tag{3.46}$$

We see that the resonance frequency does not grow without limit if we decrease the loop sizes (so that $wh \to 0$). Instead, it saturates at a limit that is determined by the plasma frequency of metal. As follows from (3.46), the limiting value of the resonant frequency of very small loops equals to

$$\lim\{\omega_0\} = \frac{\omega_p}{\sqrt{\epsilon_r + \frac{\pi}{2}\frac{l}{\delta}}}. \tag{3.47}$$

The ultimate limit, which corresponds to the situation when all the fields are only in metal so that the external inductance is zero, is given by

$$\max\{\omega_0\} = \frac{1}{\sqrt{C_{add}L_{add}}} = \omega_p. \tag{3.48}$$

This limit is achieved if $\lim\left\{\frac{l}{\delta}\right\} = 0$, assuming that $\epsilon_r \to 1$ at extremely high frequencies.

3.5 Engineering Bianisotropy

We have just learned how to control electric polarization induced by electric fields (artificial dielectrics) and magnetic polarization induced by magnetic fields (artificial magnetics). Let us now discuss how one can control magneto-electric coupling effects. Magneto-electric coupling means that applied electric fields induce some magnetic polarization and applied magnetic fields induce some electric polarization so that the material relations take the general bianisotropic form (2.1). On the level of single meta-atoms, it means that the induced electric and magnetic moments obey

$$\mathbf{p} = \bar{\bar{\alpha}}_{ee} \cdot \mathbf{E} + \bar{\bar{\alpha}}_{em} \cdot \mathbf{H}, \tag{3.49}$$

$$\mathbf{m} = \bar{\bar{\alpha}}_{mm} \cdot \mathbf{H} + \bar{\bar{\alpha}}_{me} \cdot \mathbf{E}. \tag{3.50}$$

There are two physically distinct mechanisms of magneto-electric coupling: spatial dispersion and nonreciprocal effects [33]. Spatial dispersion effects appear when the meta-atom size is small but not negligible in comparison with the wavelength. Basically, in this case, the induced current depends not only on the value of the exciting electric field at the particle location but also on its spatial derivatives. In the isotropic scenario, where there is no preferred direction in space, spatial derivatives of electric field come only in the form of multiplication by ∇, and we come to the conclusion that the induced current has a component that is proportional to $\nabla \times \mathbf{E}$, which is in view of the corresponding Maxwell's equation – tantamount to excitation by external magnetic field.

Note that the artificial magnetism of split rings and other similar particles has the same physical nature. Second-order terms in the Taylor expansion of the induced electric polarization as a function of spatially inhomogeneous electric field in isotropic media contain a second-order term:

$$\mathbf{D} = \epsilon \mathbf{E} + \cdots + \gamma \nabla \times (\nabla \times \mathbf{E}), \tag{3.51}$$

where γ is an additional material parameter. The use of such a material relation is very inconvenient due to the presence of spatial derivatives. Fortunately, this nonlocal polarization effect is equivalent to artificial magnetism, and it is possible to introduce an equivalent description in terms of effective permeability [33]. In this equivalent description, the material relations do not contain spatial derivatives, and the conventional methods for solving Maxwell's equations (using the usual boundary conditions of continuity of tangential fields) can be used.

It is important to note that there are other terms of the same order. Namely, in isotropic structures, the expansion of $\mathbf{D}$ also contains the term $\beta \nabla \nabla \cdot \mathbf{E}$ (where β is another material parameter), and effects due to the presence of this term are not possible to model by magnetic polarizability [33]. At this time, there is no complete theory of weak spatial dispersion, which would include all second-order effects. Usually, the shape of the meta-atoms is chosen so that the artificial magnetism dominates over other effects. This is the case of double split rings, for example.

Nonreciprocal mechanisms of magneto-electric coupling require presence of some external time-odd fields, like magnetic bias, and can be realized, for example, using magnetized ferrite particles [33].

Here, we will discuss only reciprocal bianisotropic coupling. The nature and strength of reciprocal coupling phenomena are determined by the particle *shape* and sizes. The coupling coefficients $\overline{\overline{\alpha}}_{\mathrm{em}}$ and $\overline{\overline{\alpha}}_{\mathrm{me}}$ are dyadics, and it is convenient to consider engineering of the symmetric and antisymmetric parts of these dyadics separately.[1] This is because the symmetric coupling basically means that the induced electric dipole is *parallel* to the exciting field, while the antisymmetric coupling means that the induced moment is *orthogonal* to the field. Obviously, different meta-atom shapes are needed to realize such different effects. Symmetric coupling is due to chirality (mirror-asymmetry) of meta-atoms. Antisymmetric coupling is called *omega coupling*, and it can exist in non-chiral structures. Also, it is important to note that symmetric and antisymmetric coupling effects manifest themselves in very different electromagnetic phenomena. Chiral metamaterials exhibit optical activity (rotation of polarization plane of linearly polarized propagating plane waves) and circular dichroism (different decay factors for right and left circularly polarized waves propagating in isotropic media). Antisymmetric (omega) coupling breaks the symmetry of reflections for illuminations of the particle from different directions.

However, it is interesting that in the microwave range, all reciprocal magneto-electric phenomena can be realized in meta-atoms of one basic shape: a short wire antenna loaded by a small loop as an inductive load; see Figure 3.20. The same particle also can be considered as a split ring loaded by a short wire antenna as a capacitive load. Changing the mutual orientation of the straight wires and the loop, both symmetric (chiral) and antisymmetric (omega) couplings can be realized.

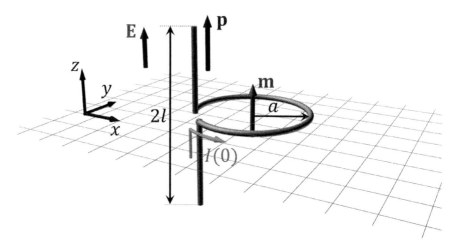

Canonical chiral particle: a connected split ring and an electric dipole. Courtesy of A. Díaz-Rubio.

Considering excitation by external electric fields polarized along the electric dipole, we can write for the current induced at the dipole center

$$I(0) = \frac{El}{Z_{\text{wire}} + Z_{\text{loop}}} \approx \frac{El}{\frac{1}{j\omega C} + R + j\omega L}, \tag{3.52}$$

which corresponds to the induced *magnetic* dipole moment $m = \mu_0 SI(0)$, because this current flows through the connected loop. This result allows us to find the amplitude of the magneto-electric coupling coefficient [34]

$$\alpha_{\text{me}} = \frac{m}{E} = \frac{\mu_0 SI}{\frac{1}{j\omega C} + R + j\omega L} = \frac{j\omega \mu_0 CSI}{1 - \omega^2 LC + j\omega CR} = -\alpha_{\text{em}}. \tag{3.53}$$

Similar consideration of excitation by external magnetic fields gives us α_{em}, which is, in fact, equal to $-\alpha_{\text{em}}$ (as a consequence of the particle reciprocity).

We see that the coupling coefficient exhibits the resonant behavior of the same type as in split rings (actually, the capacitive input impedance of the short electric dipole plays the same role as the gap capacitance of split rings). This property allows the realization of very strong coupling response even in particles of overall small sizes.

Note that when the frequency tends to zero, the coupling strength tends to zero proportionally to the frequency. Recall that the strength of the artificial magnetic response tends to zero as ω^2. This difference reflects the fact that artificial magnetism is a second-order effect of spatial dispersion while the reciprocal magneto-electric coupling is a stronger, first-order effect. In practical terms, it tells us that it may be easier to create magnetic polarization using magneto-electric coupling effects rather than the effect of artificial magnetism.

The particle shown in Figure 3.20 is a chiral particle, exhibiting symmetric magneto-electric coupling: the magnetic moment induced by external electric field is in the same direction as the field (vertical in the picture). In order to realize antisymmetric (omega) coupling, it is enough to rotate the loop 90° around the axis x so that the particle becomes planar. In that topology, the induced magnetic moment is orthogonal to the external electric field. Forming a meta-atom by two such particles, one in the xz-plane and the other in the zy-plane, we create a uniaxial omega particle with the antisymmetric coupling dyadic of the form $\overline{\overline{\alpha}}_{me} = \alpha_{me}\,\mathbf{x}_0 \times \overline{\overline{I}}$ ($\overline{\overline{I}}$ is the unit dyadic).

3.6 Modular Approach in Metamaterials: *Materiatronics*

General (linear) electromagnetic materials and metamaterials are modeled by the bianisotropic material relations

$$\mathbf{D} = \overline{\overline{\epsilon}} \cdot \mathbf{E} + \overline{\overline{a}} \cdot \mathbf{H}, \quad \mathbf{B} = \overline{\overline{\mu}} \cdot \mathbf{H} + \overline{\overline{b}} \cdot \mathbf{E}, \tag{3.54}$$

and the response of the corresponding individual meta-atoms is defined by general linear relations

$$\mathbf{p} = \overline{\overline{\alpha}}_{ee} \cdot \mathbf{E} + \overline{\overline{\alpha}}_{em} \cdot \mathbf{H}, \tag{3.55}$$

$$\mathbf{m} = \overline{\overline{\alpha}}_{mm} \cdot \mathbf{H} + \overline{\overline{\alpha}}_{me} \cdot \mathbf{E}. \tag{3.56}$$

Each of these dyadic coefficients can be represented as a 3×3 matrix in a particular coordinate system. Considering as an example the electric polarizability α_{ee}, each of the nine components of this matrix measures the dipole moment along one Cartesian axis induced by external electric fields along one of the Cartesian axes. It may seem that electric polarization processes in general include isotropic and anisotropic effects, suggesting that realizations of such general response always require rather complex-shaped inclusions. However, it is known that the permittivity of reciprocal media as well as the polarizability of electrically polarizable particles are symmetric matrices. Thus, in some coordinate system, the matrix takes a diagonal form with only three non-zero components. Moreover, each of these components measures electric dipole that is induced *in the same direction* as the applied electric field. We see that arbitrary electric properties of reciprocal inclusions can be in principle viewed as a superposition of no more than three simple canonical processes. Conceptually, it appears that the particle is "decomposed" into a set of canonical inclusions of only one type of electric dipoles polarizable only along one direction (as the meandered electric dipole or the loaded dipole shown in Figure 3.12, since in the fundamental resonance band they exhibit strong resonant response only for fields oriented along the axis). Similarly, arbitrarily anisotropic but reciprocal magnetic response can be conceptually reduced to not more than three canonical processes of one type, where external magnetic field induces magnetic moment along the same direction (conceptually, like a double split ring when its electric polarizability and bianisotropy can be neglected). Figure 3.21 illustrates these two canonical decomposition elements shown as building blocks of arbitrary meta-atoms.

Let us next consider the question: what are the basic, fundamental processes *(basic modules)* of arbitrary reciprocal magneto-electric coupling phenomena?

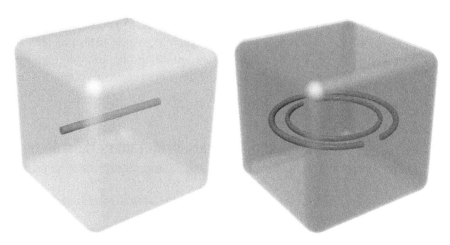

Figure 3.21 Canonical decomposition elements for modeling electric and magnetic polarization processes: electric and magnetic dipole particles. Courtesy of V. Asadchy.

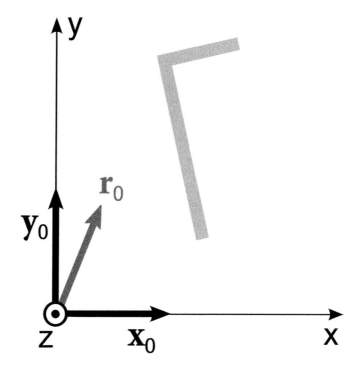

Figure 3.22 An electrically small metal particle in form the letter Γ. Unit vector $\mathbf{r}_0$ along the direction of vector $\alpha_{em}^{xz}\mathbf{x}_0 + \alpha_{em}^{yz}\mathbf{y}_0$.

To find the answer, we consider a particular example particle shape, shown in Figure 3.22, and study its magneto-electric coupling dyadics.

The particle topology determines the components of the coupling dyadics that may take non-zero values. The particle shape is planar (the conducting strips or wires are in the xy-plane). Thus, the induced magnetic moment is directed along the z-axis. On the other hand, the induced electric moment can have both x- and y-components. We conclude that the coupling dyadics have the following structure:

$$\mathbf{p} = \overline{\overline{\alpha}}_{ee} \cdot \mathbf{E} + \overline{\overline{\alpha}}_{em} \cdot \mathbf{H} = \overline{\overline{\alpha}}_{ee} \cdot \mathbf{E} + \left(\alpha_{em}^{xz}\mathbf{x}_0 + \alpha_{em}^{yz}\mathbf{y}_0\right)\mathbf{z}_0 \cdot \mathbf{H}, \qquad (3.57)$$

$$\mathbf{m} = \overline{\overline{\alpha}}_{mm} \cdot \mathbf{H} + \overline{\overline{\alpha}}_{me} \cdot \mathbf{E} = \overline{\overline{\alpha}}_{mm} \cdot \mathbf{H} + \mathbf{z}_0 \left(\alpha_{me}^{zx}\mathbf{x}_0 + \alpha_{me}^{zy}\mathbf{y}_0\right) \cdot \mathbf{E}. \qquad (3.58)$$

The reciprocity theorem imposes a relation between the two coupling dyadics [33]

$$\overline{\overline{\alpha}}_{me} = -\overline{\overline{\alpha}}_{em}^T, \qquad (3.59)$$

where T is the transpose operator.

The two vectors in the magneto-electric dyadic can be combined into one, writing

$$\overline{\overline{\alpha}}_{em} = \left(\alpha_{em}^{xz}\mathbf{x}_0 + \alpha_{em}^{yz}\mathbf{y}_0\right)\mathbf{z}_0 = K\mathbf{r}_0\mathbf{z}_0, \qquad (3.60)$$

where

$$K = \sqrt{(\alpha_{em}^{xz})^2 + (\alpha_{em}^{yz})^2}, \quad \mathbf{r}_0 = \frac{1}{K}\left(\alpha_{em}^{xz}\mathbf{x}_0 + \alpha_{em}^{yz}\mathbf{y}_0\right).$$

The trace of this dyadic is zero, $\mathrm{tr}\left(\overline{\overline{\alpha}}_{em}\right) = 0$ because for this particle shape $\mathbf{r}_0 \cdot \mathbf{z}_0 = 0$.

Splitting the magneto-electric polarizability dyadic into symmetric and antisymmetrics parts, we get

$$\overline{\overline{\alpha}}_{em} = \overline{\overline{\alpha}}_{em}^S + \overline{\overline{\alpha}}_{em}^A, \qquad (3.61)$$

where

$$\overline{\overline{\alpha}}_{em}^S = \frac{K}{2}\left(\mathbf{r}_0\mathbf{z}_0 + \mathbf{z}_0\mathbf{r}_0\right), \qquad \overline{\overline{\alpha}}_{em}^A = \frac{K}{2}\left(\mathbf{r}_0\mathbf{z}_0 - \mathbf{z}_0\mathbf{r}_0\right). \qquad (3.62)$$

The symmetric part of $\overline{\overline{\alpha}}_{em}$ can be diagonalized, writing

$$\overline{\overline{\alpha}}_{em}^S = \frac{K}{2}\left(\mathbf{v}_0\mathbf{v}_0 - \mathbf{w}_0\mathbf{w}_0\right), \qquad (3.63)$$

where

$$\mathbf{v}_0 = \frac{\sqrt{K}}{2}(\mathbf{r}_0 + \mathbf{z}_0), \qquad \mathbf{w}_0 = \frac{\sqrt{K}}{2}(\mathbf{z}_0 - \mathbf{r}_0). \qquad (3.64)$$

We see that the symmetric part of the coupling dyadic always decomposes into simple dyads of the form $\mathbf{v}_0\mathbf{v}_0$ or $\mathbf{w}_0\mathbf{w}_0$, meaning that any reciprocal magneto-electric coupling described by a symmetric coupling dyadic can be viewed as a combination of basic processes where the applied electric field induced magnetic moment along the same direction. Also, applied magnetic field induces electric dipole along the same direction. This effect is approximately realized in a thin spiral (a canonical chiral object), shown in Figure 3.23 on the left.

In this particular example, the chirality effects split into two canonical processes (the particle is "decomposed" into two canonical helices), since the decomposition contains two terms: $\mathbf{v}_0\mathbf{v}_0 - \mathbf{w}_0\mathbf{w}_0$. Note that the strength of these

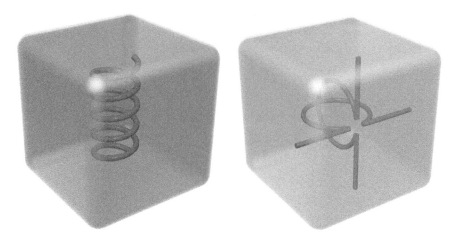

Figure 3.23 Canonical decomposition elements for modeling reciprocal magneto-electric coupling: Helix and uniaxial omega particle. Courtesy of V. Asadchy.

two processes is the same (both $\mathbf{v}_0$ and $\mathbf{w}_0$ are unit vectors), but these two terms come with the opposite signs. It means that one of the spirals is right-handed while the other one is left-handed. In the average, there is no chirality (this is called *racemic* arrangement). This fact indicates that the particle has a non-chiral shape. However, this decomposition clearly shows that, for some illumination directions, the particle exhibits chiral properties (optical activity and dichroism). This is possible because we see that these two decomposition particles are oriented orthogonally, so for waves traveling along the axis of one of the particles, the other one is not excited so that the chiral effects due to the presence of one particle are not compensated. Moreover, the orientation of the basic elements in the decomposition tells what illumination directions correspond to maximally strong chiral response. Such particles are called *pseudochiral*, since they are non-chiral but show chiral properties for some specific excitations [33, 35].

What particle topology corresponds to the antisymmetric part of the coupling dyadic, $\overline{\overline{\alpha}}_{em}^A = \frac{K}{2} (\mathbf{r}_0\mathbf{z}_0 - \mathbf{z}_0\mathbf{r}_0)$? An antisymmetric dyadic always can be presented as a vector product operation with a single (in general, complex) vector. In our example case, we can write

$$\mathbf{r}_0\mathbf{z}_0 - \mathbf{z}_0\mathbf{r}_0 = -\mathbf{u}_0 \times (\mathbf{r}_0\mathbf{r}_0 + \mathbf{z}_0\mathbf{z}_0) = -\mathbf{u}_0 \times \overline{\overline{I}}, \qquad (3.65)$$

where vector $\mathbf{u}_0 = \mathbf{r}_0 \times \mathbf{z}_0$ is the unit vector orthogonal to both $\mathbf{r}_0$ and $\mathbf{z}_0$, and $\overline{\overline{I}}$ is the unit dyadic. We see that this is an operator of 90° rotation around $-\mathbf{u}_0$. This operation can be approximately realized using a metal particle shaped as a uniaxial omega particle, whose shape resembles a hat, as shown in Figure 3.23, on the right. The presence of this particle in the decomposition tells that the particle exhibits asymmetric reflections when illuminated along the positive or negative directions of vector $\mathbf{u}_0$ (this asymmetry is the strongest for incident directions along this axis).

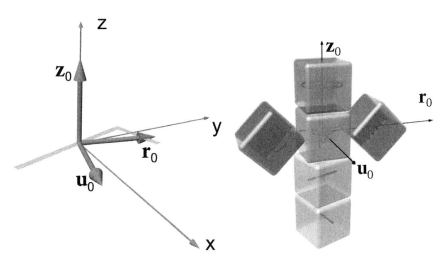

Figure 3.24   Decomposition of the Γ-shaped particle shown in Figure 3.22. Courtesy of V. Asadchy.

Finally, we see that all the polarization processes in this example particle can be viewed as a combination of processes due to a set of canonical inclusions, as shown in Figure 3.24. Due to the presence of chiral inclusions in a racemic arrangement together with a uniaxial omega inclusion, this Γ-particle belongs to the class of pseudochiral omega meta-atoms.

Although we have considered a specific example topology, from the fact that we have used the most general decomposition of all dyadics into symmetric and antisymmetric parts, we can conclude that the result holds for the most general reciprocal meta-atoms. For chiral inclusions, the trace of $\overline{\overline{\alpha}}^{S}_{em}$ is not zero, meaning that the right- and left-handed canonical chiral inclusions of the decomposition have different strengths.

This conceptual decomposition of arbitrary polarization processes into a set of fundamental processes of only four types has been named *materiatronics* [36], in analogy with electronics, where rather complex properties can be realized using sets of fundamental elements of only a few types (capacitors, inductors, resistors, and transistors). This analysis method can be extended to nonreciprocal meta-atoms, in which case four additional fundamental elements to model nonreciprocal processes are needed [36].

3.7 Tunable and Programmable Metamaterials

The metamaterial paradigm has been successfully developed and used in order to realize artificial materials with unusual, advantageous, and optimized properties. In the previous sections, we described meta-atoms with engineered electric, magnetic, and bianisotropic properties. Forming composite materials of such meta-atoms, metamaterials with desired properties can be realized. Here, we will discuss how these engineered material parameters can be made controllable and tunable. In particular, we consider how the circuit models of meta-atoms can be generalized for meta-atoms with integrated electronic components. Note that making metamaterial response tunable by external forces not only allows adjustments and reconfigurations of its functionality but also opens up possibilities to realizations of new properties [37].

At the time of writing this text, work on tunable metamaterials and meta-surfaces is very active, since it is expected that new enabling technologies can be created if engineered and optimized material parameters can be externally controlled in real time and in a programmable way. Among many envisaged applications, tunable metamaterials will hopefully enable metatronics devices, especially space-time-modulated metamaterial structures (achieving nonreciprocity through magnetless artificial materials) and creation of computational meta-materials (enhancing real-time analog signal processing capabilities of imaging devices) [38]. Most studies consider phenomena in materials with time-varying parameters (usually assuming that material permittivity or conductivity depend on time due to some external force). However, using time-domain models with time-varying permittivity $[\mathbf{D}(t) = \epsilon(t)\mathbf{E}(t)]$, one assumes instantaneous response; therefore, such an approach does not allow studies of dispersive time-modulated media. Using frequency-domain models with the permittivity dependent both on the frequency and time (due to external modulation), one assumes that the modulation is very slow at the scale of all polarization processes in the material. Only some initial studies have been made on the route toward exploring the possibilities of shaping material response by time modulations of constituent meta-atoms [39].

A fundamental tunable meta-atom should contain small elements (small "antennas") that can be excited by external fields. One can call them *sensors*. Next, it should contain polarizable elements that introduce electric and magnetic polarizations into the composite medium. One can call them *actuators*. And, most importantly, it should contain some tunable element or elements that can change the meta-atom response as desired. Within these necessary ingredients, several realization scenarios are possible.

Figure 3.25 illustrates this fundamental set of necessary ingredients in an example of an isotropic material with tunable effective permittivity and permeability [10]. Here, the tunable response is ensured by using electrically small dipole antennas loaded by bulk tunable or even programmable and computation-capable loads. In this scenario, the role of both sensors and actuators is played by the same elements: short pieces of metal wires or strips forming electric and magnetic dipoles.

Using the theory of polarizabilities of small loaded dipole scatterers [40] and the Maxwell Garnett mixing rule, it is possible to derive an analytical expression for the required load (control) impedance Z_{load}, which should be connected to each meta-atom [10] to realize the desired effective material response. The result reads

$$Z_{\text{load}} = \frac{1}{j\omega} \frac{(\epsilon_{\text{eff}} - \epsilon_0)\frac{3\epsilon_0}{(\epsilon_{\text{eff}}+2\epsilon_0)C_{\text{wire}}} - \frac{4}{3}Nl^2}{\frac{Nl^2 C_{\text{wire}}}{3} - (\epsilon_{\text{eff}} - \epsilon_0)\frac{3\epsilon_0}{\epsilon_{\text{eff}}+2\epsilon_0}}. \tag{3.66}$$

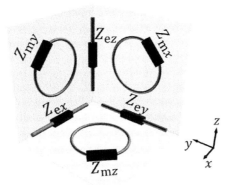

A unit-cell spatial arrangement of meta-atoms in the form of small loaded wire and loop antennas. Electric dipoles along x, y, and z provide the required electric properties for electric fields in those directions. Loops give the desired response for magnetic fields. Courtesy of A. Díaz-Rubio.

Here, ϵ_{eff} is the value of the effective permittivity of the metamaterial that we want to realize, C_{wire} is the capacitance between the two metal wires forming one meta-atom, and N is the meta-atom concentration. The value of load impedance Z_{load} of a small device in the middle of each meta-atom can be controlled by external forces using various means. Similar simple relations also can be found for the required loads of magnetically polarizable loops, necessary to realize desired values of effective permeability.

Probably for the first time, this scenario of creating metamaterials using small loaded antennas was used aiming to design broadband doubly negative metamaterials [41]. It appears that this approach in principle allows overcoming fundamental limitations on material response: the Foster theorem, which tells that, in low-loss frequency bands, the permittivity must always grow with the frequency. This fact is a good illustration of the possibilities offered by the paradigm of active and controllable metamaterials. On the other hand, we should comment that realizations of active metamaterials require care of the system stability [42], as in any active system.

Complicated electronic circuitry, including even programmable microprocessors, can be packed into electrically small volumes and connected to meta-atoms as shown in Figure 3.25. If these loads are active (contain their own power sources or can receive power from the incident light, for instance), virtually unlimited possibilities for molding electromagnetic properties of metamaterials open up. Because the meta-atoms sense the local field and we can modify them at our will, the material properties become adaptive to the environment and at the same time programmable. If the loads are connected to external excitation by some independent signal path (for instance, using radio signals at different frequency or optical means), the metamaterial response becomes fully controllable and tunable. This way, one can also modulate the material properties in time, appropriately changing the load impedance by external excitations.

In the simplest scenario shown in Figure 3.25, the roles of sensors and actuators are played by the same physical elements: short electric dipoles and small loops. This limits the realizable properties to controllable permittivity and permeability. More generally, we can return to our discussion of basic topologies of meta-atoms realizing the most general reciprocal bianisotropic response. Inspecting Figure 3.20, we see that the basic shapes contain the same simple elements – short electric dipoles and small loops – but in general, each meta-atom contains *both* elements, and they are *connected* in various ways. This consideration suggests a general approach for realization of fully controllable and tunable metamaterials with general bianisotropic response: one needs to insert controllable and possibly programmable two-port devices at the junction between the electrically and magnetically polarizable elements. For instance, it is possible to envisage a medium where *electric* field polarized along a certain direction induces *magnetic* polarization in some other, predefined direction, with a desired transformation coefficient, etc. Needless to say, a controllable response does not have to be linear.

Let us come back to our list of necessary "ingredients" of controllable meta-atoms: It should contain small elements that can be excited by external fields, polarizable elements that introduce electric and magnetic polarizations into the composite medium, and some tunable element or elements that can change the meta-atom response as desired. In the preceding discussion, we explained how all these ingredients can be realized using a small control unit embedded into a "usual," non-tunable meta-atom. However, it is not the only possibility. One can envisage a possibility to control the properties of the (usual) material from which a meta-atom is made. For instance, polarizable inclusions can be manufactured from a phase-change material. In this scenario, the whole body of the meta-atom is changed by external excitations. Still another possibility is to embed usual, non-tunable meta-atoms into a dynamic, controllable host medium – for instance, distributing metal meta-atoms in a liquid crystal host. All these opportunities deserve attention, but in our opinion, the first scenario of using small control units to adjust response of meta-atoms made of conventional materials is the most powerful and promising approach, since there are virtually unlimited possibilities in making these control units programmable and smart.

Finally, we note that this is an active field of current research and direct the reader to recent review papers for details of particular experimental realizations [43–47].

Problems and Control Questions

1. Explain the notions of isotropy, anisotropy, and bianisotropy.
2. Discuss the differences in electromagnetic response of double and single split rings. Which shape is preferable if the goal is to create an artificial magnetic material?
3. Discuss the differences of omega and chiral field coupling. Do you know any materials in nature that possess such properties?

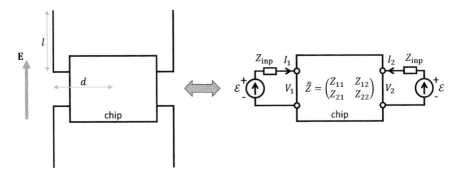

Figure 3.26 Tunable or reconfigurable unit cell. Courtesy of F. Liu.

4. Give examples of natural and artificial uniaxial media.
5. Consider excitation of an omega particle formed by a short electric dipole and a conducting loop. Explain what polarizations (electric and magnetic dipole moments) will be induced by uniform electric field directed along the electric dipole. Answer the same question for excitation by uniform magnetic field orthogonal to the loop plane. Will there be bianisotropic effects if the electric field is polarized in the loop plane orthogonally to the wire dipole?
6. Decompose electromagnetic response of T- and Π-shaped small metal particles into canonical elements. Discuss how these particles should be oriented with respect to the incidence direction and incidence plane of external plane-wave excitations in order to maximize the asymmetry in reflection for illuminations from the opposite directions.
7. Consider a metamaterial unit cell consisting of two identical small dipole antennas controlled by a chip of negligible size (Figure 3.26, left). Under excitation by uniform electric field along the dipoles, it can be modeled as a two-port network (Figure 3.26, right), in which the chip is modeled by an impedance matrix $\overline{\overline{Z}}$. The electromotive force on the antennas is $V = El$, while the input impedance of the antennas is $Z_{\mathrm{inp}} = 1/j\omega C$ (capacitive, neglecting dissipation and radiation resistance). Please answer the following questions:
   (a) Derive the formulas for the currents in the two antenna I_1 and I_2 (in terms of E, l, Z_{inp}, Z_{ij}).
   (b) At the ends of the antennas, there will be charge accumulation $Q = I/j\omega$, which will induce electric dipole moment along the antennas. What is the total electric dipole moment p in this unit cell? What is the polarizability $\alpha_{\mathrm{ee}} = p/E$?
   (c) When the currents are asymmetric in the two antennas, there is magnetic dipole moment out of the antenna plane $m = (I_1 - I_2)ld/2$. What is the magneto-electric coupling parameter $\alpha_{\mathrm{me}} = m/E$?
   (d) In what condition is only the electric dipole moment induced? What is the needed chip impedance matrix for inducing only the magnetic dipole moment?

Bibliography

[1] C. Holloway, E. F. Kuester, J. Gordon, J. O'Hara, J. Booth, and D. Smith, "An overview of the theory and applications of metasurfaces: The two-dimensional equivalents of metamaterials," *IEEE Antennas and Propagation Magazine* **54**, 10–35 (2012).

[2] S. B. Glybovski, S. A. Tretyakov, P. A. Belov, Y. S. Kivshar, and C. R. Simovski, "Metasurfaces: From microwaves to visible," *Physics Reports* **634**, 1–72 (2016).

[3] A. Safwat, S. A. Tretyakov, and A. V. Räisänen, "High-impedance wire," *IEEE Antennas and Wireless Propagation Letters* **6**, 631–634 (2007).

[4] J. C. Bose, "On the rotation of plane of polarization of electric waves by twisted structure," *Proceedings of the Royal Society* **63**, 146–152 (1898).

[5] D. R. Smith, W. J. Padilla, D. C. Vier, et al., "Composite medium with simultaneously negative permeability and permittivity," *Physical Review Letters* **84**, 4184–4187 (2000).

[6] D. R. Smith, D. C. Vier, N. Kroll, et al., "Direct calculation of permeability and permittivity for a left-handed metamaterial," *Applied Physics Letters* **77**, 2246–2248 (2000).

[7] K. F. Lindman, Über eine durch ein isotropes System von spiralförmigen Resonatoren erzeugte Rotationspolarisation der elektromagnetischen Wellen, *Annalen der Physik* **63**, 621–644 (1920).

[8] S. Tretyakov, "A personal view on the origins and developments of the metamaterial concept," *Journal of Optics* **19**, 013002 (2017).

[9] R. A. Shelby, D. R. Smith, and S. Schultz, "Experimental verification of a negative index of refraction," *Science* **292**, 77–79 (2001).

[10] S. Tretyakov, *Analytical Modeling in Applied Electromagnetics*, Artech House, 2003.

[11] N. Engheta, "An idea for thin subwavelength cavity resonators using metamaterials with negative permittivity and permeability," *IEEE Antennas and Propagation Letters* **1**, 10–13 (2002).

[12] S. A. Tretyakov, S. I. Maslovski, I. S. Nefedov, and M. K. Kärkkäinen, "Evanescent modes stored in cavity resonators with backward-wave slabs," *Microwave and Optical Technology Letters* **38**, 153–157 (2003).

[13] J. Pendry, "Negative refraction makes a perfect lens," *Physical Review Letters* **85**, 3966 (2000).

[14] V. G. Veselago, "The electrodynamics of substances with simultaneously negative values of ϵ and μ," *Soviet Physics Uspekhi* **10**, 509–514 (1968), (originally published in Russian in *Uspekhi Fizicheskikh Nauk* **92**, 517–526 (1967)).

[15] W. E. Kock, "Metallic delay lines," *Bell System Technical Journal* **27**, 58–82 (1948).

[16] A. Sihvola, *Electromagnetic Mixing Formulas and Applications*, The Institution of Electrical Engineers, 1999.

[17] R. E. Collin, *Field Theory of Guided Waves*, IEEE Press, 1991.

[18] W. Rotman, "Plasma simulations by artificial dielectrics and parallel plate media," *IRE Transactions on Antennas and Propagation* **10**, 81–96 (1962).

[19] J. Brown, "Artificial dielectrics having refractive indices less than unity," *IEE Proceedings* **100**, Part 4, 51–62 (1953).

[20] S. I. Maslovski, S. A. Tretyakov, and P. A. Belov, "Wire media with negative effective permittivity: A quasi-static model," *Microwave and Optical Technology Letters* **35**, 47–51 (2002).

[21] P. A. Belov, R. Marques, S. I. Maslovski, I. S. Nefedov, M. Silveirinha, C. R. Simovski, and S. A. Tretyakov, "Strong spatial dispersion in wire media in the very large wavelength limit," *Physical Review* **67**, 113103 (2003).

[22] P. Ikonen, C. Simovski, S. Tretyakov, P. Belov, and Y. Hao, "Magnification of subwavelength field distributions at microwave frequencies using a wire medium slab operating in the canalization regime," *Applied Physics Letters* **91**, 104102 (2007).

[23] C. R. Simovski, P. Belov, A. Atrashchenko, and Y. S. Kivshar, "Wire metamaterials: Physics and applications," *Advanced Materials* **24**, 4229–4248 (2012).

[24] S. A. Schelkunoff and H. T. Friis, *Antennas: Theory and Practice*, Wiley, 1952.

[25] M. V. Kostin and V. V. Shevchenko, "Artificial magnetics based on double circular elements," *Proceedings of Chiral'94*, 49–56, May 18–20, 1994, Périgueux, France.

[26] A. N. Lagarkov, V. N. Semenenko, V. A. Chistyaev, D. E. Ryabov, S. A. Tretyakov, and C. R. Simovski, "Resonance properties of bi-helix media at microwaves," *Electromagnetics* **17**, 213–237 (1997).

[27] J. B. Pendry, A. J. Holden, D. J. Robbins, and W. J. Stewart, "Magnetism from conductors and enhanced nonlinear phenomena," *IEEE Transactions on Microwave Theory and Techniques* **47**, 2075–2084 (1999).

[28] S. Maslovski, P. Ikonen, I. Kolmakov, S. Tretyakov, and M. Kaunisto, "Artificial magnetic materials based on the new magnetic particle: Metasolenoid, *Progress in Electromagnetics Research* **54**, 61–81 (2005).

[29] L. Jylhä, S. Maslovski, and S. Tretyakov, "High-order resonant modes of a metasolenoid," *Journal of Electromagnetic Waves and Applications* **19**, 1327–1342 (2005).

[30] J. Zhou, Th. Koschny, M. Kafesaki, et al., "Saturation of the magnetic response of split-ring resonators at optical frequencies," *Physical Review Letters* **95**, 223902 (2005).

[31] M. W. Klein, C. Enkrich, M. Wegener, C. M. Soukoulis, and S. Linden, "Single-slit split-ring resonators at optical frequencies: limits of size scaling," *Optics Letters* **31**, 1259–1261 (2006).

[32] S. Tretyakov, "On geometrical scaling of split-ring and double-bar resonators at optical frequencies," *Metamaterials* **1**, 40–43 (2007).

[33] A. N. Serdyukov, I. V. Semchenko, S. A. Tretyakov, and A. Sihvola, *Electromagnetics of Bi-Anisotropic Materials: Theory and Applications,* Gordon and Breach Science Publishers, 2001.

[34] S. A. Tretyakov, F. Mariotte, C. R. Simovski, T. G. Kharina, and J.-P. Heliot, "Analytical antenna model for chiral scatterers: Comparison with numerical and experimental data," *IEEE Transactions on Antennas and Propagation* **44**, 1006–1014 (1996).

[35] A. A. Sochava, C. R. Simovski, and S. A. Tretyakov, "Chiral effects and eigenwaves in bi-anisotropic omega structures," in *Advances in Complex Electromagnetic Materials, NATO ASI Series High Technology* **28**, Kluwer Academic Publishers, pp. 85–102, 1997.

[36] V. S. Asadchy and S. A. Tretyakov, "Modular analysis of arbitrary dipolar scatterers," *Physical Review Applied* **12**, 024059 (2019).

[37] D. Sounas and A. Alù, "Non-reciprocal photonics based on time modulation," *Nature Photonics* **11**, 774–783 (2017).

[38] N. M. Estakhri, B. Edwards, and N. Engheta, "Inverse-designed metastructures that solve equations," *Science* **363**, 1333–1338 (2019).

[39] G. Ptitcyn, M. S. Mirmoosa, and S. A. Tretyakov, "Time-modulated meta-atoms," *Physical Review Research* **1**, 023014 (2019).

[40] S. A. Tretyakov, S. I. Maslovski, and P. A. Belov, "An analytical model of metamaterials based on loaded wire dipoles," *IEEE Transactions on Antennas and Propagation* **51**, 2652–2658 (2003).

[41] S. A. Tretyakov, "Meta-materials with wideband negative permittivity and permeability," *Microwave and Optical Technology Letters* **31**, 163–165 (2001).

[42] E. Ugarte-Munoz, S. Hrabar, D. Segovia-Vargas, and A. Kiricenko, "Stability of non-Foster reactive elements for use in active metamaterials and antennas," *IEEE Transactions on Antennas and Propagation* **60**, 3490–3494 (2012).

[43] J. P. Turpin, J. A. Bossard, K. L. Morgan, D. H. Werner, and P. L. Werner, "Reconfigurable and tunable metamaterials: A Review of the theory and applications," *International Journal of Antennas and Propagation* **2014**, Article ID 429837 (2014).

[44] A. D. Boardman, V. V. Grimalsky, Y. S. Kivshar, et al., "Active and tunable metamaterials," *Laser Photonics Reviews* **5**, 287–307 (2011).

[45] G. R. Keiser, K. Fan, X. Zhang, and R. D. Averitt, "Towards dynamic, tunable, and nonlinear metamaterials via near field interactions: A review," *Journal of Infrared, Millimeter, and Terahertz Waves* **34**, 709–723 (2013).

[46] I. V. Shadrivov, P. V. Kapitanova, S. I. Maslovski, and Y. S. Kivshar, "Metamaterials controlled with light," *Physical Review Letters* **109**, 083902 (2012).

[47] F. Liu, A. Pitilakis, M. S. Mirmoosa, et al., "Programmable metasurfaces: State of the art and prospects," *IEEE International Symposium on Circuits and Systems (ISCAS2018),* May 27–30, 2018, Florence, Italy.

Note

1 In the next section, we will discuss this important decomposition in detail.

4 Metasurfaces

4.1 Introduction

Metasurface [1–4] is an electrically thin composite material layer, designed and optimized to function as a tool to control and transform electromagnetic waves (or waves of other physical nature – e.g., sound waves). The layer thickness is small and can be considered as negligible with respect to the wavelength in the surrounding space. The composite structure forming a metasurface is assumed to behave as a *material* in the electromagnetic (optical) sense, meaning that it can be homogenized on the wavelength scale, and the metasurface can be adequately characterized by its effective, surface-averaged properties. Similarly to volumetric materials, where the notions of the permittivity and permeability result from volumetric averaging of microscopic currents over volumes that are small compared to the wavelength, the metasurface parameters result from two-dimensional surface averaging of microscopic currents on the same wavelength scale. This implies that the unit-cell sizes of composite metasurfaces are reasonably small as compared to the wavelength. In terms of the optical response, this means that metasurfaces reflect and transmit plane waves as sheets of materials – in contrast to diffraction gratings, which produce multiple diffraction lobes.

Due to the presence of small but strongly polarizable inclusions, the fields inside the layer change significantly so that the effects of the layer on waves incident from surrounding space can be very strong, despite the very small thickness. To realize strong response, in the majority of designs, metasurfaces are formed by some resonant meta-atoms. In the microwave range, the material of choice is metal (most commonly, copper). Due to its high conductivity, fields are kept mostly away from the material volume, which allows the realization of low-loss structures. In optics, metal nanoparticles respond as resonant electric dipoles, but losses are high because the electric field fills the whole volume of the particle. High-permittivity dielectrics provide a route toward realization of low-loss optical metasurfaces [5–8]. Usually, nanoparticles made of high-permittivity dielectric or semiconductor materials are used, exploiting resonant electric and magnetic dipole modes.

One of the main challenges is to synthesize and realize metasurfaces that change the incident fields in desired ways. This goal means that we would like to create metasurface structures with fully adjustable and controllable reflection and transmission properties. Other applications additionally require engineering of surface waves along metasurfaces.

In antenna theory and practice, the concept of phased array antennas is widely used to control radiation patterns. Within that approach, desired radiated fields are synthesized as superpositions of fields created by many small radiating elements. Controlling the amplitudes and phases of waves radiated by individual small radiators, it is possible to realize different field patterns: fields will be strong in the directions where the waves radiated by the individual array elements interfere constructively and weak in the directions where the interference is destructive. This basic concept finds a natural generalization for metasurfaces: many important functionalities can be realized by engineering the amplitudes and phases of waves reflected from and transmitted through nonuniform meta-surfaces. Within this approach, a metasurface is treated as locally periodical, and the desired response at each point of the surface is realized by tuning the dimensions and mutual arrangements of periodically repeated unit cells.

The conventional approach to find the required metasurface response needed to realize desired reflected or transmitted wave profile is based on the phased-antenna array principle (e.g., [9]). As an example, let us consider a flat metasur-face that should perfectly reflect a plane wave incident at the incidence angle θ_i into an unusual direction, at the reflection angle $\theta_r \neq \theta_i$: The tangential component of the electric field of the incident wave is $E_{\mathrm{inc}} = E_0 e^{-jkx \sin \theta_i}$, and the desired reflected wave is $E_{\mathrm{ref}} = E_0 e^{-jkx \sin \theta_r}$. Here, x is the coordinate along the reflecting surface, in the incidence plane. Obviously, the corresponding reflection coefficient

$$R = \exp[jkx(\sin \theta_i - \sin \theta_r)] = \exp(j\Phi_r(x)) \qquad (4.1)$$

depends on the position along the surface so that

$$\sin \theta_i - \sin \theta_r = \frac{1}{k} \frac{d\Phi_r(x)}{dx}. \qquad (4.2)$$

In this design approach, at every point, one wants to have full reflection: $|R| = 1$. Thus, in order to tilt a given plane wave into a desired direction, reflection phase $\Phi_r(x)$ should linearly grow or decay along x, and its gradient determines the difference between the two angles. This is the phased-array principle, which is also called the *generalized reflection law* in recent literature [4, 10].

In fact, this approach has fundamental limitations and gives acceptable results only for a limited range of deflection angles (see a discussion in Section 4.5), but at this point, an interesting question arises: what (if any) advantages do we gain using dense arrays of electrically small unit cells as compared to the more traditional solutions of using phased arrays of half-wavelength-sized antenna elements? Indeed, if we are interested in creating a specific radiation pattern in the far zone, it is enough to make sure that reflections from all subareas of the half-wavelength size are in proper phase along the desired directions. Finer discretization of the surface and control of the reflection phase with subwave-length precision do not appear to be necessary. However, such tuning allows us to control and optimize the distributions of evanescent fields in the vicinity of the metasurface, and for some functionalities, this control is desired or even necessary. Actually, one of the examples showing advantages of metasurfaces

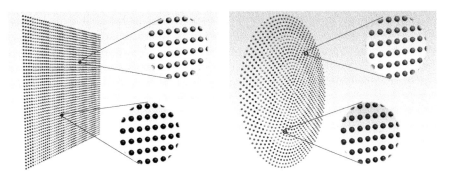

Figure 4.1 Within the physical optics approximation, the metasurface can be considered as a locally periodical structure. This way, we can synthesize metasurfaces for various transformations of incident plane waves, such as deflection of incident beams (left) or for focusing (right). Courtesy of V. Asadchy.

over conventional phased arrays is the previously discussed anomalous reflection. In fact, anomalous reflection into any desired direction becomes possible only if the metasurface supports a specific, carefully engineered distribution of surface-bound modes which is impossible in conventional phased antenna arrays (see details in Section 4.5).

On the other hand, for many applications, full control over local reflection and transmission phase for plane-wave excitation is enough to realize the desired function, and we start from discussing metasurface topology synthesis for this case. Let us outline the synthesis process for a special case of normally incident plane waves as the metasurface excitations. Actually, in many cases, this approach is appropriate even for the synthesis of nonuniform metasurfaces for rather complex field transformations, like wave focusing, thanks to the possibility of using the physical optics approximation, considering the metasurface as a locally periodical structure excited by plane waves. This concept is illustrated in Figure 4.1.

4.2 Homogenization Models

Since the period (the distance between the array inclusions) is electrically small, the reflection and transmission properties of metasurfaces are defined by the surface-averaged values of the induced surface currents. Although, in general, currents both tangential and normal to the surface can be induced, we will concentrate on the tangential components only. As is known from the Huygens principle, the knowledge of the tangential fields on the surface bounding a volume is enough to uniquely find the fields everywhere inside the volume (a discussion on the equivalence of the effects produced by tangential and normal polarizations can be found in [11]). Jumps of tangential electric and magnetic fields can be considered as equivalent electric and magnetic surface currents. Moreover, if a uniform metasurface is illuminated by a normally incident plane

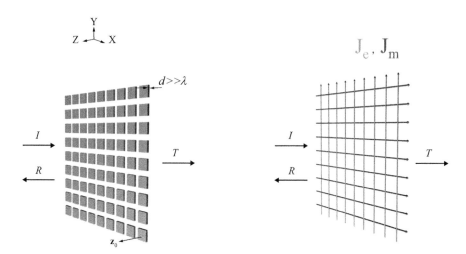

Figure 4.2   In the homogenization model, an array of electrically small unit cells is replaced by an equivalently responding set of electric and magnetic current sheets characterized by surface current densities $\mathbf{J}_e$ and $\mathbf{J}_m$. Reprinted from Y. Ra'di et al., *Physical Review Applied*, 3, 037001, 2015 with permission. © 2015 APS.

wave, normal to the surface currents do not contribute to reflected and transmitted waves. Thus, we can average the microscopic currents over a unit cell and introduce the notion of macroscopic, surface-averaged polarization densities and surface current densities; see Figure 4.2. While at small distances (comparable to or smaller than the array period), the fields of the two structures are very different, at distances much larger than the array period, the fields created by the metasurface and the model pair of two homogenized current sheets are not distinguishable.

It is clear that conventional effective parameters used to model volumetric materials (permittivity, permeability, and bianisotropy coefficients) are not applicable to metasurfaces because the very definitions of these parameters imply averaging over small *volumes* that contain many molecules or inclusions. In the case of a metasurface, this averaging volume is negligibly small because metasurfaces have negligible thickness. Thus, metasurface homogenization should imply averaging over a small *surface* area, which is small compared to the wavelength but contains many (or at least several) inclusions or unit cells. For plane-wave illumination, averaging over the area of one unit cell is sufficient. The effective models of metasurfaces are written for macroscopic quantities defined as surface-averaged values of electric and magnetic fields and surface-averaged polarizations and currents.

The surface-averaged polarization vectors can be related to the incident (external) fields [12–16]. Most commonly, these relations are written for dipole moments induced in each unit cell, $\mathbf{p}$ and $\mathbf{m}$, as

$$\mathbf{p} = \overline{\overline{\alpha}}_{ee} \cdot \mathbf{E}_{inc} + \overline{\overline{\alpha}}_{em} \cdot \mathbf{H}_{inc}, \tag{4.3}$$

$$\mathbf{m} = \overline{\overline{\alpha}}_{mm} \cdot \mathbf{H}_{inc} + \overline{\overline{\alpha}}_{me} \cdot \mathbf{E}_{inc}. \tag{4.4}$$

The polarization density vectors equal $\mathbf{P} = \mathbf{p}/S$, $\mathbf{M} = \mathbf{m}/(S\mu_0)$, where S is the unit-cell area. The coefficients in (4.3) and (4.4) are called *collective polarizability dyadics* because they model the unit-cell response when the unit cells are arranged into a periodical lattice, measuring the collective response to the incident fields. The normal (to the metasurface) components of the incident fields can be expressed in terms of the tangential components, using Maxwell's equations in the space surrounding the metasurface, and Eqs. (4.3) and (4.4) can be written with the tangential components of the incident fields $\mathbf{E}_{t\,\text{inc}}$ and $\mathbf{H}_{t\,\text{inc}}$ instead of the complete vectors.

The induced dipole moments (4.3) and (4.4) correspond to the surface-averaged electric and magnetic current sheets with the surface current densities $\mathbf{J}_e = \frac{j\omega\mathbf{p}}{S}$ and $\mathbf{J}_m = \frac{j\omega\mathbf{m}}{S}$ that radiate plane waves into surrounding medium (most commonly, free space). The reflected field is the sum of the fields radiated by these surface currents:

$$\mathbf{E}_{\text{ref}} = -\frac{j\omega}{2S}\left[\eta_0\mathbf{p} - \mathbf{z}_0 \times \mathbf{m}\right] = -\frac{1}{2}\left[\eta_0\mathbf{J}_e - \mathbf{z}_0 \times \mathbf{J}_m\right]. \tag{4.5}$$

Here, η_0 is the wave impedance of the surrounding isotropic material, and $\mathbf{z}_0$ is the unit vector normal to the metasurface plane and pointing toward the source, as in Figure 4.2. Similarly, the transmitted field is the sum of the incident field and the fields caused by the array currents:

$$\mathbf{E}_{\text{tr}} = \mathbf{E}_{\text{inc}} - \frac{j\omega}{2S}\left[\eta_0\mathbf{p} + \mathbf{z}_0 \times \mathbf{m}\right] = \mathbf{E}_{\text{inc}} - \frac{1}{2}\left[\eta_0\mathbf{J}_e + \mathbf{z}_0 \times \mathbf{J}_m\right]. \tag{4.6}$$

Note that the preceding relations for the fields created by the surface currents assume that the surface-averaged (macroscopic) currents are uniform over the surface. Thus, if the surface is infinite, these fields form plane waves propagating away from the surface in the direction normal to the surface. If the surface currents are changing over the surface (because the incident field varies or because the surface properties vary), these relations are approximately valid *locally* at every point of the surface, as long as the physical optics approximation can be used. This means the assumption of a locally periodic metasurface, applicable if the incident fields and the surface properties vary slowly with respect to the wavelength. In this situation, incident waves locally at every point experience specular reflection and partial transmission along the incidence direction, as governed by the metasurface properties at that point. The total reflected and transmitted field, found as interference of locally specularly reflected waves, can propagate in some other desired direction. In Section 4.5, we discuss how the limitations of this approximation of local periodicity can be overcome.

In this approach, the electromagnetic properties of metasurfaces are modeled by the collective polarizability dyadics $\overline{\overline{\alpha}}_{ij}(\omega)$. These values depend on the unit cell topology, dimensions, and the mutual positions of the unit cells in the arrays. Although relations (4.3) and (4.4) formally contain dipole moments of unit cells, it is important to stress that all these polarization vectors are, by definition, *surface-averaged* over an area that contains several unit cells. There exist alternative effectively homogeneous models, which relate induced surface current densities to the averaged tangential electric and magnetic fields (susceptibility model) and models based on impedance, admittance, or transmission matrices [17]. All these models are physically equivalent because they all relate surface-averaged tangential fields at the two sides of the metasurface via some effective surface parameters.

4.3 Synthesis of Metasurface Topologies

To synthesize a metasurface, we can first determine what collective polarizabilities are required to realize the desired reflection and transmission coefficients. Next, we will use the relations between the collective polarizabilities of periodically arranged unit cells and the polarizabilities of the same unit cell but measured in empty space (we call them *individual polarizabilities*). Knowing how a small object should respond to plane-wave illuminations, appropriate topologies finally can be found. The last stage is numerical optimization and fine tuning of the structure, which is needed because the analytical formulas used in the synthesis process are based on approximations.

The starting point is Eqs. (4.5) and (4.6), which define the reflected and transmitted fields in terms of the induced unit-cell electric and magnetic moments. In view of the homogenization-model relations (4.3) and (4.4), they determine the set of collective polarizabilities that we need to realize in order to get the desired reflected and transmitted fields. Indeed, substitution of (4.3) and (4.4) into (4.5) and (4.6) gives the relations between the reflected and transmitted fields in terms of the incident field and the collective polarizabilities. Let us write these expressions explicitly for metasurfaces that are isotropic in the surface plane. In this special case, dyadic coefficients in (4.3) and (4.4) are invariant with respect to rotation around the unit vector $\mathbf{z}_0$, normal to the layer plane:

$$\overline{\overline{\alpha}}_{ij} = \widehat{\alpha}_{ij}^{\mathrm{co}} \overline{\overline{I}}_\mathrm{t} + \widehat{\alpha}_{ij}^{\mathrm{cr}} \overline{\overline{J}}_\mathrm{t}, \tag{4.7}$$

where $\overline{\overline{I}}_\mathrm{t} = \overline{\overline{I}} - \mathbf{z}_0\mathbf{z}_0$ is the transverse unit dyadic and $\overline{\overline{J}}_\mathrm{t} = \mathbf{z}_0 \times \overline{\overline{I}}_\mathrm{t}$ is the vector-product operator.

Assuming the normal incidence of the incident plane wave (in which case the induced dipole moments are the same in all unit cells), we can find the amplitudes of the reflected and transmitted plane waves as [18]

$$\begin{aligned}
\mathbf{E}_{\mathrm{ref}} = -\frac{j\omega}{2S} &\left\{ \left[\eta_0 \widehat{\alpha}_{\mathrm{ee}}^{\mathrm{co}} \pm \widehat{\alpha}_{\mathrm{em}}^{\mathrm{cr}} \pm \widehat{\alpha}_{\mathrm{me}}^{\mathrm{cr}} - \frac{1}{\eta_0}\widehat{\alpha}_{\mathrm{mm}}^{\mathrm{co}} \right] \overline{\overline{I}}_\mathrm{t} \right.\\
&\left. + \left[\eta_0 \widehat{\alpha}_{\mathrm{ee}}^{\mathrm{cr}} \mp \widehat{\alpha}_{\mathrm{em}}^{\mathrm{co}} \mp \widehat{\alpha}_{\mathrm{me}}^{\mathrm{co}} - \frac{1}{\eta_0}\widehat{\alpha}_{\mathrm{mm}}^{\mathrm{cr}} \right] \overline{\overline{J}}_\mathrm{t} \right\} \cdot \mathbf{E}_{\mathrm{inc}}\\[6pt]
\mathbf{E}_{\mathrm{tr}} = &\left\{ \left[1 - \frac{j\omega}{2S} \left(\eta_0 \widehat{\alpha}_{\mathrm{ee}}^{\mathrm{co}} \pm \widehat{\alpha}_{\mathrm{em}}^{\mathrm{cr}} \mp \widehat{\alpha}_{\mathrm{me}}^{\mathrm{cr}} + \frac{1}{\eta_0}\widehat{\alpha}_{\mathrm{mm}}^{\mathrm{co}} \right) \right] \overline{\overline{I}}_\mathrm{t} \right.\\
&\left. - \frac{j\omega}{2S} \left[\eta_0 \widehat{\alpha}_{\mathrm{ee}}^{\mathrm{cr}} \mp \widehat{\alpha}_{\mathrm{em}}^{\mathrm{co}} \pm \widehat{\alpha}_{\mathrm{me}}^{\mathrm{co}} + \frac{1}{\eta_0}\widehat{\alpha}_{\mathrm{mm}}^{\mathrm{cr}} \right] \overline{\overline{J}}_\mathrm{t} \right\} \cdot \mathbf{E}_{\mathrm{inc}},
\end{aligned} \tag{4.8}$$

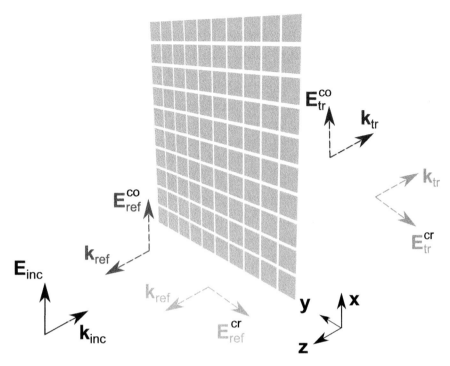

Figure 4.3   Using uniaxial metasurfaces we can engineer co- and cross-polarized reflection and transmission coefficients. Courtesy of V. Asadchy.

where η_0 is the free-space wave impedance (for simplicity and without loss of generality, we assume that the surrounding homogeneous and isotropic medium is free space). Here and thereafter, to distinguish between illuminations of the sheet from the two opposite sides, we will use double signs ($\pm$) for these two cases, where the top and bottom signs correspond to the incident plane wave propagating in the $-\mathbf{z}_0$ and $\mathbf{z}_0$ directions, respectively. The components proportional to the transverse unit dyadic $\overline{\overline{I}}_t$ define the co-polarized reflected and transmitted fields $\mathbf{E}^{\mathrm{co}}_{\mathrm{ref,tr}}$ and those proportional to the rotation operator $\overline{\overline{J}}_t$ give the values of the cross-polarized reflected and transmitted fields $\mathbf{E}^{\mathrm{cr}}_{\mathrm{ref,tr}}$; see an illustration in Figure 4.3.

Equations (4.8) also can be used for finding the effective parameters of the metasurface from calculated or measured response to probe plane waves. It can be done by solving the above equations for the collective polarizabilities if the reflected and transmitted fields created by the known incident field are known.

The next step in the synthesis procedure is selection of a set of polarizability values, which would give the desired values of the reflected and transmitted fields. Note that in the general case, the metasurface can respond differently for illuminations of its two sides, so we may need to make sure that reflection and transmission responses for illuminations of both sides of the metasurface are as required for the thought application.

There may be multiple possible choices of the polarizabilities that ensure the desired operation. We have eight (complex scalar) polarizability parameters, but the number of reflection/transmission coefficients that we want to tune can be fewer than eight. The choice is usually determined by practical considerations. For example, nonreciprocal inclusions are often expensive (such as monocrystal hexaferrites) and demand a DC-bias magnetic field. Therefore, such inclusions in metasurface design are, as a rule, not desirable. Thus, if no nonreciprocal effects are needed for the target application, it makes sense to limit the choice to reciprocal unit cells. The collective polarizability dyadics of reciprocal metasurfaces satisfy [19, 20]

$$\overline{\overline{\alpha}}_{ee} = \overline{\overline{\alpha}}_{ee}^{T}, \quad \overline{\overline{\alpha}}_{mm} = \overline{\overline{\alpha}}_{mm}^{T}, \quad \overline{\overline{\alpha}}_{em} = -\overline{\overline{\alpha}}_{me}^{T}. \tag{4.9}$$

Here, T denotes the transpose operation. For our example uniaxial metasurface, these conditions mean that

$$\widehat{\alpha}_{ee}^{cr} = \widehat{\alpha}_{mm}^{cr} = 0, \quad \widehat{\alpha}_{em}^{co} = -\widehat{\alpha}_{me}^{co}, \quad \widehat{\alpha}_{em}^{cr} = \widehat{\alpha}_{me}^{cr}, \tag{4.10}$$

which reduces the number of free parameters.

The next step is to find what are the polarizabilities of a single unit cell in free space, which would correspond to collective polarizabilities determined in the previous step. The knowledge of the individual-cell response is usually desirable because single-particle polarizabilities are easier to understand and synthesize, as there is no need to model particle interactions in infinite lattices, and for some simple topologies, even analytical models are available for polarizabilities of small scatterers in free space. The relations between the collective and individual polarizabilities can be found using the notion of the interaction factor for infinite grids. Namely, we note that each unit cell is actually excited by the local field at the unit-cell position, which is the sum of the incident field and the field created by the currents in all the other unit cells at the position of this particular reference unit cell. We express this notion in terms of the interaction constant [16, 21] β:

$$\mathbf{E}_{loc} = \mathbf{E}_{inc} + \overline{\overline{\beta}}_{e} \cdot \mathbf{p}, \tag{4.11}$$

$$\mathbf{H}_{loc} = \mathbf{H}_{inc} + \overline{\overline{\beta}}_{m} \cdot \mathbf{m}. \tag{4.12}$$

For the assumed normal-incidence plane-wave excitations, the dipole moments $\mathbf{p}$ and $\mathbf{m}$ of all unit cells are the same. Approximate analytical expressions for the interaction constants of dipolar arrays are available in the literature [16].

For a single unit cell in free space, the induced moments are proportional to the local fields (since the particle is now alone in empty infinite space, there is no difference between the local fields and the incident fields). We write these relations in terms of the individual polarizabilities as

$$\mathbf{p} = \overline{\overline{\alpha}}_{ee} \cdot \mathbf{E}_{loc} + \overline{\overline{\alpha}}_{em} \cdot \mathbf{H}_{loc}, \tag{4.13}$$

$$\mathbf{m} = \overline{\overline{\alpha}}_{mm} \cdot \mathbf{H}_{loc} + \overline{\overline{\alpha}}_{me} \cdot \mathbf{E}_{loc}. \tag{4.14}$$

Substituting the local fields from (4.11) and (4.12) into (4.3) and (4.4) and solving for the dipole moments, the individual polarizabilities can be expressed in terms of the earlier determined collective polarizabilities [18].

At this stage, we know how a single inclusion that will be used to form the metasurface with the required response should respond to external fields if tested in free space. Next, we use our knowledge on electromagnetic properties of small scatterers to find suitable realizations. Electric polarizability is easy to control. Basically, any electrically small object made of any material is electrically polarizable, and it is enough to choose the material and dimensions so that the polarizability $\overline{\overline{\alpha}}_{ee}$ takes the desired value. To access resonant regime, for microwave frequencies, metal particles of a proper shape (meander, for instance) can be used. In the infrared and visible range, plasmonic or dielectric nanoparticles offer possible realizations. To realize magnetic response in reciprocal particles, various split-ring topologies can be used at microwaves, and dielectric particles can be in the magnetic resonant mode in the visible range. Magneto-electric coupling effects (parameters $\overline{\overline{\alpha}}_{me}$ and $\overline{\overline{\alpha}}_{em}$) in reciprocal particles are realized by the proper shaping of the inclusions. For example, to control α_{em}^{co}, various spiral (mirror-asymmetric) shapes are used. Parameter α_{em}^{cr} is determined by the symmetry or asymmetry in reversal of the unit vector $\mathbf{z}_0$.

The final step is the fine tuning of the unit-cell parameters using numerical solvers for periodical arrays. This step is usually required because the analytical expressions for the interaction constants β and the analytical models for particle polarizabilities that are usually used in determining the unit-cell dimensions are approximate and the result can be improved by using accurate simulations. On the other hand, brute-force numerical optimizations without first determining the required topologies from the homogenization-model equations usually do not bring satisfactory results.

4.4 Examples

4.4.1 "Extraordinary" Reflection and Transmission

Here, we will explain why and how the amplitudes of reflection and transmission coefficients of metasurfaces can be engineered to span the whole range from nearly full reflection to full transmission. Although this is probably the simplest example of interesting and useful utilization of resonant meta-atoms, the possibility of nearly full transmission through a metal sheet with an array of small holes or full reflection from an array of small metal particles appears very unusual – "extraordinary" [22].

Let us consider a two-dimensional periodical array of small particles that respond to the exciting fields as resonant electric dipoles. The array period is smaller than the wavelength. For microwave applications, the particles can be in the form of small metal strips (or crosses if we want isotropic response). To bring them to resonance without increasing the length to $\lambda/2$, we can meander the strips or use the Jerusalem-cross shape. For optical applications, the particles can be plasmonic (Chapter 7) or high-permittivity dielectric spheres [5–8]. Considering excitation by normally incident plane waves, the dipole moments p of each unit cell are equal and proportional to the incident field:

$$p = \alpha(E_{\text{inc}} + \beta p), \qquad p = \frac{1}{\frac{1}{\alpha} - \beta} E_{\text{inc}}. \tag{4.15}$$

Here, α is the particle polarizability, and β is the interaction constant. The subwavelength array of electric dipoles creates only one secondary plane wave, also propagating in the normal direction. First, we find the averaged surface current density J by taking the time derivative of the unit-cell dipole moment and dividing by the unit-cell area S. Next, we find the electric field created by this uniform sheet of electric current and, finally, the reflection coefficient:

$$J = \frac{j\omega}{S} p = \frac{\frac{j\omega}{S}}{\frac{1}{\alpha} - \beta} E_{\text{inc}}, \qquad E_{\text{ref}} = -\frac{\eta}{2} J, \qquad R = -\frac{j\omega\eta}{2S} \frac{1}{\frac{1}{\alpha} - \beta}. \tag{4.16}$$

Here, η is the wave impedance of the surrounding space.

The interaction constant β defines the value of the interaction field $E_{\text{int}} = \beta p$, which is created at the position of a reference particle by all the other dipoles. This field equals the field of a plane wave created by the complete array (which equals $-(\eta/2)(j\omega p/S)$) minus the field of the reference dipole itself. But we consider the complete, fully regular array, and in the full expression of the denominator, the field of this "missing dipole" is compensated by the corresponding terms in the expression for the inverse polarizability of the reference dipole $1/\alpha$. Basically, the expression $(1/\alpha - \beta)$ is the inverse polarizability of the dipole in an infinite periodical array. Since there is no scattering from periodical subwavelength arrays, this plane-wave field is the only secondary field that can carry energy away from the array. Thus, assuming that there is no energy dissipation inside the particles, we find that

$$\frac{1}{\alpha} - \beta = \text{Re}\left(\frac{1}{\alpha} - \beta\right) + j\frac{\eta}{2}\frac{\omega}{S}. \tag{4.17}$$

Here, the real part of the interaction constant models near-field interactions between the cells, which usually leads to a small shift of the resonance frequency of the particles. The imaginary part is the radiation loss factor, which is due to the plane waves created by the infinite regular array.[1]

Now we can substitute this expression in (4.16) and find the reflection and transmission coefficients

$$R = -\frac{\frac{j\omega\eta}{2S}}{\text{Re}\left(\frac{1}{\alpha} - \beta\right) + \frac{j\omega\eta}{2S}}, \tag{4.18}$$

$$T = 1 + R = \frac{\text{Re}\left(\frac{1}{\alpha} - \beta\right)}{\text{Re}\left(\frac{1}{\alpha} - \beta\right) + \frac{j\omega\eta}{2S}}. \tag{4.19}$$

At the particle resonance (including the near-field interactions between unit cells), $\text{Re}\left(1/\alpha - \beta\right) \to \infty$. The preceding result tells that, at this frequency,

$R \rightarrow -1$ and $T \rightarrow 0$; the array of small resonant electric dipoles behaves as a continuous perfectly conducting wall. Note that the scattering loss, which limits the amplitude of induced dipole moments of the particles, is included in the preceding analysis; only the dissipating loss in the inclusions is neglected. Importantly, the part of the surface area that is occupied by the dipole particles can be very small, meaning that away from the resonant frequency, the array can be nearly transparent.

Next, we will consider the dual case when a continuous perfectly conducting sheet with periodically cut holes allows full transmission of normally incident plane waves. To study this case, we will use the Babinet principle. To that end, we will first reformulate the preceding results for an array of dipolar particles in terms of the effective sheet impedance of the corresponding homogenized metasurface. By definition, the sheet impedance, also called grid impedance in our case, relates the surface-averaged electric current density and the surface-averaged electric field (tangential component) at the grid plane: $Z_g J = E_{\text{tot}}$. Here,

$$E_{\text{tot}} = E_{\text{inc}} + E_{\text{ref}} = E_{\text{inc}} - \frac{\eta}{2} J, \qquad (4.20)$$

where E_{ref} is the amplitude of the reflected plane wave. We have just found how the induced electric surface current density depends on the *incident* electric field:

$$J = \frac{\frac{j\omega}{S}}{\text{Re}\left(\frac{1}{\alpha} - \beta\right) + \frac{j\omega\eta}{2S}} E_{\text{inc}}. \qquad (4.21)$$

From this result, we can express the value of the incident electric field

$$E_{\text{inc}} = \frac{S}{j\omega}\left[\text{Re}\left(\frac{1}{\alpha} - \beta\right) + \frac{j\omega\eta}{2S}\right] J \qquad (4.22)$$

and substitute it in (4.20). This way, we find the grid impedance:

$$Z_g = -j\frac{S}{\omega}\text{Re}\left(\frac{1}{\alpha} - \beta\right). \qquad (4.23)$$

In terms of Z_g, the reflection and transmission coefficients of arrays of electric dipoles take the form

$$R = -\frac{\frac{j\omega\eta}{2S}}{\text{Re}\left(\frac{1}{\alpha} - \beta\right) + \frac{j\omega\eta}{2S}} = \frac{-1}{1 + \frac{2}{\eta}Z_g}, \qquad (4.24)$$

$$T = \frac{\text{Re}\left(\frac{1}{\alpha} - \beta\right)}{\text{Re}\left(\frac{1}{\alpha} - \beta\right) + \frac{j\omega\eta}{2S}} = \frac{\frac{2}{\eta}Z_g}{1 + \frac{2}{\eta}Z_g}. \qquad (4.25)$$

We again see that at the particle resonance where $Z_g \rightarrow 0$, we get $R \rightarrow -1$ and $T \rightarrow 0$.

Now we are ready to study a continuous perfectly conducting sheet with a periodical array of subwavelength holes. The preceding expressions for the reflection and transmission coefficients in terms of the sheet impedance are also valid in this case, since the only assumption was that the metasurface supports only electric surface current, and it is periodical with a subwavelength period. Of course, the value of the impedance is different, and we denote it $Z_{\text{g holes}}$:

$$R = \frac{-1}{1 + \frac{2}{\eta} Z_{\text{g holes}}}, \qquad T = \frac{\frac{2}{\eta} Z_{\text{g holes}}}{1 + \frac{2}{\eta} Z_{\text{g holes}}}. \qquad (4.26)$$

The Babinet principle relates the grid impedance of an array of holes to the grid impedance of the complementary array of metal particles having the same shape and size as the holes (for the orthogonal polarization of the electric field):

$$Z_{\text{g holes}} Z_{\text{g}} = \frac{\eta^2}{4} \qquad (4.27)$$

(see, e.g., [23, section 5.1B]). We have just found the grid impedance of arrays of small metal patches in terms of the particle polarizability and the interaction constant, and we can use that knowledge to find the reflection and transmission coefficients for arrays of holes in a conducting sheet. In terms of the grid impedance for the complementary particle array, we get

$$R = \frac{-1}{1 + \frac{\eta}{2 Z_{\text{g}}}}, \qquad T = \frac{\frac{\eta}{2 Z_{\text{g}}}}{1 + \frac{\eta}{2 Z_{\text{g}}}}. \qquad (4.28)$$

We see that the behavior is, as expected, complementary. At the resonance of the particle polarizability, we have $Z_{\text{g}} \to 0$, $R \to 0$, $T \to 1$, and the sheet with an array of small holes becomes fully transparent. This effect of resonant tunneling is sometimes called "extraordinary transmission." Far from the resonance, the sheet with small holes is a nearly perfect reflector. Here, we have considered probably the simplest case of a sheet of negligible thickness; more elaborate models are needed to understand extraordinary transmission through thick slabs with holes.

Figures 4.4 and 4.5 present example of reflection and transmission properties of dense arrays of small lossless or low-loss resonant unit cells. In the first example, illustrated in Figure 4.4, the unit cell is a meandered metal strip. For simplicity, the array is positioned in air (no substrate). The material of the

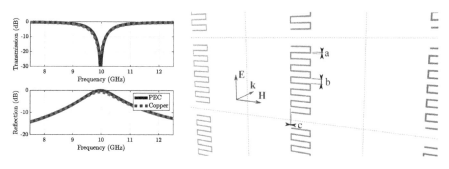

Figure 4.4 Amplitude of reflection and transmission coefficients for an electrically dense array of small resonant electric dipoles (metal strips are meandered to reduce the overall size of the elements). Simulations and graphics by F. Cuesta.

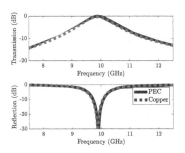 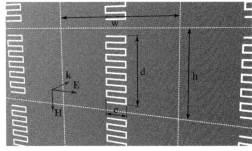

Figure 4.5 Amplitude of reflection and transmission coefficients for an electrically dense array of resonant slots in a continuous conductive sheet (complementary with respect to the topology in Figure 4.4. Simulations and graphics by F. Cuesta.

strips is either PEC (solid lines) or copper (dashed lines). The dimensions are chosen so that the unit cell sizes w and t are smaller than the wavelength (no higher-order propagating Floquet harmonics). In this example, the dimensions are the following: $a = 0.11$ mm, $b = 0.32$ mm, $c = 0.05$ mm, $d = 4.83$ mm, $e = 1.68$ mm, $w = 9.45$ mm, and $h = 5.91$ mm, which correspond to the fundamental resonance frequency at 10 GHz. We observe that, as expected from Eqs. (4.18) and (4.19), at the resonant frequency of the meander elements (slightly modified by the near-field interactions between unit cells) the array acts as a nearly perfect reflector, although only a very small portion of the array plane is occupied by metal strips. In the second example, illustrated in Figure 4.5, we consider a complementary cell with a slot of the same shape and size cut in a continuous metal sheet. As explained earlier, at the cell resonance, the first array is a nearly perfect reflector, and the second array is nearly transparent. To confirm this property, we run simulations of reflection and transmission coefficients for the orthogonal (rotated 90°) polarization plane of the incident plane wave and, indeed, see that at the same resonant frequency the metasurface is nearly fully transparent, although most of the area of the surface is filled with a highly reflective metal sheet.

Another physical mechanism for resonant full reflection or full transmission through periodical, subwavelength-structured arrays are Wood anomalies near the frequencies where the period is equal to one wavelength or several wavelengths. These effects are due to divergence of the interaction constant β at these frequencies, which happens because waves created by all the dipoles in the array come in phase at the position of the reference dipole, creating infinite interaction field. In the limit of infinite β, the grid impedance (4.23) tends to infinity. At these frequencies, periodical arrays of small (non-resonant) lossless inclusions of arbitrary size and shape become fully transparent, as is clear from Eq. (4.24). The complementary arrays of holes at these frequencies are fully reflective (Eq. (4.28)). Slightly above the frequency where the period equals one wavelength, the interaction constant is finite and large, and in Eq. (4.23), it can cancel the large value of $\mathrm{Re}(1/\alpha)$ of the inverse polarizability of small, non-resonant inclusions. At this resonance, the grid impedance (4.23) tends to zero (neglecting dissipation loss), which corresponds to the opposite behavior of the array. The array of small particles is fully reflective and the array of small holes is fully transparent [24]. However, these effects, strictly speaking, refer not to metasurfaces but to diffraction gratings because the array period is close to the wavelength.

Note that extraordinary high transmission through thin conducting sheets does not require resonance of slots. For example, it is well known that electrically dense meshes of thin metal strips are very good reflectors in the whole frequency range where the mesh period is small compared to the wavelength (the grid impedance of such meshes is proportional to the ratio of the mesh period and the wavelength [16]. Babinet-complementary sheets (continuous metal sheets with periodically cut thin slots) are highly transparent while most of the sheet area is metal.

We can conclude that, engineering the shape and size of small meta-atoms, we can tune the amplitudes of reflection and transmission coefficients in a very wide range, limited only by dissipation loss in the structure. However, if we control only electric response (allowing only electric surface currents), there are important limitations on achievable performance. Most importantly it is not possible to control reflection and transmission independently because, due to the continuity of the surface-averaged tangential electric field, they are related: $T = 1 + R$. In the following section, we will look at more interesting examples where these limitations are overcome by introducing engineered magnetic currents.

4.4.2 Matched Transmitarrays

The material of this section is based on the results of V. Asadchy et al. [25]. The functionality of transmitarrays is to shape the transmitted waves, usually by locally controlling the phase of waves passing through the metasurface. Losses due to reflections and due to absorption in the metasurface should be minimized. Typical applications are thin and (usually) flat lenses. Conventional transmitarrays based on frequency-selective surfaces [26–30] or on metasurfaces [31, 32] are multilayer structures with typically three to five layers of patch arrays. Conventional single-layer transmitarrays [33] produce considerable reflections. The metasurface concept allows the synthesis and realization of single-layer arrays with comparable or better functionalities; see an illustration in Figure 4.6.

We start from the general expressions for the reflected and transmitted fields (4.8) and demand that the reflected wave is absent,

$$\eta_0 \widehat{\alpha}_{ee}^{co} \pm \widehat{\alpha}_{em}^{cr} \pm \widehat{\alpha}_{me}^{cr} - \frac{1}{\eta_0} \widehat{\alpha}_{mm}^{co} = 0, \tag{4.29}$$

$$\eta_0 \widehat{\alpha}_{ee}^{cr} \mp \widehat{\alpha}_{em}^{co} \mp \widehat{\alpha}_{me}^{co} - \frac{1}{\eta_0} \widehat{\alpha}_{mm}^{cr} = 0, \tag{4.30}$$

that the transmitted wave has the same polarization as the incident wave,

$$\eta_0 \widehat{\alpha}_{ee}^{cr} \mp \widehat{\alpha}_{em}^{co} \pm \widehat{\alpha}_{me}^{co} + \frac{1}{\eta_0} \widehat{\alpha}_{mm}^{cr} = 0, \tag{4.31}$$

A metasurface as a device to shape transmitted fields. Courtesy of V. Asadchy.

and that, at each point of the metasurface, the transmitted wave has the same amplitude as the incident wave, and its phase is shifted by the desired angle ϕ:

$$1 - \frac{j\omega}{2S}\left(\eta_0\widehat{\alpha}_{ee}^{co} \pm \widehat{\alpha}_{em}^{cr} \mp \widehat{\alpha}_{me}^{cr} + \frac{1}{\eta_0}\widehat{\alpha}_{mm}^{co}\right) = e^{j\phi}. \qquad (4.32)$$

The $\pm$ signs in the preceding relations refer to the incident waves illuminating the two sides of the metasurface (waves propagating along $\mp z_0$). If we want to design a transmitarray that works the same way when illuminated from any side, we should make sure that the preceding relations are satisfied when we take either the top or bottom sign in the preceding equations. Under this restriction, we should choose unit cells whose parameters obey

$$\widehat{\alpha}_{em}^{cr} + \widehat{\alpha}_{me}^{cr} = 0, \qquad \widehat{\alpha}_{em}^{cr} - \widehat{\alpha}_{me}^{cr} = 0, \qquad (4.33)$$

$$\widehat{\alpha}_{em}^{co} + \widehat{\alpha}_{me}^{co} = 0, \qquad \widehat{\alpha}_{em}^{co} - \widehat{\alpha}_{me}^{co} = 0, \qquad (4.34)$$

which means that all these coupling coefficients must equal zero, and the unit cells should not exhibit any bianisotropic effects.

Thus, for symmetric transmitarrays, the preceding requirements (4.29)–(4.32) simplify to

$$\eta_0\widehat{\alpha}_{ee}^{co} - \frac{1}{\eta_0}\widehat{\alpha}_{mm}^{co} = 0 \qquad (4.35)$$

$$\eta_0\widehat{\alpha}_{ee}^{cr} - \frac{1}{\eta_0}\widehat{\alpha}_{mm}^{cr} = 0, \qquad \eta_0\widehat{\alpha}_{ee}^{cr} + \frac{1}{\eta_0}\widehat{\alpha}_{mm}^{cr} = 0 \qquad (4.36)$$

$$1 - \frac{j\omega}{2S}\left(\eta_0\widehat{\alpha}_{ee}^{co} + \frac{1}{\eta_0}\widehat{\alpha}_{mm}^{co}\right) = e^{j\phi}. \qquad (4.37)$$

Equations (4.36) mean that $\widehat{\alpha}^{\mathrm{cr}}_{\mathrm{ee}} = \widehat{\alpha}^{\mathrm{cr}}_{\mathrm{mm}} = 0$, the electric and magnetic polarizabilities are symmetric dyadics, and, thus, the unit cells must be reciprocal (see (4.10)). The remaining two simple equations (4.35) and (4.37) have a unique solution for the required electric and magnetic collective polarizabilities:

$$\eta_0 \widehat{\alpha}^{\mathrm{co}}_{\mathrm{ee}} = \frac{1}{\eta_0} \widehat{\alpha}^{\mathrm{co}}_{\mathrm{mm}} = \frac{S}{j\omega} \left(1 - e^{j\phi}\right). \tag{4.38}$$

Next, we use the relations [18] between the collective polarizabilities of the particles in infinite arrays $\widehat{\alpha}^{\mathrm{co}}_{\mathrm{ee}}, \widehat{\alpha}^{\mathrm{co}}_{\mathrm{mm}}$ and the polarizabilities of the same particles considered as individual, single scatterers in free space. For this simple case of reciprocal non-bianisotropic unit cells, these relations read [18, 25]

$$\frac{1}{\eta_0 \alpha_{\mathrm{ee}}} = \frac{1}{\eta_0 \widehat{\alpha}^{\mathrm{co}}_{\mathrm{ee}}} + \frac{\beta_{\mathrm{e}}}{\eta_0}, \qquad \frac{1}{\alpha_{\mathrm{mm}}/\eta_0} = \frac{1}{\widehat{\alpha}^{\mathrm{co}}_{\mathrm{mm}}/\eta_0} + \frac{\beta_{\mathrm{e}}}{\eta_0}. \tag{4.39}$$

From here, we find the required polarizabilities of individual unit cells in free space:

$$\frac{1}{\eta_0 \alpha_{\mathrm{ee}}} = \frac{1}{\alpha_{\mathrm{mm}}/\eta_0} = \frac{1}{\eta_0} \mathrm{Re}(\beta_{\mathrm{e}}) - \frac{\omega}{2S} \frac{\sin \phi}{1 - \cos \phi} + j \frac{k^3}{6\pi \sqrt{\epsilon_0 \mu_0}}. \tag{4.40}$$

This result corresponds to a lossless dipole scatterer because the imaginary part is only due to radiation damping. The real part of the interaction constant can be estimated analytically for moderate values of the unit cell sizes and small dipolar particles [16]. For electrically small unit cells, the following quasi-static approximation is valid:

$$\frac{1}{\eta_0} \mathrm{Re}(\beta_{\mathrm{e}}) \approx \frac{0.36}{\sqrt{\epsilon_0 \mu_0}\, a^3}, \tag{4.41}$$

where a is the array period (considering square unit cells).

To realize such a transmitarray, we need to design electrically small particles with the polarizabilities given by (4.40). The particle should be polarizable by both electric and magnetic fields, have no bianisotropy, and be made from a material with negligible losses. A good candidate for realizing low-loss particles with balanced electric and magnetic polarizabilities at microwave frequencies is a metal spiral; see the previous section. However, spirals are chiral objects, but for this application, chirality is not allowed (Eqs. (4.34)). A possible solution is to use unit cells containing racemic combinations of two or more balanced spirals. Such arrays behave as non-chiral metasurfaces because the number of right- and left-handed spirals in each unit cell is the same, and the chirality of individual spirals is compensated by other spirals in the same unit cell.

Note that the same approach can be used to realize single-layer absorbers, which are transparent outside of the absorption band [25] because for that application also, one needs to ensure balanced electric and magnetic response, but chirality is not allowed.

Figure 4.7   Single-layer metamirrors can independently control reflections from their two sides, while the transmission through the metasurface is negligible. Courtesy of V. Asadchy.

4.4.3 Metamirrors

The material of this section is based on the results of Y. Ra'di, V. Asadchy, et al. [34, 35]. Let us consider synthesis of metasurfaces that fully reflect incident waves allowing full control over the reflection phase. Often, it is desirable to utilize both sides of the metasurface, so we will seek for possibilities to independently control the reflection phase for illuminations of both sides of the metasurface. Let us also demand that the polarization state does change upon reflection. Such a metamirror can be used, for example, as reflectarray antennas, replacing large and heavy parabolic reflectors. Thought realizations as single arrays of small scatterers offer additional application possibilities since metamirrors (in contrast to conventional reflectarrays) have no ground plane and are transparent outside of the operational frequency band. Possible functionalities of metamirrors [34] are illustrated on Figure 4.7.

Similarly to the previous examples, we start from the general expressions for the reflection and transmission coefficients (4.8) in terms of the collective polarizabilities of unit cells. Because the desired properties should hold for any polarization of the incident fields, the use of uniaxial structures (with the only preferred direction being the direction normal to the surface) is the only possibility. For the thought application, we demand that

$$1 - \frac{j\omega}{2S}\left(\eta_0\widehat{\alpha}_{\mathrm{ee}}^{\mathrm{co}} \pm \widehat{\alpha}_{\mathrm{em}}^{\mathrm{cr}} \mp \widehat{\alpha}_{\mathrm{me}}^{\mathrm{cr}} + \frac{1}{\eta_0}\widehat{\alpha}_{\mathrm{mm}}^{\mathrm{co}}\right) = 0 \qquad (4.42)$$

(co-polarized transmission coefficient is zero),

$$\eta_0 \widehat{\alpha}_{\text{ee}}^{\text{cr}} \mp \widehat{\alpha}_{\text{em}}^{\text{co}} \pm \widehat{\alpha}_{\text{me}}^{\text{co}} + \frac{1}{\eta_0} \widehat{\alpha}_{\text{mm}}^{\text{cr}} = 0 \qquad (4.43)$$

(cross-polarized transmission coefficient is zero),

$$\eta_0 \widehat{\alpha}_{\text{ee}}^{\text{cr}} \mp \widehat{\alpha}_{\text{em}}^{\text{co}} \mp \widehat{\alpha}_{\text{me}}^{\text{co}} - \frac{1}{\eta_0} \widehat{\alpha}_{\text{mm}}^{\text{cr}} = 0 \qquad (4.44)$$

(cross-polarized reflection coefficient is zero),

$$-\frac{j\omega}{2S} \left(\eta_0 \widehat{\alpha}_{\text{ee}}^{\text{co}} + \widehat{\alpha}_{\text{em}}^{\text{cr}} + \widehat{\alpha}_{\text{me}}^{\text{cr}} - \frac{1}{\eta_0} \widehat{\alpha}_{\text{mm}}^{\text{co}} \right) = e^{j\phi} \qquad (4.45)$$

(co-polarized reflected field has the same amplitude as the incident field and the desired phase shift ϕ if the incident wave propagates along $-\mathbf{z}_0$),

$$-\frac{j\omega}{2S} \left(\eta_0 \widehat{\alpha}_{\text{ee}}^{\text{co}} - \widehat{\alpha}_{\text{em}}^{\text{cr}} - \widehat{\alpha}_{\text{me}}^{\text{cr}} - \frac{1}{\eta_0} \widehat{\alpha}_{\text{mm}}^{\text{co}} \right) = e^{j\theta} \qquad (4.46)$$

(co-polarized reflected field has the same amplitude as the incident field and the desired phase shift θ if the incident wave propagates along $+\mathbf{z}_0$).

We see that there are several alternative possibilities to realize metamirrors, using reciprocal or nonreciprocal unit cells. However, in most practical situations, the physical properties of inclusions that we can use can be restricted by various considerations. For example, let us consider condition (4.43), which ensures that there is no cross-polarized (with respect to the polarization of the incident wave) transmitted field. Physically, this condition means that the strengths of all physical effects that result in creation of cross-polarized fields behind the metasurface must be balanced so that the total cross-polarized transmission is zero. From (4.43) we see that the cross-polarized transmission can appear if at least one of the following is true:

1. $\widehat{\alpha}_{\text{ee}}^{\text{cr}} \neq 0$
2. $\widehat{\alpha}_{\text{mm}}^{\text{cr}} \neq 0$
3. $\widehat{\alpha}_{\text{em}}^{\text{co}} - \widehat{\alpha}_{\text{me}}^{\text{co}} \neq 0$

In case (1), when $\widehat{\alpha}_{\text{ee}}^{\text{cr}} \neq 0$, the electric response is nonreciprocal, for example, we have some magnetized plasma filling our unit cells. In case (2), when $\widehat{\alpha}_{\text{mm}}^{\text{cr}} \neq 0$, the magnetic response is nonreciprocal, meaning that we have, for example, some magnetized ferrite materials in the unit cells (there are other possibilities to realize nonreciprocal unit cells). In both cases, the polarization transformation in transmission is due to the Faraday effect. In case (3), if the cells are reciprocal, we have chiral (mirror-asymmetric) unit cells, and there is cross-polarizated transmission due to the optical activity of the metasurface.

We see from (4.43) that if we are going to use nonreciprocal unit cells (for example, to have more flexibility in shaping reflections), either *both* electric and magnetic response must be nonreciprocal or, if only one of the parameters $\widehat{\alpha}_{\text{ee}}^{\text{cr}}$ and $\widehat{\alpha}_{\text{mm}}^{\text{cr}}$ is non-zero, we must use chiral unit cells and carefully balance the strength of both effects so that (4.43) is satisfied.

In fact, it is clear that the required functionality (full control over the reflection phase) does not require any nonreciprocal phenomena. Thus, it is most reasonable to use reciprocal structures for realizing metamirrors. Based on these considerations, we can discard possible nonreciprocal realizations, which means that $\widehat{\alpha}_{ee}^{cr} = \widehat{\alpha}_{mm}^{cr} = 0$ (see the reciprocity conditions (4.10)). In this case, to satisfy (4.43), we must ensure that $\widehat{\alpha}_{em}^{co} - \widehat{\alpha}_{me}^{co} = 0$. But the reciprocity condition tells that $\widehat{\alpha}_{em}^{co} = -\widehat{\alpha}_{me}^{co}$; thus, both these coupling coefficients must vanish. Physically, this means that if a metamirror is reciprocal, it must be non-chiral. Similar considerations can be applied to other requirements (4.42)–(4.46). For reciprocal metamirrors, the solution of (4.42)–(4.46) is unique, and it reads [34]

$$\eta_0 \widehat{\alpha}_{ee}^{co} = \frac{S}{j\omega} \left[1 - \frac{e^{j\phi} + e^{j\theta}}{2} \right],$$

$$\widehat{\alpha}_{em}^{cr} = \widehat{\alpha}_{me}^{cr} = \frac{-S}{j\omega} \left[\frac{e^{j\phi} - e^{j\theta}}{2} \right], \tag{4.47}$$

$$\frac{1}{\eta_0} \widehat{\alpha}_{mm}^{co} = \frac{S}{j\omega} \left[1 + \frac{e^{j\phi} + e^{j\theta}}{2} \right].$$

The next step is to find what are the required individual polarizabilities of a single unit cell in free space (not interacting with the other particles in the array); see (4.13) and (4.14). This can be done using the concept of the interaction constant $\overline{\overline{\beta}}$, as explained before. The result reads [34]

$$\eta_0 \alpha_{ee}^{co} = \frac{1 - e^{j(\theta+\phi)} + \frac{j\omega\eta_0}{\beta_e S} \left[-\frac{1}{2} \left(e^{j\phi} + e^{j\theta} \right) + 1 \right]}{-e^{j(\theta+\phi)} + \left(1 + j\frac{\omega\eta_0}{\beta_e S} \right)^2} \frac{\eta_0}{\beta_e},$$

$$\alpha_{em}^{cr} = \alpha_{me}^{cr} = \frac{-\frac{j\omega\eta_0}{2\beta_e S} \left(e^{j\phi} - e^{j\theta} \right)}{-e^{j(\theta+\phi)} + \left(1 + j\frac{\omega\eta_0}{\beta_e S} \right)^2} \frac{\eta_0}{\beta_e}, \tag{4.48}$$

$$\frac{1}{\eta_0} \alpha_{mm}^{co} = \frac{1 - e^{j(\theta+\phi)} + \frac{j\omega\eta_0}{\beta_e S} \left[\frac{1}{2} \left(e^{j\phi} + e^{j\theta} \right) + 1 \right]}{-e^{j(\theta+\phi)} + \left(1 + j\frac{\omega\eta_0}{\beta_e S} \right)^2} \frac{\eta_0}{\beta_e}.$$

Approximate analytical formulas for calculations of the electrical interaction coefficient β_e can be found in the literature [16, 34]; see also Eq. (4.41).

The result shows that the unit cells should be lossless bianisotropic omega particles. The cells should be polarizable electrically ($\alpha_{ee}^{co} \neq 0$), magnetically ($\alpha_{mm}^{co} \neq 0$), and there should be bianisotropic omega coupling ($\alpha_{em}^{cr} = \alpha_{me}^{cr} \neq 0$). The last condition means that electric field applied to the particle in the direction orthogonal to $\mathbf{z}_0$ should induce magnetic moment in the direction orthogonal to the applied electric field and also orthogonal to $\mathbf{z}_0$. The typical topology of such particles is an Ω-shaped piece of a metal wire (appropriate for microwave-frequency realizations). Knowing the required values of the polarizabilities, we can now use the known analytical models of polarizabilities of omega particles [20, 36] and find the particle dimensions for which the polarizabilities take the required values. At the final stage, the dimensions can be optimized using full-wave simulations of the designed particles in the infinite array.

Reports on numerical and experimental studies of metamirrors realized as arrays of metal Ω-shaped particles can be found in the literature [34, 35, 37]. Specially shaped dielectric particles can be used for realization of metamirrors for optical applications [38].

4.5 Perfect Control of Transmission and Reflection

The design methodology explained in the previous sections is based on the assumption that the metasurface introduces a phase shift in transmission or reflection without any modification in the wave amplitude, following the same design principle as used in conventional phased arrays. Recently, it has been demonstrated that this simplistic method does not allow us to achieve perfect performance due to the inevitable scattering of some energy in undesirable directions [39–41]. These studies show that for realizing perfectly performing metasurfaces, which redirect all the energy in the desired direction or directions, the amplitudes of the incident and transmitted/reflected waves have to be different. Here, we explain recent advances in the design of perfect metasurfaces for shaping transmitted and reflected waves. We will go into detail about the power considerations in both transmission and reflection scenarios. Moreover, methods for calculating the required collective polarizabilities in both scenarios will be discussed.

4.5.1 Perfect Transmitarrays

In order to realize perfect control of transmission through metasurfaces, an important condition that ensures complete suppression of the energy radiated in undesirable directions should be satisfied. Here, we present this condition and an approach for calculating the collective polarizabilities that ensure perfect anomalous refraction into any desired direction.

First, we define the electric and magnetic fields when a TE-polarized wave, with the amplitude $\mathbf{E}_i$, illuminates the metasurface from medium 1 at a certain angle θ_i (see Figure 4.8(a)). We refer to this case of incidence as *forward illumination*. Assuming ideally zero reflection and anomalous refraction into a certain direction, the tangential components of the electric and magnetic fields on both sides of the metasurface can be written as

$$\mathbf{E}_{t1} = \mathbf{E}_i e^{-jk_1 \sin\theta_i z}, \quad \mathbf{n} \times \mathbf{H}_{t1} = \frac{\cos\theta_i}{\eta_1} \mathbf{E}_i e^{-jk_1 \sin\theta_i z}, \tag{4.49}$$

$$\mathbf{E}_{t2} = t_{\mathrm{TE}} \, \mathbf{E}_i e^{-jk_1 \sin\theta_i z}, \quad \mathbf{n} \times \mathbf{H}_{t2} = t_{\mathrm{TE}} \, \frac{\cos\theta_t}{\eta_2} \mathbf{E}_i e^{-jk_1 \sin\theta_i z}, \tag{4.50}$$

where t_{TE} is the complex transmission coefficient

$$t_{\mathrm{TE}} = |t_{\mathrm{TE}}|e^{j\Phi_t}, \quad \Phi_t = (k_1 \sin\theta_i - k_2 \sin\theta_t)z + \phi_t, \tag{4.51}$$

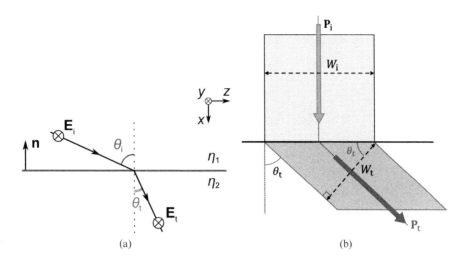

Figure 4.8 Illustration of the performance desired in an ideal transmitarray. (a) Coordinate system and field definitions. Reprinted from [39] with permission. © 2016 APS. (b) Schematic representation of the power conservation in the transmitarray.

and $\eta_{1,2}$ are wave impedances of the two half-spaces. ϕ_t is the desired phase shift in transmission through the metasurface.

Perfect transmission can be obtained by ensuring that all the incident energy goes into the desirable direction and there is no dissipation in the metasurface; thus, the normal component of the Poynting vector has to be the same at both sides of the metasurface:

$$\frac{1}{2}\mathrm{Re}\left(\mathbf{E}_{t1} \times \mathbf{H}_{t1}^{*}\right) = \frac{1}{2}\mathrm{Re}\left(\mathbf{E}_{t2} \times \mathbf{H}_{t2}^{*}\right). \tag{4.52}$$

This condition defines a relation between the incident and the reflected field amplitudes, which reads

$$\mathbf{E}_t = \mathbf{E}_i\sqrt{\frac{\cos\theta_i}{\cos\theta_t}}\sqrt{\frac{\eta_1}{\eta_2}} = \mathbf{E}_i|t_{\mathrm{TE}}|. \tag{4.53}$$

Equation (4.53) shows that the amplitudes of the incident and transmitted waves are different. This condition can be easily understood intuitively by considering the example of a metasurface which steers a beam of a finite width W_i. Figure 4.8(b) conceptually illustrates this scenario, when the incident wave impinges normally into the metasurface (for simplicity). We can see that the transmitted beam is tilted by θ_t, and it has a smaller width W_t. Using simple trigonometrical analysis, we can obtain the relation between both widths, $W_i = W_t/\cos\theta_t$, supporting the conclusion extracted from Eq. (4.53).

To find the required collective polarizabilities, we need to study the reciprocal case when the metasurface is illuminated by the incident wave with the electric field $\mathbf{E}_t$ from medium 2 at the angle θ_t. We refer to this incidence scenario as *backward illumination.* In this case, the tangential components of the electric and magnetic fields read

$$\mathbf{E}_{t1} = \mathbf{E}_i e^{jk_1 \sin \theta_i z}, \quad \mathbf{n} \times \mathbf{H}_{t1} = -\frac{\cos \theta_i}{\eta_1} \mathbf{E}_i e^{jk_1 \sin \theta_i z}, \tag{4.54}$$

$$\mathbf{E}_{t2} = t_{\text{TE}}^* \mathbf{E}_i e^{jk_1 \sin \theta_i z}, \quad \mathbf{n} \times \mathbf{H}_{t2} = -t_{\text{TE}}^* \frac{\cos \theta_t}{\eta_2} \mathbf{E}_i e^{jk_1 \sin \theta_i z}, \tag{4.55}$$

where the asterisk (*) denotes the complex conjugate operator.

Once we have defined the functionality of the metasurface ensuring the reciprocal response, the next step in the design methodology is to obtain analytical formulas for the collective polarizabilities of the ideal refractive metasurface. Using Eqs. (4.3) and (4.4) and the definition of the fields in the forward illumination (f) scenario, we can find an expression for the tangentially oriented dipole moments of unit cells $\mathbf{p}_t^f$ and $\mathbf{m}_t^f$:

$$\mathbf{p}_t^f = \left(\widehat{\alpha}_{\text{ee}}^{yy} + \widehat{\alpha}_{\text{em}}^{yz} \frac{\cos \theta_i}{\eta_1} \right) \mathbf{E}_i e^{-jk_1 \sin \theta_i z},$$

$$\mathbf{n} \times \mathbf{m}_t^f = \left(\widehat{\alpha}_{\text{me}}^{zy} + \widehat{\alpha}_{\text{mm}}^{zz} \frac{\cos \theta_i}{\eta_1} \right) \mathbf{E}_i e^{-jk_1 \sin \theta_i z}. \tag{4.56}$$

Following the same procedure for the case of backward illumination (b), the same polarizabilities relate another set of dipole moments $\mathbf{p}_t^b$ and $\mathbf{m}_t^b$ to the fields incident from medium 2 (that is, to $t_{\text{TE}}^* \mathbf{E}_i e^{jk_1 \sin \theta_i z}$):

$$\mathbf{p}_t^b = \left(\widehat{\alpha}_{\text{ee}}^{yy} - \widehat{\alpha}_{\text{em}}^{yz} \frac{\cos \theta_t}{\eta_2} \right) t_{\text{TE}}^* \mathbf{E}_i e^{jk_1 \sin \theta_i z},$$

$$\mathbf{n} \times \mathbf{m}_t^b = \left(\widehat{\alpha}_{\text{me}}^{zy} - \widehat{\alpha}_{\text{mm}}^{zz} \frac{\cos \theta_t}{\eta_2} \right) t_{\text{TE}}^* \mathbf{E}_i e^{jk_1 \sin \theta_i z}. \tag{4.57}$$

In the next step, we apply the boundary conditions that link the tangential fields on the two sides of the metasurface to the electric $\mathbf{p}_t/S$ and magnetic $\mathbf{m}_t/S$ surface polarization densities induced in the metasurface. These classical boundary conditions simply express the fact that the jumps of the tangential field components at the interface equal to the corresponding surface current densities:

$$\mathbf{E}_{t1} - \mathbf{E}_{t2} = \frac{j\omega}{S} \mathbf{n} \times \mathbf{m}_t^{f,b},$$

$$\mathbf{n} \times \mathbf{H}_{t1} - \mathbf{n} \times \mathbf{H}_{t2} = \frac{j\omega}{S} \mathbf{p}_t^{f,b}. \tag{4.58}$$

Substituting (4.49)–(4.55) and (4.56)–(4.57) in (4.58) for the two cases of illumination into the boundary conditions, one can obtain the following system of linear equations

$$\frac{\cos\theta_i}{\eta_1} - \frac{\cos\theta_t}{\eta_2} t_{\text{TE}} = \frac{j\omega}{S}\left(\widehat{\alpha}_{\text{ee}}^{yy} + \widehat{\alpha}_{\text{em}}^{yz}\frac{\cos\theta_i}{\eta_1}\right),$$

$$1 - t_{\text{TE}} = \frac{j\omega}{S}\left(\widehat{\alpha}_{\text{me}}^{zy} + \widehat{\alpha}_{\text{mm}}^{zz}\frac{\cos\theta_i}{\eta_1}\right),$$

$$-\frac{\cos\theta_i}{\eta_1} + \frac{\cos\theta_t}{\eta_2} t_{\text{TE}}^* = \frac{j\omega}{S}\left(\widehat{\alpha}_{\text{ee}}^{yy} - \widehat{\alpha}_{\text{em}}^{yz}\frac{\cos\theta_t}{\eta_2}\right) t_{\text{TE}}^*,$$

$$1 - t_{\text{TE}}^* = \frac{j\omega}{S}\left(\widehat{\alpha}_{\text{me}}^{zy} - \widehat{\alpha}_{\text{mm}}^{zz}\frac{\cos\theta_t}{\eta_2}\right) t_{\text{TE}}^*. \tag{4.59}$$

Solving this system, we find the required polarizabilities

$$\widehat{\alpha}_{\text{ee}}^{yy} = \frac{S}{j\omega}\frac{\cos\theta_i\cos\theta_t}{\eta_1\cos\theta_t + \eta_2\cos\theta_i}\left[2 - \left(\sqrt{\frac{\eta_1\cos\theta_t}{\eta_2\cos\theta_i}} + \sqrt{\frac{\eta_2\cos\theta_i}{\eta_1\cos\theta_t}}\right)e^{j\Phi_t(z)}\right], \tag{4.60}$$

$$\widehat{\alpha}_{\text{mm}}^{zz} = \frac{S}{j\omega}\frac{\eta_1\eta_2}{\eta_1\cos\theta_t + \eta_2\cos\theta_i}\left[2 - \left(\sqrt{\frac{\eta_1\cos\theta_t}{\eta_2\cos\theta_i}} + \sqrt{\frac{\eta_2\cos\theta_i}{\eta_1\cos\theta_t}}\right)e^{j\Phi_t(z)}\right], \tag{4.61}$$

$$\widehat{\alpha}_{\text{em}}^{yz} = -\widehat{\alpha}_{\text{me}}^{zy} = \frac{S}{j\omega}\frac{\eta_2\cos\theta_i - \eta_1\cos\theta_t}{\eta_1\cos\theta_t + \eta_2\cos\theta_i}. \tag{4.62}$$

Equations (4.60) and (4.61) show that the electric and magnetic polarizabilities depend on the coordinate z. In other words, zero reflection at any point of the metasurface is required for obtaining a perfect refractive metasurface, which demands the balance of the induced electric and magnetic surface currents at any point (the Huygens condition). On the other hand, the omega coupling coefficient in (4.62) is constant with respect to z and depends only on the impedances and the angles. This result reflects the fact that bianisotropic coupling of the omega type is necessary to ensure that the waves incident on both sides of the metasurface at different angles θ_i and θ_t see the same surface impedance so that reciprocal full transmission is realized. It is important to notice that when the impedances of the incident and transmitted waves are the same – that is, $\frac{\eta_1}{\cos\theta_i} = \frac{\eta_2}{\cos\theta_t}$ – the required coupling coefficient vanishes. Basically, the bianisotropic coupling in the metasurface is needed for matching the impedances of plane waves at the opposite sides of the metasurface.

4.5.2 Perfect Anomalous Reflection

Perfect realizations of anomalous reflections are also not possible just tuning the reflection phase over the metasurface area. Let us assume that an incident plane wave with the amplitude $\mathbf{E}_i$ and the incident angle θ_i is reflected into another plane wave with the amplitude $\mathbf{E}_r$ and the direction defined by the reflection angle

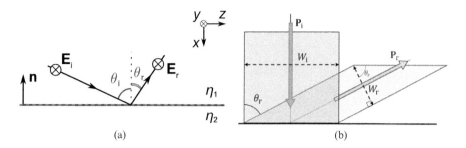

Figure 4.9 Illustration of the performance desired in a perfect reflectarray. (a) Coordinate system and definitions of the fields. Reprinted from [39] with permission. © 2016 APS. (b) Schematic representation of the power conservation in the transmitarray.

θ_r (see Figure 4.9(a)). Considering reflective metasurfaces, we demand that the tangential fields behind the metasurface are zero:

$$\mathbf{E}_{t2} = 0, \quad \mathbf{n} \times \mathbf{H}_{t2} = 0. \tag{4.63}$$

Let us define the desired performance. The incident plane wave is reflected at the angle θ_r with the phase delay ϕ_r. Tangential components of the electric and magnetic fields in medium 1 read

$$\mathbf{E}_{t1} = \mathbf{E}_i e^{-jk_1 \sin \theta_i z} + \mathbf{E}_r e^{-jk_1 \sin \theta_r z + j\phi_r},$$

$$\mathbf{n} \times \mathbf{H}_{t1} = \mathbf{E}_i \frac{1}{\eta_1} \cos \theta_i e^{-jk_1 \sin \theta_i z} - \mathbf{E}_r \frac{1}{\eta_1} \cos \theta_r e^{-jk_1 \sin \theta_r z + j\phi_r}. \tag{4.64}$$

This field structure corresponds to the position-dependent reflection coefficient for the TE incidence given by

$$r_{\mathrm{TE}} = \frac{|\mathbf{E}_r| e^{-jk_1 \sin \theta_r z + j\phi_r}}{|\mathbf{E}_i| e^{-jk_1 \sin \theta_i z}} = |r_{\mathrm{TE}}| e^{j\Phi_r}, \quad \Phi_r = k_1 (\sin \theta_i - \sin \theta_r)z + \phi_r. \tag{4.65}$$

The condition for perfect reflection can be found, similarly to the transmission case, by analyzing the normal component of the Poynting vector. In this case, we have to ensure that all the energy illuminating the metasurface is reflected only into the desired direction. The power carried by the incident plane wave is $P_i = \frac{|\mathbf{E}_i|^2}{2\eta_1} \cos \theta_i$, while the power carried by the reflected wave is $P_r = \frac{|\mathbf{E}_r|^2}{2\eta_1} \cos \theta_r$. Equating these two expressions, we can find the relation between the incident and reflected field amplitudes. This relation reads

$$\mathbf{E}_r = \mathbf{E}_i \sqrt{\frac{\cos \theta_i}{\cos \theta_r}} \sqrt{\frac{\eta_1}{\eta_2}} = \mathbf{E}_i |r_{\mathrm{TE}}|. \tag{4.66}$$

Next, we use the same boundary conditions as in the case of perfect refractive metasurfaces – i.e., boundary conditions (4.58) – however, only for the forward illumination direction (since the metamirror metasurface behaves as a boundary disconnecting medium 1 from medium 2). Substituting the fields from (4.64) and the surface polarizations (4.56) in the boundary conditions (4.58), one can obtain the following system of equations:

$$1 + r_{\mathrm{TE}} = \frac{j\omega}{S}\left(\widehat{\alpha}_{\mathrm{ee}}^{yy} + \widehat{\alpha}_{\mathrm{em}}^{yz}\frac{\cos\theta_{\mathrm{i}}}{\eta_1}\right),$$

$$\frac{\cos\theta_{\mathrm{i}}}{\eta_1} - \frac{\cos\theta_{\mathrm{r}}}{\eta_1}r_{\mathrm{TE}} = \frac{j\omega}{S}\left(\widehat{\alpha}_{\mathrm{me}}^{zy} + \widehat{\alpha}_{\mathrm{mm}}^{zz}\frac{\cos\theta_{\mathrm{i}}}{\eta_1}\right).$$

(4.67)

Obviously, these equations have infinitely many solutions for polarizabilities that realize the desired response. The metasurface either can be bianisotropic (omega coupling) or it can be a non-bianisotropic pair of electric and magnetic current sheets. For the non-bianisotropic realization, we set

$$\widehat{\alpha}_{\mathrm{em}}^{yz} = \widehat{\alpha}_{\mathrm{me}}^{zy} = 0 \tag{4.68}$$

and find the unique solution

$$\widehat{\alpha}_{\mathrm{ee}}^{yy} = \frac{S}{j\omega}\left(1 + \sqrt{\frac{\cos\theta_{\mathrm{i}}}{\cos\theta_{\mathrm{r}}}}\sqrt{\frac{\eta_1}{\eta_2}}e^{\Phi_{\mathrm{r}}}\right), \tag{4.69}$$

$$\widehat{\alpha}_{\mathrm{mm}}^{zz} = \frac{S}{j\omega}\left(1 - \sqrt{\frac{\cos\theta_{\mathrm{r}}}{\cos\theta_{\mathrm{i}}}}\sqrt{\frac{\eta_1}{\eta_2}}e^{\Phi_{\mathrm{r}}}\right). \tag{4.70}$$

These expressions show the collective polarizabilities required for the desired performance when the bianisotropy is absent and we can immediately see what are the appropriate topologies of unit cells for implementing them. Since we need both electric and magnetic polarizations, the physical thickness of the reflecting layer must be different from zero to allow the formation of tangential magnetic moments in unit cells. For example, it is not possible to realize the desired performance by any patterning of a single, infinitesimally thin sheet of a perfect conductor.

If we relax the requirement of non-bianisotropic particles, the equations suggest the use of a single array of small particles that are polarizable both electrically and magnetically, such as small metal spirals as in [25]. A typical realization based on the bianisotropic route is a high-impedance surface with a PEC ground plane (such as "mushroom layers" [42]). This implementations based on bianisotropic particles present the advantage of stronger magnetic excitation via bianisotropic coupling. Especially for optical applications, it is easier to realize strong bianisotropy (which is a first-order dispersion effect) as compared to the artificial magnetism (which is a weaker, second-order effect) [20].

However, direct realizations as lossless metasurfaces with these collective polarizabilities are not possible because the surface impedance of the metasurface producing the reflected field as in (4.64) is not purely reactive. It is given by the expression

$$Z_{\mathrm{s}}(x) = \frac{\eta_1}{\sqrt{\cos\theta_{\mathrm{i}}\cos\theta_{\mathrm{r}}}}\frac{\sqrt{\cos\theta_{\mathrm{r}}} + \sqrt{\cos\theta_{\mathrm{i}}}\,e^{j\Phi_{\mathrm{r}}(x)}}{\sqrt{\cos\theta_{\mathrm{i}}} - \sqrt{\cos\theta_{\mathrm{r}}}\,e^{j\Phi_{\mathrm{r}}(x)}}, \tag{4.71}$$

and it is easy to check that the real part of this input impedance (the input resistance) is non-zero and takes both positive and negative values. Although on average the surface is lossless, for perfect operation as an anomalous reflector, it must exhibit nonlocal properties. In some areas, some energy "enters" the metasurface (where the real part of Z_s is positive), and it is then launched back into space (where the real part of Z_s is negative). Although the basic topology of an omega-bianisotropic boundary is appropriate, design of actual structures needs numerical optimization of interacting unit cells. Experimental realization of a nearly perfect anomalously reflecting metasurfaces as an inhomogeneous high-impedance surface has been reported in [43].

Metasurfaces for anomalous reflection of plane waves are periodical surfaces where the period depends on the incidence and reflection angles as well as the wavelength. For a small tilt of reflected waves, the period is very large, and the conventional design approach based on engineering the reflection phase in the physical optics (locally periodical) approximation works very well. On the other hand, when the needed period of the metasurface allows propagation of only one or two unwanted Floquet harmonics, theoretically perfect performance can be achieved using diffraction gratings, called metagratings in recent literature. However, it appears that the only approach to sending reflected waves into arbitrary directions is the use of carefully engineered nonlocal metasurfaces, which support wave propagation along the surface.

Research on metasurfaces for various applications is very active, since there are many more possibilities than discussed in the preceding paragraphs; see, e.g., a roadmap paper [44].

Problems and Control Questions

1. Define the notion of *metasurface*. Discuss similarities and differences of metasurfaces, frequency-selective surfaces, and phased arrays (reflectarrays and transmitarrays).

2. Discuss what functionalities are possible if the metasurface maintains only electric surface currents. What additional functionalities become possible if the surface is also magnetically polarizable? What functionalities are not possible if there is no magneto-electric coupling?

3. Consider a metasurface in the yz-plane excited by a normally incident plane wave (the electric field $\mathbf{E}_{\text{inc}}$ is in the z-direction), propagating along the x-direction). The electric dipole moment of each unit cell is $\mathbf{p} = \hat{\alpha}_{\text{ee}} E_{\text{inc}} \mathbf{z}_0$, where $\frac{1}{\hat{\alpha}_{\text{ee}}} = \frac{1}{\alpha} - \beta$ is a complex number and $\mathbf{z}_0$ is the unit vector in the z-direction. The electric surface current density is $\mathbf{J} = \frac{j\omega}{S}\mathbf{p}$, where S is the area of the unit cell. What are the scattered electric field and the scattered magnetic field at each side of the metasurface? What is the radiated power by one unit cell at each side of the metasurface? What is the total radiated power by one unit cell? Hint: $P^{\text{rad}} = 2 \int \frac{1}{2}\text{Re}[\mathbf{E} \times \mathbf{H}^*]dS = S\,\text{Re}[\mathbf{E} \times \mathbf{H}^*]$, where the asterisk

denotes the complex conjugate operation. The factor 2 takes into account that the metasurface currents radiate waves on both sides of the metasurface plane.

4. Based on the results of the previous exercise, prove that $\mathrm{Im}[1/\hat{\alpha}_{\mathrm{ee}}] = \frac{\omega\eta}{2S}$, where η is the wave impedance of the surrounding space. Hint: the power consumption on the excitation of one unit cell can be calculated by $P^{\mathrm{ext}} = \frac{S}{2}\mathrm{Re}[\mathbf{J}^* \cdot \mathbf{E}_{\mathrm{inc}}]$. Express the surface current density in terms of the unit-cell polarizability and the incident field: $P^{\mathrm{ext}} = -\frac{\omega}{2}\,\mathrm{Im}\,[\hat{\alpha}_{\mathrm{ee}}]\,|\mathbf{E}_{\mathrm{inc}}|^2$. Use the energy conservation: if there is no absorption in the metasurface, we have $P^{\mathrm{rad}} = P^{\mathrm{ext}}$.

5. Consider a metasurface formed by a two-dimensional array of magnetically polarizable particles (e.g., double split-ring resonators) in the yz-plane, as shown in Figure 4.10. The periods in both z- and y-directions are small compared to the wavelength. When the array is under a plane-wave excitation with magnetic field $\mathbf{H}_{\mathrm{inc}}$ along the z-direction and propagation is along the x-direction, the magnetic dipole moment of each unit cell is $\mathbf{m} = \hat{\alpha}_{\mathrm{mm}}H_{\mathrm{inc}}\mathbf{z}_0$, where $\frac{1}{\hat{\alpha}_{\mathrm{mm}}} = \frac{1}{\alpha_{\mathrm{mm}}} - \beta_{\mathrm{m}}$ is a complex number (inverse of the magnetic collective polarizability) and $\mathbf{z}_0$ is the unit vector in the z-direction. The magnetic surface current density is $\mathbf{J}_{\mathrm{m}} = \frac{j\omega}{S}\mathbf{m}$, where S is the area of the unit cell. What are

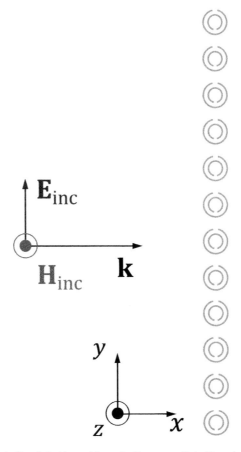

An array of small magnetically polarizable particles excited by a normally incident plane wave. Courtesy of F. Liu.

the scattered magnetic field and the scattered electric field at each side of the metasurface? Hint: use Maxwell's equation $\nabla \times \mathbf{E} = -\mathbf{J}_\mathrm{m} - \frac{\partial \mathbf{B}}{\partial t}$.

Bibliography

[1] S. B. Glybovski, S. A. Tretyakov, P. A. Belov, Y. S. Kivshar, and C. R. Simovski, "Metasurfaces: From microwaves to visible," *Physics Reports* **634**, 1–72 (2016).

[2] C. L. Holloway, E. F. Kuester, J. A Gordon, et al., "An overview of the theory and applications of metasurfaces: The two-dimensional equivalents of metamaterials," *IEEE Antennas and Propagation Magazine* **54**, 10–35 (2012).

[3] A. V. Kildishev, A. Boltasseva, and V. M. Shalaev, "Planar photonics with metasurfaces," *Science* **339**, 6125 (2013).

[4] N. Yu and F. Capasso, "Flat optics with designer metasurfaces," *Nature Materials* **13**, 139–150 (2014).

[5] A. I. Kuznetsov, A. E. Miroshnichenko, M. L. Brongersma, Y. S. Kivshar, and B. Luk'yanchuk, "Optically resonant dielectric nanostructures," Science **354**, aag2472 (2016).

[6] S. Jahani and Z. Jacob, "All-dielectric metamaterials," *Nature Nanotechnology* **11**, 23–36 (2016).

[7] D. G. Baranov, D. A. Zuev, S. I. Lepeshov, et al., "All-dielectric nanophotonics: the quest for better materials and fabrication techniques," *Optica* **4**, 814 (2017).

[8] Z. J. Yang, R. Jiang, X. Zhuo, et al., "Dielectric nanoresonators for light manipulation," *Physics Reports* **701**, 1–50 (2017).

[9] J. Huang and J. A. Encinar, *Reflectarray Antennas*, Wiley, 2008.

[10] N. Yu, P. Genevet, M. A. Kats, et al., "Light propagation with phase discontinuities: Generalized laws of reflection and refraction," *Science* **334**, 333 (2011).

[11] M. Albooyeh, D.-H. Kwon, F. Capolino, and S. A. Tretyakov, "Equivalent realizations of reciprocal metasurfaces: Role of tangential and normal polarization," *Physical Review B* **95**, 115435 (2017).

[12] C. R. Simovski, M. S. Kondratjev, P. A. Belov, and S. A. Tretyakov, "Interaction effects in two-dimensional bianisotropic arrays," *IEEE Transactions on Antennas and Propagation* **47**, 1429–1439 (1999).

[13] S. I. Maslovski and S. A. Tretyakov, "Full-wave interaction field in two-dimensional arrays of dipole scatterers," *International Journal of Electronics & Communication Technology Archiv für Elektronik und Übertragungstechnik: AEÜ* **53**, 135–139 (1999).

[14] S. A. Tretyakov, A. J. Viitanen, S. I. Maslovski, and I. E. Saarela, "Impedance boundary conditions for regular dense arrays of dipole scatterers," *IEEE Transactions on Antennas and Propagation* **51**, 2073–2078 (2003).

[15] V. V. Yatsenko, S. I. Maslovski, S. A. Tretyakov, et al., "Plane-wave reflection from double arrays of small magnetoelectric scatterers," *IEEE Transactions on Antennas and Propagation* **51**, 2–11 (2003).

[16] S. Tretyakov, *Analytical Modeling in Applied Electromagnetics,* Artech House, 2003.

[17] V. Asadchy, A. Díaz-Rubio, D.-H. Kwon, and S. Tretyakov, "Analytical modeling of electromagnetic surfaces," in *Surface Electromagnetics, With Applications in Antenna, Microwave, and Optical Engineering,* ed. F. Yang and Y. Rahmat-Samii, Cambridge University Press, pp. 30–65, 2019.

[18] T. Niemi, A. Karilainen, and S. Tretyakov, "Synthesis of polarization transformers," *IEEE Transactions on Antennas and Propagation* **61**, 3102–3111 (2013).

[19] L. D. Landau and E. M. Lifshitz, *Statistical Physics*, Butterworth-Heinemann, vol. 5, sec. 125, Course of Theoretical Physics, pt. 1, 1997.

[20] A. N. Serdyukov, I. V. Semchenko, S. A. Tretyakov, and A. Sihvola, *Electromagnetics of Bi-Anisotropic Materials: Theory and Applications*, Gordon and Breach Science Publishers, 2001.

[21] V. V. Yatsenko, S. I. Maslovski, S. A. Tretyakov, S. L. Prosvirnin, and S. Zouhdi, "Plane-wave reflection from double arrays of small magnetoelectric scatterers," *IEEE Transactions on Antennas and Propagation* **51**, 2–11 (2003).

[22] T. W. Ebbesen, H. J. Lezec, H. F. Ghaemi, T. Thio, and P. A. Wolff, "Extraordinary optical transmission through sub-wavelength hole arrays," *Nature* **391**, 667 (1998).

[23] J. A. Kong, *Electromagnetic Wave Theory,* EMW Publishing, 2005.

[24] F. J. García de Abajo, R. Gómez-Medina, and J. J. Sáenz, "Full transmission through perfect-conductor subwavelength hole arrays," *Physical Review* **72**, 016608 (2005).

[25] V. S. Asadchy, I. A. Faniayeu, Y. Ra'di, S. A. Khakhomov, I. V. Semchenko, and S. A. Tretyakov, "Broadband reflectionless metasheets: Frequency-selective transmission and perfect absorption," *Physical Review X* **5**, 031005 (2015).

[26] D. M. Pozar, "Flat lens antenna concept using aperture coupled microstrip patches," *Electronics Letters* **32**, 2109–2111 (1996).

[27] N. Gagnon, A. Petosa, and D. A. McNamara, "Research and development on phase-shifting surfaces (PSSs)," *IEEE Antennas and Propagation Magazine* **55**, 29–48 (2013).

[28] B. Rahmati and H. R. Hassani, "High-efficient wideband slot transmitarray antenna," *IEEE Transactions on Antennas and Propagation* **63**, 5149–5155, 2015.

[29] A. H. Abdelrahman, A. Z. Elsherbeni, and F. Yang, "Transmission phase limit of multilayer frequency-selective surfaces for transmitarray designs," *IEEE Transactions on Antennas and Propagation* **62**, 690–697 (2014).

[30] C. G. M. Ryan, M. R. Chaharmir, J. Shaker, et al., "A wideband transmitarray using dual-resonant double square rings," *IEEE Transactions on Antennas and Propagation* **58**, 1486–1493 (2010).

[31] C. Pfeiffer and A. Grbic, "Millimeter-wave transmitarrays for wavefront and polarization control," *IEEE Transactions on Antennas and Propagation* **61**, 4407–4417 (2013).

[32] F. Monticone, N. M. Estakhri, and A. Alù, "Full control of nanoscale optical transmission with a composite metascreen," *Physical Review Letters* **110**, 203903 (2013).

[33] R. Milne, "Dipole array lens antenna," *IEEE Transactions on Antennas and Propagation* **30**, 704–712 (1982).

[34] Y. Ra'di, V. S. Asadchy, and S. A. Tretyakov, "Tailoring reflections from thin composite metamirrors," *IEEE Transactions on Antennas and Propagation* **62**, 3749–3760 (2014).

[35] V. S. Asadchy, Y. Ra'di, J. Vehmas, and S. A. Tretyakov, "Functional metamirrors using bianisotropic elements," *Physical Review Letters* **114**, 095503 (2015).

[36] C. R. Simovski, S. A. Tretyakov, A. A. Sochava, et al., "Antenna model for conductive omega particles," *Journal of Electromagnetic Waves and Applications* **11**, 1509–1530 (1997).

[37] S. N. Tcvetkova, V. S. Asadchy, and S. A. Tretyakov, "Scanning characteristics of metamirror antennas with sub-wavelength focal distance," *IEEE Transactions on Antennas and Propagation* **64**, 3656–3660, (2016).

[38] V. Asadchy, M. Albooyeh, and S. Tretyakov, "Optical metamirror: All-dielectric frequency-selective mirror with fully controllable reflection phase," *Journal of the Optical Society of America* **33**, A16–A20, (2016).

[39] V. S. Asadchy, M. Albooyeh, S. N. Tcvetkova, et al., "Perfect control of reflection and refraction using spatially dispersive metasurfaces," *Physical Review B* **94**, 075142 (2016).

[40] J. Wong, A. Epstein, and G. Eleftheriades, "Reflectionless wide-angle refracting metasurfaces," *IEEE Antennas and Wireless Propagation Letters* **15**, 1293 (2016).

[41] N. M. Estakhri and A. Alù, "Wavefront transformation with gradient metasurfaces," *Physical Review X* **6**, 041008 (2016).

[42] D. Sievenpiper, L. Zhang, R. F. J. Broas, N. G. Alexópolous, and E. Yablonovitch, "High-impedance electromagnetic surfaces with a forbidden frequency band," *IEEE Transactions on Microwave Theory and Techniques* **47**, 2059 (1999).

[43] A. Díaz-Rubio, V. Asadchy, A. Elsakka, and S. Tretyakov, "From the generalized reflection law to the realization of perfect anomalous reflectors," *Science Advances* **3**, e1602714 (2017).

[44] O. Quevedo-Teruel, H. Chen, A. Díaz-Rubio, et al., "Roadmap on metasurfaces," *Journal of Optics,* **21** 073002 (2019).

Note

1 More details about scattering cancellations in periodical arrays can be found in [16].

5 Photonic Crystals

5.1 Introduction

5.1.1 What Are Photonic Crystals?

Photonic crystals can be defined as composites with a space-periodical internal structure operating at frequencies where the wavelength is comparable to the period of the structure. This is probably the most general definition, encompassing most complex structures operating in various frequency ranges. Alternative names are *electromagnetic crystals* and *bandgap structures* [1–3]. The latter name is given because the frequency bands in which the electromagnetic waves cannot propagate (stopbands) are called bandgaps if they refer to photonic crystals. When such regular composites are designed to operate at optical frequencies (from the far infrared range to the visible and ultraviolet light), they are usually called photonic crystals [1–4]. Building blocks of photonic crystals are made of usual (effectively homogeneous) materials, and since homogeneous materials in the optical range lose magnetic susceptibility, the internal periodicity of photonic crystals is often defined as a periodic coordinate dependence of the permittivity $\varepsilon(\mathbf{r}) = \varepsilon(\mathbf{r+a})$, where $\mathbf{a} = (a_x, a_y, a_z)$ is the lattice unit vector [1, 3, 4]. For isotropic dielectric media, the relative permittivity ε and the refractive index n are related as $\varepsilon = n^2$; thus, in this case, photonic crystals can be defined as artificial media with a space-periodic contrast of refractive index [2].

However, it is important to remember that the definitions in terms of the permittivity or refractive index are limited to specific (although very wide) classes of electromagnetic crystals. For example, periodical arrangements of magnetic or magneto-optical inclusions, the so-called *magnetophotonic* crystals, are very important nonreciprocal photonic crystals (see, e.g., [5–8]). Lattices of magnetically biased ferrimagnetic crystals (like yttrium iron garnet) or self-biased hexaferrite inclusions are conventionally used. In the microwave and millimeter ranges these materials have tensorial magnetic permittivity with a non-zero antisymmetric part, responsible for nonreciprocal effects. In the optical range, these materials exhibit nonreciprocal magneto-optical effects due non-zero antisymmetric part of the permittivity tensor. In photonic crystals made of biased magnetic and magneto-optical materials, the nonreciprocal effects can be dramatically enhanced due to light localization [5, 7, 8]. Stopband (bandgap) frequencies for the two orthogonal polarizations can be electrically controlled [6].

The wavenumber in magnetic and magneto-optical media depends not only on the frequency but also on the propagation direction. The refractive index for such material is simply meaningless, and the definition of [2] is not applicable. Furthermore, the definition in terms of refractive index is not suitable for metallic photonic crystals that are also quite important for applications (see, e.g., [9]). Though metals in optics are reciprocal and isotropic materials, they do not possess refractive properties because $\text{Re}(\varepsilon_{\text{metal}}) < 0$.

Photonic or electromagnetic crystals can be alternatively considered as meta-materials with periodically arranged meta-atoms in cases when the metamaterial unit cell becomes comparable to the wavelength and spatial dispersion effects become very significant. These regular metamaterials are not effectively continuous media.

Similarly, metasurfaces in the frequency ranges where the period is comparable to λ behave as two-dimensional analogues of photonic crystals. These two-dimensional arrays exhibit stopbands for surface waves, similarly to stopbands for plane waves in volumetric photonic crystals. Obviously, the definition in terms of periodically varying permittivity does not apply to these planar stopband structures. Interestingly, it is possible to combine the notions of photonic crystals and metamaterials, considering photonic crystals formed by complex inclusions performed of a metamaterial. Such photonic crystals are sometimes called *meta-metamaterials* [10, 11].

In this book, we will consider photonic crystals in the form of inhomogeneous media whose complex permittivity tensor is spatially periodic. This class covers all nonmagnetic classes of photonic crystals operating at optical frequencies, including anisotropic and metal lattices. Within this definition, we do not exclude the low-frequency region where the period is small on the wavelength scale so that the photonic crystal behaves as an effectively homogeneous medium. However, in this chapter, we do not consider interesting phenomena that may arise at low frequencies. The low-frequency region is included into our general consideration mainly to show a complete *dispersion diagram* of a photonic crystal – a plot whose ordinate axis starts from zero frequency.

Geometrically, volumetric photonic crystals split onto three classes: one-dimensional (single-period lattices of layers), two-dimensional (doubly periodic crystals of cylinders, not necessarily circular ones), and three-dimensional (triply periodic lattices). Examples of such lattices are shown in Figure 5.1. In fact,

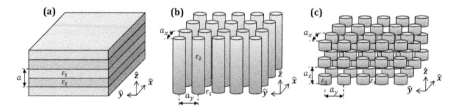

Figure 5.1 Schematic illustration of the idea of (a) one-dimensional, (b) two-dimensional, and (c) three-dimensional photonic crystals. Courtesy of Ana Díaz-Rubio.

one-dimensional and even two-dimensional photonic crystals had been known before this terminology appeared. The term "photonic crystal" arose about 30 years ago, when the interest in spatially periodic optical composites sharply grew. The reason of this keen interest was the publication of two pioneering works in the same month of 1987: that by E. Yablonovitch [12] and that by S. John [13]. These experimental papers reported two exciting effects: an amazing local enhancement of light intensity in photonic crystals with broken periodicity and inhibited spontaneous emission of quantum emitters in strictly periodic three-dimensional photonic crystals. Both these effects were theoretically predicted earlier [14, 15]. However, these theoretical works have not produced a break-through. What has increased both theoretical and experimental investigation of photonic crystals was their fabrication and experimental confirmation of the effects predicted long ago.

5.1.2 Natural Photonic Crystals

Among natural crystalline media, the only known photonic crystal is opal. Opal has a larger lattice unit cell than those of other solid crystals. Though in the visible and IR ranges of frequencies it also behaves as a continuous medium, it becomes a photonic crystal in the ultraviolet light [14]. However, natural photonic crystals operating in the visible range are also known. They are not natural solid crystals but parts of living systems. Two most famous examples of living photonic crystals are illustrated by Figure 5.2. The first one is the sea mouse hair in Figure 5.2, left panel. The sea mouse is a marine worm with a 15–20 cm long body living on the sea shelf at rather shallow depths. It is covered with thin, internally structured hairs also called tubular, spines. Each hair is a two-dimensional photonic crystal of submicron cylindrical holes oriented along the hair axis in the matrix of tissue whose refractive index is close to 1.5. When light is incident perpendicular to the axis of a hair, it exhibits red coloration. For off-axis incidences, green and blue shades appear. This optical effect is due to the so-called *directional photonic bandgap*. For the normal incidence, the bandgap corresponds to the red light. The term bandgap means that the light whose frequencies lie in a certain band (in the present case, red light) cannot propagate inside the structure and, therefore, is totally reflected from it. The *directional bandgap* means that the band of frequencies for which this total reflection holds varies versus the incident light direction. When the incidence angle increases, the band of light totally reflected from the hair becomes yellow, green, and finally – for large angles – blue. This effect visually manifests in the angle-dependent coloration of the hair observed in the sunlight.

Another example of a photonic crystal in a living system is the moth eye. The photonic crystal (a hexagonal lattice) is formed by identical dielectric protrusions with lengths about 600 nm and the lattice step 400 nm (Figure 5.2, right panel). It operates as a superprism – a device with extremely high dispersion of light.[1] This

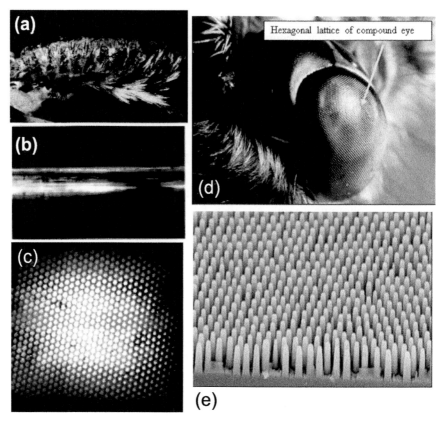

Figure 5.2   Sea mouse: (a) general view, (b) its hair illuminated by an obliquely incident light, (c) micrograph of the photonic-crystal structure. Reprinted from [16] with permission of the authors. Moth eye: (d) general view, (e) scanning electron microscope (SEM) picture of the analog of the moth eye fabricated in a laboratory on a flat substrate. Reprinted from [17] with permission. Copyright (2016) of ISIS.

is an important nanophotonic device, and we discuss it in the next chapter. Here, we only mention that this superprism in the eye allows a butterfly to distinguish any more colors and shades that humans can do. Descriptions of some other examples of living photonic crystals can be found in [18].

5.2 Photonic Band Structures

5.2.1 Bragg Phenomenon as a Reason of Bandgaps

The key phenomenon inherent to every photonic crystal is the so-called *Bragg phenomenon* or Bragg's scattering. This phenomenon results from the in-phase interference of waves, reflected from a set of parallel crystal planes. The phenomenon was explained theoretically by Lord Rayleigh in [19] after W. L. Bragg and W. H. Bragg – had revealed complete reflection of X-rays impinging

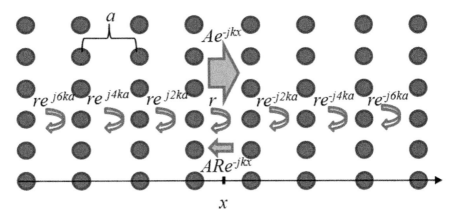

Figure 5.3 Illustration of the Bragg phenomenon. In the uniform space between any two crystal planes, the propagating field of a lattice eigenmode can be considered as a sum of two waves traveling in opposite directions: one with the amplitude A propagating along x with the wavevector $\mathbf{k} = k\mathbf{x}_0$, and the other one propagating in the opposite direction with the amplitude RA. At any point x, this "reflected" wave is a sum of partial waves reflected by all crystal planes located behind x.

natural crystals at a set frequencies specific for a given crystal [20]. The theory by Rayleigh not only predicted this phenomenon for any periodic arrays of electromagnetic scatterers; it was further used in solid-state physics, where it was developed by F. Bloch and other scientists for electron wave functions in dielectric and semiconductor crystals. In natural crystals, an electron propagates through the crystal as a package of the so-called *de Broglie waves* [21], and these particles are atoms (ions). In photonic crystals, constitutive particles are macroscopic inclusions in the host matrix, and propagating waves are electromagnetic – light – waves.

In Figure 5.3, we show the formation of the reverse-directed partial wave in the crystal as a summation of partial waves reflected from planar arrays of scattering particles. These planar arrays form a set of partially reflecting grids. If the grid period is sufficiently small, these grids can be thought as effective semitransparent sheets. The reflection coefficient R at the reference plane x between two arbitrary selected adjacent sheets results from multiple reflections that occur with the reflection coefficient r at each plane located behind x. Here, the reflection coefficient r is that of a single effective sheet, and $|r| < 1$ since the scatterers are passive objects. The path of the wave reflected from an arbitrary crystal plane is evidently larger by $2a$ than the path of the wave reflected from the previous crystal plane. It implies the phase shift $2ka$ for two waves reflected by adjacent crystal planes. Therefore, the result for R comprises a geometric progression with a factor $\exp(-j2ka)$:

$$R = re^{-jkx} + re^{-jkx}e^{-j2ka} + re^{-jkx}e^{-j4ka} + \cdots = re^{-jkx}\frac{1}{1-e^{-j2ka}}. \qquad (5.1)$$

The right-hand side of (5.1) diverges when $\exp(-2jka) = 1$, i.e., when $k \equiv \omega/v = m\pi/a$, where m is an integer number and v is the speed of light in the host medium. This physically meaningless divergence implies that the

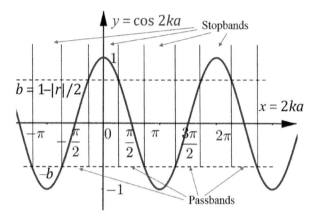

Figure 5.4 Illustration of the concept of stopbands in a set of parallel equidistant crystal planes shown in Figure 5.3. If the absolute value of the cosine function exceeds b the absolute value of R exceeds unity.

propagation of waves along x at frequencies $\omega_n = mv\pi/a$ is impossible. Formula $\omega_m = mv\pi/a$ (or its equivalent in terms of the wavelengths $\lambda_m = 2a/m$) is called the Bragg condition,[2] which describes the frequencies (wavelengths) at which the crystal completely reflects normally incident plane waves. For the oblique propagation, the differential path $2a$ should be replaced by $2a \sin \theta$, and the set of frequencies for which $R \to \infty$ changes respectively.

The Bragg condition $R \to \infty$ gives a discrete set of frequencies (or wavelengths) at which the propagation is forbidden for the given direction of propagation. However, finite values of $|R|$ larger than unity are also prohibited. Really, the propagation of a wave with finite energy along x in a passive structure such as a lattice of scatterers cannot give birth to a wave with higher energy. Imposing the condition $|R| < 1$ to the right-hand side of (5.1), we obtain after a simple algebra an inequality describing the passbands:

$$|\cos 2ka| < 1 - \frac{|r|^2}{2}. \tag{5.2}$$

At these frequencies (recall that $k \equiv \omega/v$), the propagation along x is allowed, whereas at frequencies for which (5.2) is not satisfied, the wave cannot propagate along x.

In Figure 5.4, we show the graphic solution of (5.2) for these frequencies and see that the lattice passbands and stopbands periodically alternate. In this example, r does not depend on ka, and our set of bandgaps includes the range of low frequencies, where $ka < \arcsin |r|/2$. This quasi-static bandgap does not arise for lattices of three-dimensional scatterers. Since the optical size of the scatterer vanishes in the quasi-static limit, for any planar array of finite-size scatterers, we have $|r| \to 0$ when $ka \to 0$. It is possible to prove that in this limit the absolute value of R in (5.1) keeps lower than unity. The same refers to the case when the lattice is formed by planar grids of dielectric cylinders. However, a lattice formed by perfectly conducting wires (metal wires at radio frequencies can be thought perfectly conducting if we do not aim to study finer effects and only calculate the stopbands and passbands) has the low-frequency passband for its eigenmode whose electric field is parallel to the axes of the wires. In this case, $|r|$ does not vanish in the quasi-static limit, and formula (5.2) stays relevant. If the electric field is orthogonal to the wires, $|r|$ tends to 0 when ka tends to 0. The same refers to the grids of dielectric cylinders and to the grids of three-dimensional scatterers. For this mode, the low-frequency bandgap in a lattice of parallel wires does not exist, either.

If we consider an oblique propagation in the same lattice, we obtain a set of passbands and stopbands that depend on the propagation direction. By specifying a frequency and a propagation direction, we specify the wavevector; therefore, generalizing the Rayleigh formula (5.1) for arbitrary direction allows us to relate the eigenfrequencies in the passbands to the wavevector. This relationship, which is different from that established for continuous media, manifests *spatial dispersion* of photonic crystals. A directional bandgap is the most spectacular manifestation of spatial dispersion. It means that in a certain sheer of directions, a signal of a given frequency cannot propagate, whereas in all other directions, it can propagate. This is impossible in continuous media, even in anisotropic media.

In realistic photonic crystals, the reflection coefficient r is itself dependent on the frequency. Therefore, the width of passbands and stopbands varies over the frequency axis. Moreover, many photonic crystals have a lattice more complex than the orthorhombic (rectangular) lattice that is depicted in Figure 5.3. This makes the band structure quite complicated. Therefore, the Bragg conditions $\lambda_m = 2a/m$ (for normal incidence) and $\lambda_m = 2a \sin \theta/m$ (for oblique incidence), which are very important for natural crystals, are not so useful for photonic crystals. For natural crystals, the Bragg conditions determines the frequencies ω_m of X-rays that completely reflect from the crystal for the given incidence direction.

Neglecting electromagnetic (optical) losses (as we did earlier), we have to prohibit the propagation of X-rays not only at frequencies $\omega_m = mc\pi/a$ (except that with $m = 0$, for which $r = 0$) but also in the stopbands centering these frequencies. However, high optical losses of natural crystals in the range of X-rays reduce the impact of the stopbands and make the absolute value of the reflection coefficient smaller than unity at all frequencies. It maximally approaches unity at Bragg frequencies, where the impact of the optical loss turns negligible compared to that of the interference.

It is not so for photonic crystals. Here, the constitutive particles are usually substantial in size, as compared to the lattice period. The reflection coefficient r of a crystal plane in a photonic crystal strongly depends on both frequency and incidence angle. It results in a more complicated spatial dispersion than that of X-rays in solids. Also, photonic crystals operate at the visible and infrared light frequencies, where optical losses – at least for dielectric photonic crystals – are lower than those inherent to solids in the range of X-rays. The reflection of the electromagnetic wave from such photonic crystals is practically complete over the whole stopband [2]. Therefore, the Bragg frequencies for photonic crystals are not very relevant (moreover, it is difficult to estimate them due to the strong frequency dispersion of r). Instead of Bragg frequencies, for photonic crystals, one calculates the set of bandgaps and passbands. The Rayleigh approach to this calculation expressed by formula (5.2) is not very helpful, even though we know the frequency dispersion of r. In many photonic crystals, this value is modified by interaction between adjacent crystal planes and cannot be calculated separately. Moreover, calculating the positions of bandgaps on the frequency axis is insufficient to describe electromagnetic properties of photonic crystals. In this chapter, we will discuss which characteristics of photonic crystals need to be calculated and how to calculate them.

5.2.2 Bandgap Structures in General

Since there are similarities in the theories of electromagnetic waves in photonic crystals and theories of electron waves in crystalline semiconductors [1, 2, 21], and solid-state physics was developed earlier, the terminology of photonic crystals is largely borrowed from solid-state physics. Stopbands are usually called bandgaps, and one distinguishes *complete* and *directional* bandgaps. In a complete bandgap, the propagation is forbidden in all directions and for all polarizations of waves. An optical emitter embedded into a three-dimensional photonic crystal cannot radiate photons with the frequencies inside a complete bandgap. Respectively, a complete bandgap is sometimes called *photonic bandgap*. If the propagation is forbidden for some directions and allowed for the other ones the corresponding frequency band is called directional bandgap.

Most of three-dimensional photonic crystals have only directional bandgaps. Complete bandgaps are possible in diamond-like lattices and in so-called *inverted opal* lattices. As to simple lattices, such as the cubic or orthorhombic ones, a complete bandgap is possible if the inclusions are resonant. In the optical range, they can be, for example, metallic nanoparticles. Even simple-shape nanoparticles such as nanospheres of some metals experience the so-called *plasmon resonances* in the range of the visible light; see Chapter 7. From the applications point of view, plasmonic structures have the disadvantages of larger dissipation losses. Here, we do not consider photonic crystals formed by resonant inclusions. We concentrate on photonic crystals of non-resonant constituents.

For one-dimensional photonic crystals, the relevant propagation direction is usually fixed in the direction normal to the layers, and one does not distinguish complete and directional bandgaps. For two-dimensional photonic crystals – lattices of cylindrical inclusions parallel to a certain axis – the complete bandgap as a rule implies prohibited propagation in the plane perpendicular to the axis. The exceptions are so-called *photonic-crystal fibers* in which only the propagation along the axes of cylinders is allowed (these cylinders are practically air voids in the fiber glass). Thus, not only the propagation across the fiber but also the oblique propagation should be prohibited.

A typical sample of a two-dimensional photonic crystal represents an optically thick dielectric layer with optically contrast (another dielectric or void) cylindrical inclusions stretched across the layer. In this situation, the finite length of the cylindrical inclusions plays no role: cylinders occupy the whole volume between the boundaries of the photonic crystal slab. The photonic bandgap prevents leakage of electromagnetic energy through the layer perimeter. The waves may propagate along the axes of the cylinders and obliquely to them but not orthogonally. This means that leakage of waves is possible through the top and bottom boundaries of the layer – which, in fact, can be useful for applications. One interface is usually mirrored, and the light is radiated by the second interface. In the next chapter, we will discuss this application of two-dimensional photonic crystals.

Notice that two-dimensional photonic crystals of metal rods in a dielectric host are also known and used in some applications. However, they are more practically important in the low-frequency regime than in the photonic-crystal regime. In the low-frequency regime – i.e., when the period is much smaller than the wavelength – such arrays are called *wire media*. Wire media refer to spatially dispersive metamaterials and deserve a separate study. We refer the reader to an overview [22]. As to the frequency range where a wire medium becomes a metallic photonic crystal described by a set of passbands and stopbands, it found applications only at microwaves (see, e.g., in [23]). In microwave electromagnetic bandgap structures, the wire diameters are in the millimeter and submillimeter ranges, whereas in the optical wire media, they are *nanowires*. Since metals in the optical range are rather lossy, metallic photonic crystal are usually inefficient. In both passbands and stopbands, waves propagate with decay, and the difference of decay rates in a passband and in a stopband is not very large. In optics, dielectric photonic crystals are practically more important than metallic ones.

As to the analogy of photonic crystals to natural ones, it is straightforward only for three-dimensional photonic crystals, because natural ones are bulk media of atoms (ions). Energy zone structures for all three solid states – conducting, semiconducting, and insulating – are illustrated in Figure 5.5. This band diagram shows only the *main bandgap* – that between the valence band and the conduction band. Three-dimensional photonic crystals without a complete

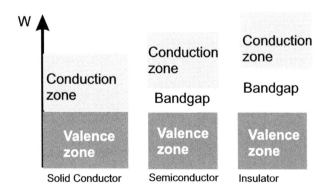

Figure 5.5 Band structures of electron states in solid conductors, semiconductors, and insulators. Potential energy W_c of an electron in the conductive state is related to the frequency of optical transition ω as $W_c = \hbar\omega + W_v$, where W_v is the electron potential energy in the valence zone.

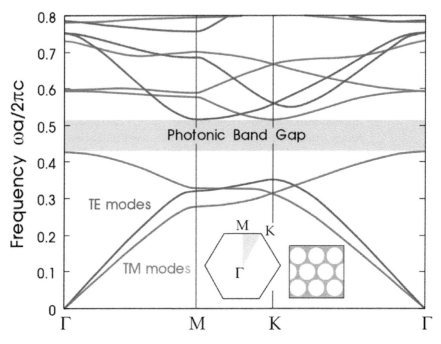

Figure 5.6   Dispersion diagram and bandgap structure of a honeycomb lattice of holes in a dielectric matrix. The bandgap is shown in the dispersion diagram by yellow color, and the lattice structure is depicted in an inset together with the so-called *Brillouin zone*. Reprinted from [1] with permission of the editors.

bandgap are analogous to metals, those with a narrow complete bandgap are analogous to semiconductors, and those with a broad complete bandgap – to insulators.

This simplified band diagram does not allow us to see the directional bandgaps and, in general, does not tell anything about spatial dispersion (relations between eigenfrequencies and wavevectors). The spatial dispersion and the band structure of photonic crystals are visualized in so-called *Brillouin dispersion diagrams* – plots of the frequency (or photon energy) versus the wavevector of the propagating eigenmode. An example of the dispersion diagram is shown in Figure 5.6. In the following paragraphs, we learn this graphic tool and relate it with another graphic tool – constant-frequency surfaces and contours, also called *isofrequencies*.

We have briefly discussed these tools in Chapter 2. Now, let us recall that the dispersion diagram of an isotropic dielectric space corresponds to the dispersion equation $\omega = vk$, where $k = \sqrt{k_x^2 + k_y^2 + k_z^2}$ is the wavenumber and $v = c/n = c/\sqrt{\varepsilon}$ is the phase velocity in the dielectric medium (ε is its relative permittivity, and n is the refractive index). For $k_z = 0$ (in-plane propagation), the dispersion diagram is a light cone. A section of this light cone by a plane $\omega =$ const is a circle of radius $k = \omega/v$ in the coordinates (k_x–k_y), and this circle is an isofrequency contour of the uniform isotropic space. Higher frequencies correspond to larger circles in the plane (k_x–k_y). Thus, isofrequency contours in the plane ($k_x - k_y$) are horizontal sections of the dispersion plot $\omega(k_x, k_y)$ by the planes of constant ω. This link of the dispersion diagram and the isofrequency contours holds for photonic crystals as well.

5.2.3 Bloch's Theorem

In order to understand the dispersion diagram of an infinite periodic structure, we need to know a very important theorem concerning eigenwaves of a periodic structure. It was formulated and proven by F. Bloch for wave functions of electrons in natural crystals and further generalized to all periodic structures supporting electromagnetic waves, acoustic waves, plasma waves, etc. This became possible because electron wave functions obey the Schrödinger equation, which in the linear regime reduces to the wave equation like Maxwell's equations do in linear media (see in [24–26]). The same refers to the linearized dynamic equations for gases, plasmas, and liquids: they all reduce to the wave equation.

For simplicity, let us consider a one-dimensional photonic crystal with the period a, shown in Figure 5.7. In accordance with the *Bloch theorem*, complex amplitudes of each monochromatic eigenmode of this structure (in our case, the eigenmode is formed by electric **E** and magnetic **H** fields) can be presented as a sum of the following waves:

$$\begin{cases} \mathbf{E}(\mathbf{r}) \\ \mathbf{H}(\mathbf{r}) \end{cases} = \begin{cases} \mathbf{E}_0(x,y) \\ \mathbf{H}_0(x,y) \end{cases} e^{-jqz} + \sum_{n\neq0} \begin{cases} \mathbf{E}_n(x,y) \\ \mathbf{H}_n(x,y) \end{cases} e^{-j(qz+G_nz)}, \tag{5.3}$$

where $G_n = 2\pi n/a$, n is an integer, and the time dependence of fields is assumed in form $\exp(j\omega t)$. Parameter q is called the *Bloch wavenumber*. The Bloch theorem evidently generalizes to two-dimensional and three-dimensional photonic crystals. For a three-dimensional photonic crystal with the periods (a_x, a_y, a_z), we write $\mathbf{n} = (n_x, n_y, n_z)$ and $\mathbf{G_n} \equiv (G_x, G_y, G_z) = \left(2\pi n_x/a_x, 2\pi n_y/a_y, 2\pi n_z/a_z\right)$ and arrive at the Bloch plane-wave expansion for the phasors of electric and magnetic field vectors:

$$\begin{cases} \mathbf{E}(\mathbf{r}) \\ \mathbf{H}(\mathbf{r}) \end{cases} = \begin{cases} \mathbf{E}_0 \\ \mathbf{H}_0 \end{cases} e^{-j(\mathbf{q}\cdot\mathbf{r})} + \sum_{n\neq0} \begin{cases} \mathbf{E_n} \\ \mathbf{H_n} \end{cases} e^{-j(\mathbf{q}\cdot\mathbf{r}+\mathbf{G_n}\cdot\mathbf{r})}. \tag{5.4}$$

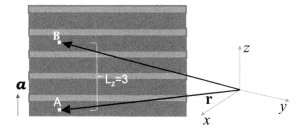

Figure 5.7   Let point *B* in an infinite photonic crystal with period *a* in the *z*-direction be obtained from point *A* shifted by L_z periods along the coordinate axis. Then the electromagnetic field of the eigenmode at these two points differs only by the phase shift $L_z\mathbf{q} \cdot \mathbf{a}$, where L_z is an integer and $\mathbf{q}$ is called the Bloch vector of the eigenmode.

In formulas (5.3) and (5.4), the fundamental ($n = 0$) spatial harmonic whose wavevector is equal to $\mathbf{q}$ is shared out from the infinite plane-wave expansion. This plane wave is called the *Bloch wave*. In the one-dimensional case, the value of this fundamental *Bloch wavenumber* $q \equiv q_z$ is restricted by the condition $-\pi/a < q < \pi/a$. All spatial harmonics with $n \neq 0$ oscillate with smaller spatial periods. If we assume that in the Bloch expansion the absolute value of the Bloch wavenumber is higher (e.g., $\pi/a < q < 2\pi/a$), this expansion term will be not distinguishable from that with the wavenumber $(q - 2\pi/a)$, which enters the second term of expression (5.3): the series of higher-order harmonics. Then, the replacement $q_{new} \to (q - 2\pi/a)$ will return the correct Bloch wavenumber with $-\pi/a < q_{new} < 0 < \pi/a$. Similarly, in the three-dimensional case, the components of the *Bloch wavevector* satisfy $-\pi/a_{x,y,z} < q_{x,y,z} < \pi/a_{x,y,z}$.

Bloch's theorem has two important implications. First, the problem of the infinite lattice can be reduced to the so-called cell problem. Really, from (5.3) the so-called *Bloch quasi-periodicity conditions* follow:

$$\begin{cases} \mathbf{E}(\mathbf{r} + L_z\mathbf{a}) \\ \mathbf{H}(\mathbf{r} + L_z\mathbf{a}) \end{cases} = \begin{cases} \mathbf{E}(\mathbf{r}) \\ \mathbf{H}(\mathbf{r}) \end{cases} e^{-jL_z\mathbf{q}\cdot\mathbf{a}}. \qquad (5.5)$$

For three-dimensional crystals, $L_z\mathbf{q}\cdot\mathbf{a}$ reads $L_x q_x a_x + L_y q_y a_y + L_z q_z a_z$. The special case $L_{x,y,z} = 1$ shows that the fields at the opposite faces of the unit cell differ only by the phase shift $\mathbf{q}\cdot\mathbf{a}$. The boundary problem for the wave equation in a spatial domain of the lattice unit cell in the case when the a priori unknown wave functions on the front and back sides of the cell are related to one another in a known way is called in mathematical physics the *Hamilton*[3] *cell problem* [27]. The Hamiltonian cell problem makes sense only in the case when the cell comprises inhomogeneities with known boundary conditions on their boundaries. In the case of photonic crystals, Maxwell's boundary conditions are satisfied at the surfaces of inclusions and the quasi-periodicity condition relates the values of the wave function at the cell sides. In the general case, the Hamiltonian cell problem is solved using the so-called *Hamilton operator*, often simply called the *Hamiltonian*. We refer the readers to [27, 28] for the proof that for lossless cells the solutions are discrete sets of real-valued eigenfrequencies ω corresponding to a given Bloch wavevector $\mathbf{q}$ $(-\pi/a_{x,y,z} < q_{x,y,z} < \pi/a_{x,y,z})$.

We see that Bloch's theorem is the possibility to find all eigenmodes of the photonic crystal considering one isolated unit cell of the lattice and varying $\mathbf{q}$. The second implication is a possibility to restrict the wavevector axis in the dispersion diagram by the fundamental wavevector $\mathbf{q}$. This allowed L. Brillouin to replace the dispersion plot $\omega(\mathbf{q})$, where $\mathbf{q}$ is the wavevector with an arbitrary absolute value $(-\infty < q_{x,y,z} < \infty)$ inherent to the unbounded space by the Brillouin's dispersion diagram [26].[4] The idea by Brillouin is illustrated by Figure 5.8 for a one-dimensional crystal. In this figure, the Brillouin diagram is shown for a photonic crystal formed by two dielectric layers of equal thicknesses $a/2$ and relative permittivities $\varepsilon_1 = 6$ and $\varepsilon_2 = 7$.

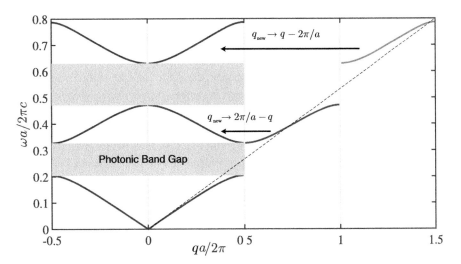

Figure 5.8 Brillouin's band diagram of an infinite stack of dielectric bilayers oriented orthogonally to axis z. Layers of equal thicknesses have relative permittivities 6 and 7, respectively. In the first passband the wavenumber is equal to q. In the second passband q is replaced by $q_{new} = 2\pi/a - q$. In the third passband $q_{new} = q - 2\pi/a$.

Let us explain the physical meaning of this procedure. For a uniform medium with permittivity $\varepsilon = (\varepsilon_1 + \varepsilon_2)/2 = 6.5$, the dispersion equation is as simple as $\omega = v|k|$, where $v = c/n = c/\sqrt{\varepsilon}$ is the phase velocity. The dispersion curve of this homogeneous medium is shown in Figure 5.8 for positive k as a straight dashed line. Next, if we solve the Hamiltonian cell problem for this one-dimensional photonic crystal, we obtain the dispersion curve that in the band of normalized frequencies $\omega a/2\pi c = [0.0{-}0.8]$ comprises three passbands separated by two bandgaps. In the passbands, in the intervals $\omega a/2\pi c = [0.0{-}0.2]$, $[0.32{-}0.48]$, $[0.63{-}0.79]$, the eigenmode is quite similar to the wave in the homogeneous medium with the averaged permittivity ε, because the contrast of the two permittivity is not high. The dispersion plot $\omega(q)$ in the second passband is the thick dark curve in Figure 5.8, and that in the third passband is the lighter curve. This dispersion is also not very different from the dispersion of the homogeneous medium. However, the fundamental wavenumber q of the Bloch plane-wave expansion is restricted by π/a. Therefore, for the second passband, we replace $q \rightarrow 2\pi/a - q$. Due to this shift, the dispersion branch inverts, being transposed to the Brillouin zone. In the Brillouin diagram presented in Figure 5.8, the curves for all even-numbered passband, seemingly appear to correspond to backward waves because here, $\partial\omega/\partial q < 0$. However, we should remember that these branches have been inverted when transposed to the Brillouin zone. Thus, in this example, all waves in the full dispersion diagram are forward waves with co-directed phase and group velocities. On the other hand, if one measures the field phase at discrete points separated by the period, the wave will indeed appear as a backward wave. This property is exploited in some devices, such as backward-wave amplifiers and generators.

Notice that the dispersion diagram for one-dimensional photonic crystals with reciprocal constituents is symmetric: identical eigenmodes can propagate in opposite directions. In Figure 5.8, we have shown the part of the diagram where $-\pi/a < q < 0$, but usually the part with $q < 0$ for such photonic crystals is omitted. It is worth noting that in a standard Brillouin diagram, the wavenumber and frequency axes are normalized, as we have done in Figure 5.8.

To conclude this discussion, let us compare the optical properties of photonic crystals in their passbands with optical properties of natural anisotropic crystals (see Chapter 2). The differences follow from the fact that in natural crystals at optical frequencies, the period is small as compared to the wavelength while, in photonic crystals, these two quantities are comparable. If the incident light wave is polarized so that it does not feel the anisotropy of a natural crystal, it excites only an ordinary wave in the crystal. For such incidence, the interface of a crystal behaves as that of an isotropic dielectric. For general excitation, both ordinary and extraordinary waves are excited. It results in double refraction, also called the ray birefringence. In a photonic crystal, analogous response is allowed only for the first passband where the absolute majority of photonic crystals behave as effectively continuous media. Indeed, at low frequencies, the wavelength of incident light is large compared to the lattice period and the quickly varying higher-order terms of the Bloch expansion ray birefringence (with $n \neq 0$) are evanescent waves which exist near the inhomogeneities of the internal structure. The propagation direction inside the crystal is determined by the fundamental spatial harmonics of the Bloch expansion – the Bloch waves of two eigenpolarizations. For two-dimensional crystals – most often, TE- and TM-polarized – waves propagate in the crystal bulk. Thus, in the first passband, a photonic crystal refracts incident waves like a natural crystal does. If the internal geometry is isotropic, two polarizations degenerate, and there is one refracted wave. If is it anisotropic, birefringence of refracted waves is observed. These properties mean that the photonic crystal behaves similarly to an effectively homogeneous natural crystal.

In the other extreme of high frequencies of external excitation, when $2\pi n/a < k_0$ with $n \neq 0$ (k_0 is the wavenumber in the medium from where the incident wave comes), one or more higher-order harmonics can correspond to propagating waves in external space and also in the crystal bulk. Several propagating reflected waves can be created as in a diffraction grating. These higher-order modes also propagate inside the photonic crystal body, corresponding to several refracted waves. Thus, at high frequencies, the photonic crystal either completely reflects the incident wave (in stopbands) or splits it into a number of transmitted waves (in passbands). This splitting is not surprising for an optician. A photonic crystal slab is the same as a multilayer diffraction grating. For a finite-thickness photonic crystal, the wavevectors of dominating spatial harmonics in the Bloch expansion are wavevectors of Fraunhofer's diffraction lobes in the theory of diffraction gratings.

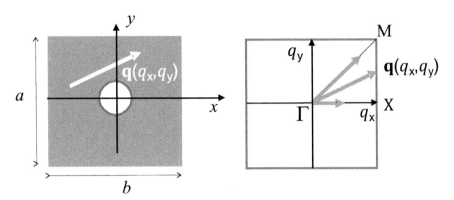

Figure 5.9 Geometry of a unit cell of a rectangular lattice of holes with periods a and b where the eigenmode with the Bloch vector **q** propagates obliquely (left panel) and the Brillouin zone of this lattice (right panel). The Brillouin dispersion diagram is obtained by the variation of **q** from the characteristic point $\Gamma(q_x = 0, q_y = 0)$ to point $X(q_x = \pi/a, q_y = 0)$, then to point $M(q_x = \pi/a, q_y = \pi/b)$ and again to Γ.

5.2.4 Brillouin Diagrams of Two-Dimensional Photonic Crystals

A reciprocal lattice with the coordinate axes (q_x, q_y) that are the components of the wavevector **q** can be introduced for a two-dimensional lattice with periods (a_x, a_y). As usual, only the Brillouin zone $-\pi/a_{x,y} < q_{x,y} < \pi/a_{x,y}$ is considered. The Brillouin zone of a rectangular lattice ($a_x \equiv a, a_y \equiv b$) is illustrated by Figure 5.9. Since the Brillouin diagram is a plot $\omega(\mathbf{q})$, where **q** is a vector, the horizontal axis splits onto intervals in which different components of the Bloch vector vary separately. In order to make this diagram more compact, L. Brillouin has suggested the following algorithm.

First, for reciprocal lattices with a unit cell symmetric along axes x and y, the negative part of the Brillouin zone $-\pi/a < q_{x,y} < 0$ is removed, as containing no new information; only one quadrant of the Brillouin zone is relevant. Second, there is no need to find all ω for all possible values of the components of **q** within this quadrant. It is enough to vary this vector over a triangle formed by the characteristic points of the Brillouin zone: points Γ, X, and M. The variation of the vector **q** on the Brillouin diagram starts from and ends at point Γ: the origin of the reciprocal space ($q_x = q_y = 0$). In the first part of the band diagram, the wavevector **q** is directed along x, and $q = q_x$ varies from zero to π/a (point X). In the second part, the wavevector **q** keeps the same x-component $q_x = \pi/a$, and its y-component varies from zero to π/b (point M). Finally, keeping the same diagonal direction of the wavevector one decreases its absolute value from $\sqrt{(\pi/a_x)^2 + (\pi/a_y)^2}$ to zero, returning in this way to Γ.

Wavevectors varying over any other contour onto which the Brillouin zone may split will not reveal new eigenfrequencies. The sufficiency of the triangle Γ-X-M results from the so-called *group symmetry* of the rectangular cell problem [1, 2, 27]. In the case of more complicated lattices, such as honeycomb lattices,

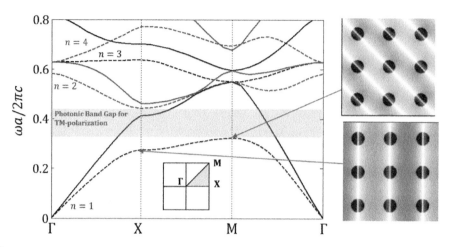

Figure 5.10 The Brillouin diagram of a square lattice of cylinders with the period a, the relative permittivity $\varepsilon = 10$, and the diameter $a/2$ hosted by free space. Dispersion curves of the TE-polarized eigenmodes are solid; those of the TM-polarized ones are dashed. The photonic bandgap holds only for the TM-polarized modes. On the inset, the electric field phase maps are shown for two Bragg modes, corresponding to points X and M on the boundary of the Brillouin zone. Simulations are done by Ana Díaz-Rubio.

the Brillouine zone and the corresponding path of $\mathbf{q}$ become more complicated. We will discuss this issue in the following paragraphs.

The Brillouin diagram for a square $(a = b)$ lattice of dielectric cylinders of diameter $a/2$ with permittivity $\varepsilon = 10$ located in free space is shown in Figure 5.10. From this diagram, it is clear why the photonic bandgap term is preferable for two-dimensional photonic crystals rather than the complete bandgap. Really, the stopband for all propagation directions (in the plane orthogonal to the cylinders axes) is observed only for waves whose magnetic field is polarized along the cylinders: for TM-polarized eigenwaves. For TE-polarized waves whose dispersion curves are shown as solid lines, only directional stopbands are present. For example, for propagation along x, there is a gap $\omega a/2\pi c = 0.41\dots0.44$ between the first and second dispersion curves for TE-waves. For the bisectorial propagation there is a gap $\omega a/2\pi c = 0.56\dots0.59$, etc.

On the insets of Figure 5.10 the color maps of simulated electric field are shown for two eigenmodes: that with $q_x = \pi/a$, $q_y = 0$ (corresponding to point X) and that with $q_x = q_y = \pi/a$ (point M). At these boundary points of the Brillouin zone any pair of inclusions neighboring along the propagation direction is polarized in the opposite phase. Strictly speaking, such propagating waves cannot exist because for them, the Bragg condition (5.1) is satisfied. Therefore, the modes whose $\mathbf{q}$ corresponds to the intersections of the dispersion curve with points X and M are called the Bragg modes or (more often) the *staggered modes* of the photonic crystal. It is seen in Figure 5.10 that the Bragg mode has zero group velocity $(\partial\omega/\partial q = 0)$, which means absence of energy propagation. These points on the dispersion curve are assumed to be punctured.

In fact, a Bragg mode on the dispersion curve is truly prohibited shallow in a purely lossless crystal. Due to finite (even very small) optical losses, the group velocity in photonic crystals is not exactly equal to $\partial\omega/\partial q$, and Bloch's vector $\mathbf{q}$ depicted in the horizontal axis of the Brillouin diagram represents the real part of the complex wavevector $\mathbf{k} = \mathbf{q} + j\mathrm{Im}(\mathbf{k})$. This imaginary part is negative and responsible for wave decay. Therefore, in realistic photonic crystals, the Bragg mode exponentially attenuates versus the distance from the source. If it is excited by an incident plane wave at the crystal boundary, it practically decays over a certain path that may be longer than the lattice period. In the present simulations, $\mathrm{Im}(\varepsilon) = -10^{-7}$), and in this case, the decay length of the Bragg mode equals to about a dozen of lattice periods.

Modes that have $\partial\omega/\partial q = 0$ inside the Brillouin zone can exist, too. In the present case, this point is on the dispersion curve of the $2d$ passband between M and Γ points. Due to non-zero losses, this property does not mean that the light is "frozen" (as it is sometimes erroneously thought). This simply means that, at this point, the imaginary part of the wavevector prevails, and the group velocity has no physical meaning. Waves with $\mathrm{Re}(q) \neq 0$ and $|\mathrm{Im}(q)| > \mathrm{Re}(q)$ practically do not propagate into the bulk and rapidly decay. Such modes are sometimes called complex modes or staggered modes. However, most often they are called *polaritons* [2, 3]. In realistic photonic crystals, polaritons (including the Bragg modes) can be excited either by an embedded source or by a plane wave. In the last case, they are excited at the interface of the crystal. The effect of polaritons can be used in some practical applications involving relatively thin photonic crystal layers. In a thin structure of a few periods across the layer, polariton fields reach the rear interface before it decays and may even experience multiple reflections on both interfaces (see, e.g., in [29]).

5.2.5 Brillouin Diagrams of Complex Three-Dimensional Photonic Crystals

For two-dimensional photonic crystals whose lattice is different from the rectangular one and for three-dimensional photonic crystals whose lattice is different from the orthorhombic one (i.e., the unit cell is not a parallelepiped) determination of the Brillouin zone is not so simple. The answer for any explicit lattice can be found applying the theory of groups. We do not explain this theory here. Recipes for building the Brillouin zone for various photonic crystals are available in books [1–4]. Deeper understanding can be gained by reading [30].

In Figure 5.11, we show two typical band diagrams for two typical photonic crystals with three-dimensional honeycomb (opal) lattices, also called body-centered cubic lattices. The difference between Figure 5.11(a) and Figure 5.11(b) is the content of the inclusions and the matrix. For a usual opal, the spherical inclusions have a higher permittivity than that of the matrix. For the inverted opal, the matrix has a higher permittivity than the inclusions. In the inset showing the Brillouin zone, we can see that the number of characteristic points (six) is larger than that for a simple cubic lattice (four).

Figure 5.11 is instructive for comparison of directional and complete bandgaps. The appearance of a directional bandgap is seen in Figure 5.11(a). Between points Γ and X, light is propagating along the x-axis. Between points X and W, q_x is fixed ($q_x = \pi/a$), and q_y grows from zero – i.e., light is obliquely

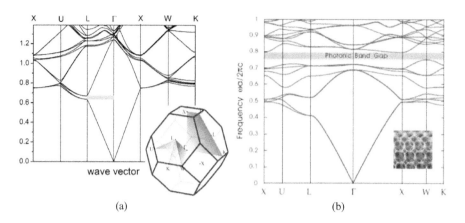

Figure 5.11 The Brillouin band diagrams for two three-dimensional photonic crystals with complex lattices: (a) that of a three-dimensional honeycomb lattice of spherical inclusions called an artificial opal, (b) a honeycomb lattice of voids (inverted opal). (a) Gray rectangle points out a stopband along the vertical axis of the lattice. This stopband (directional bandgap) shifts upward when the propagation direction deviates from the vertical axis. The inset shows the Brillouin zone of this lattice. Reprinted from [31] with permission of the author. (b) A honeycomb lattice of voids (inverted opal) manifests the complete bandgap. The inset shows a micrograph of a sample of inverted opal. Reprinted from [1] with permission of the editors.

propagating with an increasing angle in the xy-plane. The central frequency of the first directional bandgap (between the first and second passbands) clearly increases versus the propagation angle. For a sea mouse hair, this prohibited frequencies correspond to red light at point X and to blue light at point W. Similarly, the directional bandgap effect holds when the propagation angle changes in other planes, between other characteristic points of the diagram. For the inverse opal whose inclusions are voids, one observes in Figure 5.11(b) a complete bandgap.

A wider complete bandgap than that of opals is observed for artificial diamond lattices. Diamond-type photonic crystal with voids (called yablonovite [1–3] since the technique of its fabrication was suggested by the group of E. Yablonovich) was theoretically and experimentally studied in works [12, 32, 33], which manifested the start of the era of nanophotonics in modern optics and electrodynamics.

5.3 Isofrequencies of Photonic Crystals

The use of isofrequency contours is a powerful graphical tool for qualitative solving boundary problems for plane-wave reflection and transmission. When a spatial harmonic – a propagating plane wave or even an evanescent wave – impinges a flat boundary of two media, the isofrequency drawn for the medium of incidence compared with that drawn for the medium of transmission allows us to determine the main parameters of the transmitted wave and the conditions of transmission.

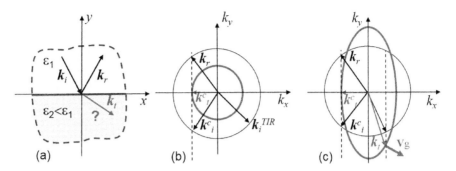

Figure 5.12   Illustration of the application of isofrequencies for finding the critical angle in the problem of the plane-wave transmission into a less refractive dielectric: (a) problem sketch, (b) graphic solution of the problem for isotropic media, (c) graphical solution for a TM-polarized incident wave if the medium of incidence is isotropic and the medium of refraction is anisotropic.

5.3.1 Isofrequencies in Boundary Problems

Consider as our first example the problem of finding the critical angle of total internal reflection for two isotropic media. Let the incident wave with wavevector $\mathbf{k}_i$ reflect from the boundary of two dielectrics modeled by real-valued permittivities $\varepsilon_{1,2}$. Let us assume that $\varepsilon_1 > \varepsilon_2$ as in Figure 5.12(a) and ask ourselves if this reflection is total or not? If it is partial reflection, there is a transmitted wave with the wavevector $\mathbf{k}_t$ in the medium with ε_2. If reflection is total, there is only a reflected wave with the wavevector $\mathbf{k}_r$ in the medium of incidence. In the incidence plane $(x-y)$, the dispersion equation of both media is trivial: $k_0^2 \varepsilon_{1,2} = k_x^2 + k_y^2$, where k_x is the same is both media ($k_{ix} = k_{rx} = k_{tx}$), as Maxwell's boundary conditions demand a plane wave. The isofrequency of the incidence medium in Figure 5.12(b) is shown as the black circle of radius $k_0 \sqrt{\varepsilon_1}$, and the isofrequency of the refracting medium is the thin outer of radius $k_0 \sqrt{\varepsilon_2}$. The incidence angle corresponding to $\mathbf{k}_i^c$ shown in this figure is the critical angle. It corresponds to a refracted wave propagating horizontally along the interface, whereas the wave with $\mathbf{k}_i^{TIR}$ cannot excite a propagating wave in the lower medium and is totally reflected from it (TIR means total internal reflection).

Next, in Figure 5.12(c), the same method is applied to the case when the lower medium is anisotropic (uniaxial). In the uniaxial medium, the permittivity tensor $\overline{\overline{\varepsilon}}$ written in the coordinate frame defined by its eigenvectors has two different components – axial (in the present coordinate frame, it is $\varepsilon_{yy} \equiv \varepsilon_n$) and transverse ($\varepsilon_{xx} = \varepsilon_{zz} \equiv \varepsilon_t$). The dispersion equations are different for the ordinary wave polarized in the transverse plane and for the extraordinary wave polarized in the plane comprising the optical axis. These two equations are as follows (respectively):

$$k_0^2 = \frac{k_y^2}{\varepsilon_t} + \frac{k_x^2}{\varepsilon_t}, \quad k_0^2 = \frac{k_y^2}{\varepsilon_t} + \frac{k_x^2}{\varepsilon_n}. \tag{5.6}$$

The first equation coincides with that for isotropic media. If the polarization of the incident wave is TE (electric field is along z), the isofrequencies are circular, as in Figure 5.12(b). Figure 5.12(c) corresponds to TM polarization, when the refracted wave is extraordinary. Since the second equation in (5.6) is the equation of an ellipse,[5] we have to juxtapose the circle of the upper medium with the ellipse of the lower medium, as it is done in Figure 5.12(c). Again, the condition $k_{ix} = k_{rx} = k_{tx}$ allows us to graphically find the critical regime when $k_t = k_{tx} = k_t^c$.

It is worth recalling that the group velocity vector $\mathbf{v}_g$ in lossless media is equal to $\partial \omega / \partial \mathbf{k}$. Therefore, $\mathbf{v}_g$ is directed orthogonally to the isofrequency surface for the same reason the electrostatic field is directed orthogonally to equipotential surfaces. Thus, if the dissipation is negligible, the normal to the isofrequency contour shows the direction of $\mathbf{v}_g$ that is also the direction of the power transport, i.e., that of the refracted wave. When the interface is illuminated at the critical angle, the refracted wave direction as well as the wavevector (phase velocity) direction are horizontal. When the incidence angle is smaller than the critical one, as it is also depicted in Figure 5.12(c), the wave energy velocity (the group velocity $\mathbf{v}_g$) and the wave phase velocity (the wavevector $\mathbf{k}_t$) of the transmitted wave point into different directions. Such waves, well known in optics of crystals, are not purely transversal – vector $\mathbf{E}$ is tilted with respect to $\mathbf{k}_t$.

5.3.2 Isofrequencies of Photonic Crystals

For natural anisotropic crystals, isofrequency surfaces for different frequencies form a family of ellipsoids whose sizes grow versus frequency. Respectively, the isofrequency contours – sections of these ellipsoids by propagation planes – are ellipses. Isofrequency contours of a photonic crystal in the range where the spatial dispersion is present are much more complex than the family of ellipses. Only at low frequencies, where the photonic crystal can be homogenized, its isofrequency surfaces are as simple as those of a natural medium.

However, there are some photonic crystals that are not effectively continuous even at very low frequencies. In Figure 5.13(a), two unit cells of such a photonic crystal are depicted. If such cylindrical inclusions are perfectly conducting (that implies the limit $\sigma \to \infty$ for the complex permittivity $\varepsilon = \mathrm{Re}(\varepsilon) - j\sigma/\omega\varepsilon_0$ of inclusions) the dispersion diagram is as complex as it is shown in Figure 5.13(b).

This photonic crystal is called *crossed-wire medium* (see, e.g., [22]). It practically represents a parallel set of alternating mutually crossing (but not touching) grids of parallel metal wires. This type of wire media is used in the frequency ranges where the approximation $\sigma \to \infty$ is adequate – at frequencies of the far infrared light and lower frequencies. At far infrared frequencies, the period a can be of the order of several microns, at microwaves – of the order of a few millimeters. Eigenwaves of the crossed-wire medium have hybrid polarizations: the dispersion curves and isofrequencies cannot be split onto two separate families corresponding to TE- and TM-waves.

For simplicity, we restrict the consideration by waves propagating in the plane $(y-z)$ only. In this case ($q_x \equiv 0$), the point $q_y = \pi/a$, $q_z = 0$ is denoted as point Z, and the point $q_y = q_z = \pi/a$ as point L. Let us choose in the dispersion diagram

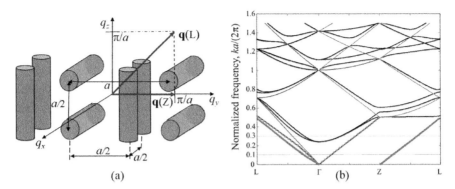

(a)   (b)

Figure 5.13 A crossed-wire medium and its dispersion diagram. (a) Internal geometry (two unit cells along the axis y are shown). (b) Dispersion diagram of the crossed-wire medium for waves propagating in the plane $(y-z)$. Thin straight lines corresponding to free-space dispersion are shown for comparison. Calculations are done by Pavel Belov.

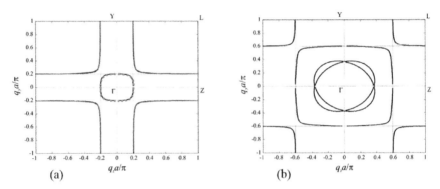

(a)   (b)

Figure 5.14 Isofrequencies of a crossed-wire medium in the plane $(q_y - q_z)$: (a) contours corresponding to $ka/2\pi = 0.1$; (b) contours corresponding to $ka/2\pi = 0.3$. Calculations are done by Pavel Belov.

two frequencies $\omega a/2\pi c = 0.1$ and $\omega a/2\pi c = 0.3$. Thin lines corresponding to the dispersion of free space and called light lines are shown in this diagram for comparison (Figure 5.13(b)). We see that two lines highlighted by dashes rather closely mimic the dispersion of free space. The right dashed line (**q** varies between Γ and Z) visually coincides with the light line. The left dashed line (**q** varies between L and Γ) is also close to the corresponding light line. Therefore, the isofrequency contour corresponding to the dashed dispersion branch is close to the circle corresponding to free space.

However, a small deviation of the left dashed line from the light line visible in Figure 5.13(b) makes this contour slightly different from the circle. Numerical simulations show that it is a superquadric contour, which is depicted with dashes in Figure 5.14(a). Here, we only explain the physical meaning of the obtained isofrequencies and how to use them for the graphical analysis. Methods of their calculation as well as calculations of the dispersion diagram will be considered in the following paragraphs.

A four-branch hyperbola in the isofrequency plot corresponds to the dispersion curves highlighted (and thicker) in Figure 5.13(b). Such dispersion – when the propagation along y is absent but the diagonal propagation with a long wavevector $\mathbf{q}(L)$ is allowed – is not observed in any continuous medium. And recall that this is the same frequency range where there is an eigenwave described by the superquadric contour. In the direction from Γ to L, two eigenmodes with different $\mathbf{q}$ propagate. For this bisectorial propagation, the frequency $\omega a/2\pi c = 0.1$ corresponds to one eigenmode with $q_y = q_z \approx 0.15\pi/a$ (dashed isofrequency) and to another one with $q_y = q_z \approx 0.25\pi/a$ (hyperbolic isofrequency). Definitely, the properties of this crystal are dramatically different from those of continuous media even at such low frequencies.

For $\omega a/2\pi c = 0.3$, we observe eight intersections with the dispersion curves in Figure 5.13(b) and, respectively, obtain four isofrequency contours in Figure 5.14(b): a four-branch hyperbola, a superquadric contour, and two tilted ellipses. It is clear from the isofrequency plot that in the same direction from Γ to L, four different eigenmodes propagate. In the direction from Γ to Z (i.e., along the axis z), three eigenmodes propagate. More exactly, along z travel similarly two waves – one with the wavevector $q_z a/\pi \approx 0.4$ and another one with $q_z a/\pi \approx 0.6$. However, the same wavevector $q_z a/\pi \approx 0.4$ corresponds to two different modes with group velocities vectors tilted nearly by $\pm 30°$ with respect to the z-axis.

Crossed-wire media refer to arguably most complicated photonic crystals (see, e.g., [22]), and here we do not aim to further describe its electromagnetic properties, but we hope that this example clearly shows drastic difference between photonic-crystal isofrequencies and those of continuous media (isotropic or anisotropic).

For many three-dimensional photonic and two-dimensional photonic crystals, the isofrequency plots are simpler because waves can be classified as TE and TM modes. Although the dispersion of TE-polarized waves is different from that of TM-waves, as we could observe in Figure 5.10, solving the boundary problems for incident waves from an isotropic space one can separately consider two different families of isofrequencies for these two polarizations, and superquadric isofrequency contours rarely coexist with hyperbolic ones at the same frequencies.

5.3.3 Isofrequencies of a Photonic Crystal in a Boundary Problem

Here, we apply the isofrequency technique to a problem of transmission of a plane wave into low-loss photonic crystals. The problem is: to find the refracted wave direction (the same as the direction of the group velocity $\mathbf{v}_g$ and the Poynting vector $\mathbf{S}_t$). An explicit example is illustrated by Figure 5.15. A photonic crystal of cylindrical air holes in a high-permittivity dielectric (in the present example, $\varepsilon = 10$) is illuminated by a TE-polarized plane wave with the wavevector $\mathbf{k}_i$ forming the incidence angle θ with the normal axis Y. The wave is incident from free space; its Poynting vector $\mathbf{S}_i$ is parallel to $\mathbf{k}_i$. The lattice axes x and y are

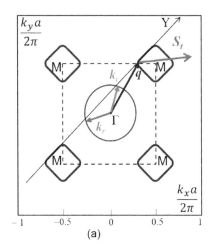
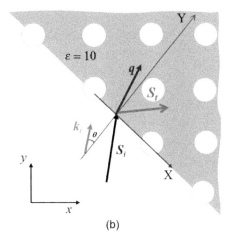

(a)   (b)

Figure 5.15 The use of isofrequency technique for finding the direction of the Poynting vector of a wave refracted into a photonic crystal. (a) A photonic crystal is illuminated by a plane wave having the wavevector k_i at the normalized frequency $k_0 a/2\pi = 0.19$. The corresponding isofrequency of free space (circle) is centered at point Γ of the crystal Brillouin zone. The isofrequency contours of the crystal at $\omega a/2\pi c = 0.19$ are superquadric and centered at points M. The refracted wave is an eigenmode with the wavevector $\mathbf{q}$, and the normal to the isofrequency at the point $\mathbf{q}$ shows the direction of the Poynting vector $\mathbf{S}_t$ of the refracted wave. (b) The same vectors shown in the real space.

tilted by $\pi/4$ with respect to the normal axis Y and tangential axis X, as depicted in Figure 5.15(b).

At the normalized frequency $k_0 a/2\pi = 0.19$, the isofrequency of the photonic crystal with cylindrical voids inside the Brillouin zone represents a quarter of a superquadric contour. It is relevant to extend the isofrequency plot beyond the Brillouin zone as it is done in Figure 5.15(a). Here, the area in the coordinates $(k_x - k_y)$ is twofold of the Brillouin zone ($|k_{x,y}a/2\pi| \leq 1$), and in this area, we see four complete superquadric isofrequency contours. The only relevant contour is centered at the top-right corner of the Brillouin zone. Really, the incident wavevector $\mathbf{k}_i$ defines the plane-wave incidence in the real space that is shown in Figure 5.15(b). Three other superquadric isofrequencies refer to different cases of wave incidence. Comparing the relevant isofrequency contour with that of free space, the direction of the surface normal (axis Y) and the boundary condition $k_{iX} = q_X$ graphically define the refracted wavevector $\mathbf{q}$, similarly to the cases of continuous media analyzed above. In both Figure 5.15(a) and Figure 5.15(b), we show the wavevector $\mathbf{q}$ found in this way. The group velocity $\mathbf{v}_g$ and Poynting vector $\mathbf{S}_t$ of the refracted wave are normal to the isofrequency contour. $\mathbf{S}_t$ is also shown in Figure 5.15.

When the normalized frequency grows from $k_0 a/2\pi = 0.19$ to $k_0 a/2\pi = 0.22$, the superquadric isofrequency contour squeezes to point M. Therefore, the normal to the contour corresponding to $\mathbf{v}_g$ is directed inside the contour and $\mathbf{S}_t$ in Figure 5.15 is tilted to the right with respect to Y. Inspecting in Figure 5.15(b) the direction of both $\mathbf{q}$ and $\mathbf{S}_t$ with respect to the interface normal Y, we see that refraction of the wavevector is positive, whereas the refraction of the Poynting vector is negative. Thus, using the isofrequency technique, we have theoretically revealed a very important phenomenon of negative refraction observed in many photonic crystals (see, e.g., [9] and Chapter 3).

Isofrequencies are also helpful for clarifying the issue of total reflection from a photonic crystal. Unlike the case of interfaces between two continuous low-dispersion dielectrics, the critical angle of total reflection in this case depends on the frequency. And total reflection within a passband does not imply that the incident wave comes from a higher-permittivity dielectric. In the last example, the incidence angle θ, as it is seen in Figure 5.15(a), is very close to the critical one. The condition $k_{iX} = q_X$ cannot be fulfilled if $\theta > \theta_c \approx 40°$. Total reflection into free space is observed, though the photonic crystal matrix is a high-permittivity dielectric and air holes are relatively small. Notice that the frequency in this example corresponds to the first passband, where the absolute majority of photonic crystals behaves similarly to continuous media. This crystal is not an exception. In what concerns reflection, this example photonic crystal mimics a continuous medium whose effective refractive index at the given frequency $k_0a/2\pi = 0.19$ is smaller than unity. In what concerns refraction, it mimics an anisotropic medium: in anisotropic media, $\mathbf{q}$ and $\mathbf{S}_l$ can be directed differently, as it was shown in Figure 5.12(c). However, there are no natural continuous media offering a possibility for negative refraction for waves with small angles of incidence and total reflection for large angles (at the same frequency). Therefore, this photonic crystal at frequencies around $k_0a/2\pi = 0.19$ (namely at $k_0a/2\pi = 0.17$–0.22) is an *artificial* effectively continuous medium, having to natural analogs in what concerns its electromagnetic properties. Such media by definition refer to metamaterials, discussed in Chapter 3. In other words, this photonic crystal is a metamaterial at frequencies $k_0a/2\pi = 0.17$–0.22.

5.4 Numerical Simulations of Photonic Crystals and Their Scalability

The theory of the cell problem formulated via the Hamiltonian operator is instructive and allows one to understand the main features of the lattice dispersion. However, in practical cases, one needs to calculate the parameters of eigenmodes of the photonic crystal numerically. For this purpose, one uses the so-called governing equation that provides information on both Bloch wavevector $\mathbf{q}$ and relations between the amplitudes of the spatial harmonics in the Bloch plane-wave expansion of the eigenmode. Let us derive this equation.

5.4.1 Numerical Solution via Plane-Wave Expansion

If a photonic crystal is built from isotropic nonmagnetic materials (dielectrics or metals), the photonic crystal properties are determined by a scalar periodically varying permittivity $\varepsilon(\mathbf{r})$. From Maxwell's equations

$$\nabla \times \mathbf{H}(\mathbf{r}) + j\omega\varepsilon(\mathbf{r})\varepsilon_0\mathbf{E}(\mathbf{r}) = 0,$$

$$\nabla \times \mathbf{E}(\mathbf{r}) - j\omega\mu_0\mathbf{H}(\mathbf{r}) = 0, \tag{5.7}$$

we obtain the governing equation of a photonic crystal:

$$\nabla \times \left[\varepsilon^{-1}(\mathbf{r})\nabla \times \mathbf{H}(\mathbf{r}) \right] - k_0^2\mathbf{H}(\mathbf{r}) = 0. \tag{5.8}$$

Let us restrict our consideration by the case of simple (orthorhombic) lattices, where the natural lattice axes are Cartesian ones. In this case, we have $\varepsilon(x, y, z) = \varepsilon(x + a_x, y + a_y, z + a_z)$, and the Bloch plane-wave expansion takes the form

$$\mathbf{H} = \sum_{mps} \mathbf{H}_{mps} e^{-j\left(q_x + \frac{2\pi m}{a_x}x\right) - j\left(q_y + \frac{2\pi p}{a_y}y\right) - j\left(q_z + \frac{2\pi s}{a_z}z\right)}, \tag{5.9}$$

where $G_x = 2\pi m/a_x$, $G_y = 2\pi p/a_y$, $G_z = 2\pi s/a_z$ are the components of the lattice vector $\mathbf{G}$ (m, p, and s are integers). The Bloch expansion allows us to replace the operator $\nabla\times$ in the governing equation by the vector product $-j(\mathbf{q} + \mathbf{G})\times$.

Since $\varepsilon(\mathbf{r})$ is a periodic function of coordinates, the inverse permittivity $\varepsilon^{-1}(\mathbf{r})$ is also a periodic function, and it can be expanded into a Fourier series,

$$\varepsilon^{-1}(\mathbf{r}) = \sum_{MPS} b_{MPS} e^{-j\left(\frac{2\pi M}{a_x}x + \frac{2\pi P}{a_y}y + \frac{2\pi S}{a_z}z\right)}. \tag{5.10}$$

The Fourier coefficients of the inverse permittivity b_{MPS} can be found via the standard integration procedure since the function $\varepsilon(x, y, z)$ is known. Substituting (5.9) and (5.10) into (5.8), we obtain an infinite system of algebraic equations for unknown coefficients $\mathbf{H}_{mnp}$ with the unknown vector $\mathbf{q}$ having, in the general case, all three components q_x, q_y, and q_z. This system is source-free; i.e., to allow for nontrivial solutions, its determinant must be equal to zero. This fact allows us to find $\mathbf{q}$ for a given frequency $\omega \equiv k_0 c$ without solving the cell problem. Assume that the wave propagates along x – i.e., $q_y = q_z = 0$. In this case we obtain the system

$$\sum_{mnp} \xi_{MmPpSs} \mathbf{H}_{mps} e^{-j\left(q_x + \frac{2\pi m}{a_x}x\right) - j\left(\frac{2\pi p}{a_y}y + \frac{2\pi s}{a_z}z\right)} = 0, \quad (M, P, S) = 0, \pm 1, \pm 2, \ldots, \tag{5.11}$$

where coefficients ξ_{MmPpSs} can be easily expressed through q_x and b_{MPS} using (5.8) with the substitution of (5.10) and $\nabla = -j(q_x + 2\pi m/a_x)\mathbf{x}_0 - j2\pi p/a_y\mathbf{y}_0 - j2\pi s/a_z\mathbf{z}_0$. Then we truncate the system (5.11), choosing finite $M_{\max}$, $P_{\max}$, and $S_{\max}$ so that the finite-sum approximation of the inverse permittivity initially described by the infinite sum (5.10) provides required accuracy. Most often, numbers $M_{\max}$, $P_{\max}$, and $S_{\max}$ corresponding to system (5.11) are smaller than similar numbers that would arise if we were using the alternative governing equation

$$\nabla \times \nabla \times \mathbf{E}(\mathbf{r}) - k_0^2\varepsilon(\mathbf{r})\mathbf{E}(\mathbf{r}) = 0, \tag{5.12}$$

which also follows from Maxwell's equations, and applied the Fourier expansion for $\varepsilon(\mathbf{r})$ [1]. After this truncation, we obtain a linear source-free system of equations (5.11) for a finite number of unknowns $H_{x,mps}$, $H_{y,mps}$, and $H_{z,mps}$. Equating the determinant of the system to zero, we come to an algebraic equation for q_x that can be numerically solved, for example, using MATLAB.

Similarly, one can find q_y and q_z for the cases when the wave propagates along these axes. Next, one fixes the relationship between the components of the Bloch wavevector **q** that corresponds to a selected direction of oblique propagation. The needed propagation directions are determined by the corners of the Brillouin zone. For oblique propagation, we again obtain a finite source-free linear system with a scalar q whose absolute value is found equating the determinant to zero.

Once we know q for a given propagation direction, we may set the amplitude of the component of $\mathbf{H}_{000}$ orthogonal to this propagation direction equal to unity and find the relative amplitudes of all spatial harmonics. It is what we need because, in general, the eigenmode amplitude is not defined – this is all that we can know about an eigenmode propagating in a given direction at a given frequency.

As we have already discussed, the geometry of majority of photonic crystals allows separate existence of the TE- and TM-polarized modes along all relevant directions. This property simplifies the eigenvalue problem for $\mathbf{H}_{mps}$. In this case, it becomes a scalar problem for two field components. Using Maxwell's equations, the calculation of the other components of the modal electromagnetic field is straightforward.

We see that, using the governing equation and the Bloch theorem, we can find the relative amplitudes of all spatial harmonics of eigenmodes propagating in the infinite lattice at a given frequency in a given direction. The knowledge of q for all propagation directions at a given frequency ω gives us the isofrequency surface or contour, and the same data for a sufficient set of frequencies provides a family of isofrequencies and the dispersion diagram. The amplitude $\mathbf{H}_{000}$ is defined by excitation of waves by external sources (most often, by plane waves illuminating crystal surfaces). Here we do not discuss methods for numerical solutions of boundary problems for photonic crystals; we refer the reader to book [2].

5.4.2 Numerical Solution of the Cell Problem

Numerical simulations of the band diagram and isofrequencies of photonic crystals can be done using commercial software packages such as HFSS, CST Studio, and COMSOL Multiphysics. These packages comprise solvers of cell problems using Bloch's quasi-periodicity conditions applied for a unit cell having six faces and possessing translation symmetry. Such cells are called *monoclinic prisms*.

In the HFSS solver, the option for solving such cell problems is called *Optimetric Analysis*. Choosing this option, the user draws the unit cell of the lattice with its internal geometry in an HFSS project, defines in the descriptive part of the project the material parameters of all the internal parts of the cell, and selects the option *Eigenmode solution*. Next, the user suggests varying from zero to π the phase shift of the eigenwave traveling across the unit cell, determining in the solver the so-called *master and slave faces* of the cell. Master faces are the cell sides "illuminated by the eigenwave." Slave faces are opposite to the master faces – i.e., located at the back side of the monoclinic prism. The solver varies the propagation directions and the absolute value of **q** so that **q** varies from the center (Γ) to the boundary of the Brillouin zone. Then, for the all passible values of **q** (varying with a very small step), the solver delivers the set of eigenfrequencies. If there are optical losses (i.e., the permittivity of the inclusion or the host medium is complex), the eigenfrequencies corresponding to given real-valued **q** are complex numbers.

Only real parts of these complex numbers are used in the dispersion diagram that is calculated within the framework of the optimetric analysis. Of course, photonic crystals are never prepared of very lossy constituents since their practical applications are not absorption and they must be sufficiently transparent in their passbands. Therefore, eigenfrequencies of the propagating waves are in the practical cases sufficiently close to real values; otherwise, either the solution is incorrect or losses are too high.

Similar general procedures correspond to other commercial solvers of the cell problem for photonic crystals. However, the numerical solution can be done in different ways. The HFSS solver uses the finite-element method for Maxwell's equations together with the Bloch quasi-periodicity conditions, whereas the CST Studio Suite involves two techniques: volume integral equations in the frequency domain and the finite-difference-time-domain (FDTD) approach. For reliability, one often simulates the same photonic crystal using two or more numerical techniques. Beyond commercial software, there is homemade software applying different techniques for solving the cell problem. Besides the plane-wave expansion method that we have already discussed, people also use the so-called *transfer-matrix approach* [34] and FDTD approach [35]. In some special cases, it is possible to solve the cell problem for simple photonic crystals analytically in closed form. It was done in [36] for one-dimensional photonic crystals and for a two-dimensional photonic crystal (a square lattice of circular cylinders) in [37]. However, these closed-form solutions are very involved, even for these basic cases, and full-wave numerical simulations remain the main tool for the analysis of photonic crystals.

5.4.3 Scalability of Photonic Crystals

Here we discuss the scalability of photonic crystals – the invariance of the solution with respect to the magnification of the unit cell accompanied by the corresponding reduction of the operation frequency.

For the dispersion equation, this scalability follows from the fact that both axes in the dispersion diagram are dimensionless: $\omega a/2\pi c$ and $qa/2\pi$. For the electromagnetic field vectors, the scalability is not so evident, and we need to prove it. Denote the radius vector in the enlarged photonic crystal as **R**, and in the original photonic crystal, let it be **r**. Then $\nabla_r = s\nabla_R$, and we may rewrite (5.8) as

$$s^2 \nabla_R \times \left[\varepsilon^{-1}(\mathbf{R})\nabla_R \times \mathbf{H}(\mathbf{R}) \right] = k_0^2 \mathbf{H}(\mathbf{R}). \tag{5.13}$$

Dividing both parts of this equation by s^2, we obtain $k_0^{new} = k_0/s$, i.e., $\lambda_{new} = s\lambda$. This means that optical phenomena of nanostructured photonic crystals can be modeled at microwave frequencies where the unit cell is scaled to millimeters or even centimeters. Scalability offers a great tool for experimental study and optimization of photonic crystals. The main restriction of this approach refers to photonic crystals with metal inclusions. Metals in the optical range have complex permittivity whose negative real part prevails over the imaginary part. Such materials at microwaves and other radio frequency ranges are absent, except some kinds of plasma. An inclusion of plasma needs an encapsulation. It is difficult to prepare a matrix with voids filled with plasma, and it is even more difficult to make plasmas identical in all these voids. It would be easier and cheaper to fabricate an original photonic crystal with submicron metal inclusions.

Problems and Control Questions

Problems

1. Let an eigenmode propagate in a two-dimensional lattice of optically thin cylinders whose axes are orthogonal to the propagation plane, as it is shown in Figure 5.16(a), but the propagation is oblique with respect to the lattice axes. Describe how the set of stopbands varies versus θ, assuming that $a \gg b$ and the partial reflection coefficient r of the grid does not depend on both ω and θ?

2. The geometry of a two-dimensional photonic crystal of cylinders with square cross sections is shown in Figure 5.16(b). In order to solve the general Eq. (5.8), one expands the inverse permittivity $1/\varepsilon(\mathbf{r})$ of the structure into the Fourier series (5.10) that, in the present case, takes the form

$$\varepsilon^{-1}(\mathbf{r}) = \sum_{M=-L}^{L} \sum_{P=-L}^{L} b_{MP} e^{-j\left(\frac{2\pi M}{a_x} x + \frac{2\pi P}{a_y} y\right)},$$

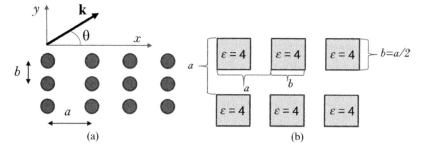

(a)   (b)

Figure 5.16 Oblique propagation in a two-dimensional photonic crystal with rectangular lattice (a). A two-dimensional photonic crystal of free-standing cylinders with square cross sections and permittivity $\varepsilon = 4$ (b).

where L is called truncation order of the infinite series (5.10). Assuming that the host medium is free space and using standard formulas of the Fourier theory, calculate b_{MP} analytically via double integration. Calculate these coefficients numerically – e.g., in, assuming $L = 5$. Is the truncation order $L = 5$ for the present case sufficient?

3. Assume that a TM-polarized mode ($H = H_z$) propagates in the crystal depicted in Figure 5.16(b). Write explicitly in matrix or index form the linear system of equations for complex amplitudes of Bloch's spatial harmonics H_{mn} entering the two-dimensional Bloch plane-wave expansion for the magnetic field. Here, the three-dimensional expansion (5.9) reduces to

$$ H = \sum_{m=-L}^{L} \sum_{n=-L}^{L} H_{mp} e^{-j\left(q_x + \frac{2\pi m}{a_x} x\right) - j\left(q_y + \frac{2\pi p}{a_y} y\right)}, $$

as follows from Eq. (5.8), assuming that you know coefficients b_{MP} (here, the fundamental term is not shared). What is the order of this system for a given L (i.e., how many equations are in it)? What is the power order of the dispersion equation?

Questions

1. Is the explanation given by Lord Rayleigh to the phenomenon of Bragg valid for lattices of particles of substantial size? Can the reflection by a planar grid of large particles be treated as a reflection by an infinitesimally thin crystal plane? Explain your opinion.

2. Why are Bloch's quasi-periodicity conditions called quasi-periodic? When they become periodic conditions, what is the limit case in terms of the Bloch wavevector?

3. What is the advantage of using the Brillouin zone in the dispersion diagram of lattices?

4. Let an isotropic medium with the relative permittivity $\varepsilon_1 = 4$ form a flat interface with a uniaxial medium having tangential component of the permittivity tensor $\varepsilon_{2t} = 4$ and the normal component $\varepsilon_{2n} = 1$. Can the effect of total internal reflection be observed for waves obliquely incident from the isotropic medium? Use the isofrequency technique for finding the answer.

5. Can the effect of total internal reflection be observed for waves incident from the uniaxial medium? Again, use the isofrequency technique.

6. How do the answers of the two previous questions change if $\varepsilon_{2n} = 4$ and the normal component $\varepsilon_{2t} = 1$?

7. Why is the governing equation of the photonic crystal formulated for the magnetic field of an eigenmode and not for the electric field?

8. Why is the scalability of photonic crystals not used by researchers to simplify the experimental investigations of lattices of metal nanoparticles and optical wire media?

Bibliography

[1] J. D. Joannopoulos, R. D. Mead, and J. N. Winn, *Photonic Crystals: Molding the Flow of Light*, Princeton University Press, 1995.

[2] K. Sakoda, *Optical Properties of Photonic Crystals*, Springer-Verlag, 2005.

[3] Special Volume on Electromagnetic Applications of Photonic Band Gap Materials and Structures, *Progress in Electromagnetic Research (PIER)* **41** (2003).

[4] Focus Issue: Photonic Bandgap Calculations, *Optics Express* **8**, no. 3 (2001).

[5] M. Inoue and K. Arai, "Magneto-optical properties of one-dimensional photonic crystals composed of magnetic and dielectric layers," *Journal of Applied Physics* **83**, 6768–6770 (1998).

[6] P. A. Belov, S. A. Tretyakov, and A. J. Viitanen, "Nonreciprocal microwave band-gap structures," *Physical Review E* **66**, 016608 (2002).

[7] A. P. Vinogradov, A. V. Dorofeenko, S. G. Erokhin, et al., "Surface-state peculiarities in 1D photonic crystal interfaces," *Physical Review B* **74**, 045128 (2006).

[8] D. P. Belozorov, M. K. Khodzitsky, and S. I. Tarapov, "Tamm states in magnetophotonic crystals and permittivity of the wire medium," *Journal of Physics D: Applied Physics* **42**, 055003 (2009).

[9] C. Luo, S. G. Johnson, and J. D. Joannopoulos, "Negative refraction without negative index in metallic photonic crystals," *Optics Express* **11**, 746–754 (2003).

[10] C. Rockstuhl, F. Lederer, C. Etrich, T. Pertsch, and T. Scharf, "Design of an artificial three-dimensional composite metamaterial with magnetic resonances in the visible range of the electromagnetic spectrum," *Physical Review Letters* **99**, 017401 (2007).

[11] A. Sihvola, "Metamaterials: A personal view," *Radioengineering* **18**, 90–94 (2009).

[12] E. Yablonovich, "Inhibited spontaneous emission in solid-state physics and electronics," *Physical Review Letters* **58**, 2059–2062 (1987).

[13] S. John, "Strong localization of photons in certain disordered dielectric superlattices," *Physical Review Letters* **58**, 2486–2489 (1987).

[14] V. P. Bykov, "Spontaneous emission in a periodic structure," *Soviet Physics Journal of Experimental and Theoretical Physics* **35**, 269–273 (1972). V. P. Bykov, "Spontaneous emission from a medium with a band spectrum," *Soviet Journal of Quantum Electronics* **4**, 861–871 (1975).

[15] P. W. Anderson, "Absence of diffusion in certain random lattices," *Physical Review* **109**, 1492–1505 (1958).

[16] R. C. McPhedran, N. A. Nicorovici, D. R. McKenzie, L. C. Botten, A. R. Parker, and G. W. Rouse, "The sea mouse and the photonic crystal," *Australian Journal of Chemistry* **54**, 241–244 (2001).

[17] K. H. Sabry, "Insect's photonic crystals and their applications," *Bioscience Research* **13**, 15–20 (2016).

[18] P. Fratzl, J. W. C. Dunlop, and R. Weinkamer, *Materials Design Inspired by Nature*, RSC Publishing, 2013.

[19] Lord Rayleigh, "On the reflection of light from a regularly stratified medium," *Proceedings of the Royal Society of London* **93**, 565–577 (1917).

[20] W. H. Bragg and W. L. Bragg, "The reflection of X-rays by crystals," *Proceedings of the Royal Society of London* **88**, 428–438 (1913).

[21] C. Kittel, *Introduction to Solid State Physics*, 8th ed., Wiley, 2005.

[22] C. R. Simovski, P. A. Belov, A. V. Atraschenko, and Y. S. Kivshar, "Wire metamaterials: Physics and applications," *Advanced Materials* **24**, 4229–4248 (2012).

[23] C. Jin, B. Cheng, B. Man, D. Zhang, S. Ban, B. Sun, L. Li, X. Zhang, and Z. Zhang, "Two-dimensional metallodielectric photonic crystal with a large band gap," *Applied Physics Letters* **75**, 1201–1205 (1999).

[24] A. Sommerfeld, "Elektronentheorie der Metalle," *Zeitschrift für Physik* **47**, 1–60 (1928).

[25] F. Bloch, "Uber die Quantenmechanik der Electronen in Kristallgittern," *Zeitschrift für Physik* **52**, 555–600 (1928).

[26] L. Brillouin, "Les électrons dans les métaux et le classement des ondes de de-Broglie correspondantes," *Comptes Rendus Hebdomadaires des Séances de l'Académie des Sciences* **191**, 292–299 (1930).

[27] H. Sagan, *Boundary and Eigenvalue Problems in Mathematical Physics*, 2nd ed., Dover Publications Inc. (1989).

[28] J. M. Sanz-Serna and M. P. Calvo, *Numerical Hamiltonian Problems*, Dover Publications Inc., 2018.

[29] S. K. S. Rahman, T. Klein, J. Gutowski, S. Klembt, and K. Sebald, "Tunable Bragg polaritons and nonlinear emission from a hybrid metal-unfolded ZnSe-based microcavity," *Scientific Reports* **7**, 767 (2017).

[30] E. Wigner, *Group Theory and Its Application to the Quantum Mechanics of Atomic Spectra*, Academic Press, 1959.

[31] J. F. G. Lopez, An optical study of opal based photonic crystals, PhD thesis, Deptartment of Material Physics, Autonomous University of Madrid, 2005.

[32] E. Yablonovitch and T. J. Gmitter, "Photonic band structure: The face-centered-cubic case," *Physical Review Letters* **63**, 1950–1953 (1989).

[33] E. Yablonovitch, T. J. Gmitter, and K. M. Leung, "Photonic band structure: The face-centered-cubic lattice employing nonspherical atoms," *Physical Review Letters* **67**, 2295–2298 (1991).

[34] J. B. Pendry, "Calculating photonic band structure," *Journal of Physics: Condensed Matter* **8**, 1085–1108 (1996).

[35] U. Andersson, M. Qiu, and Z. Zhang, "Parallel-power computation for photonic crystal devices," *Methods and Applications of Analysis* **13**, 149–156 (2006).

[36] S. M. Rytov, "Electromagnetic properties of a finely stratified medium," *Soviet Physics Journal of Experimental and Theoretical Physics* **2**, 466–475 (1956).

[37] H. Xie and Y. Y. Lu, "Modeling two-dimensional anisotropic photonic crystals by Dirichlet-to-Neumann maps," *Journal of the Optical Society of America A* **26**, 1606–1612 (2009).

Notes

1 Superprism operation corresponds to the light propagation across the nanorods. This is achieved in the moth eye because the lattice of nanorods covers a spherical surface, and these spherical nanostructured surfaces are triply repeated in the moth eye, being symmetrically tilted. Therefore, for whatever direction of the incident light, there is obviously a part of the moth eye operating as a superprism.
2 The Bragg condition corresponds to $m = 1, 2, 3, \ldots$ The quasi-static limit case when $m = 0$ and $\lambda_0 \to \infty$ will be discussed separately.
3 W. R. Hamilton – one of the fathers of mathematical physics – formulated, in addition to the cell problem of mathematical physics, also the path problem and the cycle problem [28].
4 This procedure was initially done by Brillouin in studies of dispersion of electron waves in solids.
5 For natural solid media when $\varepsilon_t > 0$ and $\varepsilon_n > 0$. If $\varepsilon_t > 0$, $\varepsilon_n < 0$, or $\varepsilon_t < 0$, $\varepsilon_n > 0$ which is possible for some metamaterials, the isofrequency contour is hyperbolic.

Nanophotonic Applications of Photonic Crystals

6.1 Conventional Applications of Photonic Crystals

In this section, we discuss conventional (already commercialized) applications of photonic crystals that leave no doubt that this branch of nanophotonics is fully practical today.

6.1.1 Photonic-Crystal Defect States

The most straightforward application of photonic crystals is related to the use of defect states in order to confine light energy. For example, waves can be guided along a line defect called the *defect channel* [1–3]. Such *defect states* (dispersion curves of the guided light inside the bandgap) are also called *mini-bands* because propagation is allowed not at just a single frequency but within a finite-width frequency band. A mini-band occupies a part of the bandgap, as shown in Figure 6.1(a) for the case of a two-dimensional photonic crystal of free-standing rods with the period a. Since the channel mode can propagate only along the channel, the defect state exists only between the points Γ and M of the *Brillouin zone*. This waveguide can be a single-mode one: if $a \approx \lambda/2$, and the wavelength is inside the bandgap, the air channel allows propagation of the fundamental TE-mode while all the higher-order modes are below cutoff. Since the wave at this frequency cannot propagate in the photonic crystal, the channel interfaces operate as walls of an equivalent parallel-plate waveguide.

However, a defect waveguide cannot be directly compared to the metallic parallel-plate waveguide, because in the optical range, metal walls are not practical, unlike in its microwave analog. Metals are lossy plasmas in the optical range, and metal-tube waveguides as well as single-mode TEM waveguides in optics are impossible. This conclusion is true also for photonic-crystal defect channels. The channel interfaces are not perfectly electrically or magnetically conducting. They reflect light with the reflection phase θ equal to neither 0 nor π. In a two-dimensional, waveguide, the guided mode of a single-mode waveguide is a superposition of two plane waves (called in the theory of waveguides the *Brillouin waves*) internally reflected from the channel interfaces with the reflection coefficient $R = \exp(j\theta)$. Like in a dielectric slab waveguide, the reflection phase determines the characteristic impedance of the guided mode and the phase velocity at a given frequency. The last one evidently varies versus frequency. This variation determines dispersion of the mode in the mini-band, shown in Figure 6.1(a) as a nearly parabolic curve.

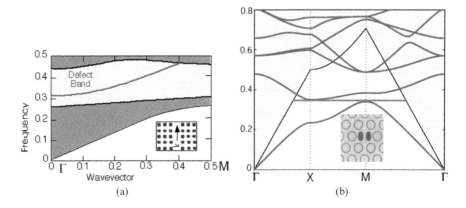

Figure 6.1 (a) Defect state (mini-passband) of a two-dimensional photonic crystal corresponding to a channel with a guided mode in the bandgap. Reprinted from [1] with permission of the editors. (b) Defect state of a two-dimensional photonic crystal corresponding to a resonator with one eigenfrequency within the bandgap. Reprinted from [4] with permission. Copyright 2004 American Physical Society.

Other conventional applications of photonic crystals utilize localized defect states. In this case, the defect forms a resonator, as we have already discussed. The defect can be created, for example, by a single missing inclusion in a photonic-crystal ambient. This case for a lattice of cylindrical air holes in a dielectric matrix of permittivity $\varepsilon = 7$ is illustrated by Figure 6.1(b). The defect state defines the resonator eigenfrequency that can be formally attributed to any propagation direction. As already mentioned, such resonators have high-quality factors that obviously correspond to high field concentration. Once excited by an embedded or externally coupled source, this resonator confines light energy in a quite small area of the order of $\lambda \times \lambda$. The density of the stored energy is evidently very high as compared to the case when the same energy is radiated into the whole space.

Photonic-crystal defects split into two groups – so-called *air-type defect* and *dielectric-type defect*. If the crystal is formed by optically dense inclusions in a less refractive matrix, the air defect may imply a reduced inclusion or a completely absent inclusion. A dielectric defect is as a rule an enlarged inclusion, rarely an inclusion with a higher refractive index than the others [1, 2]. On the contrary, in a hollow photonic crystal (an optically dense matrix hosting a regular array of holes), the air defect is an enlarged hole, and the reduced or absent hole is a dielectric defect.

6.1.2 Peculiarities of Photonic-Crystal Defect Waveguides

Photonic-crystal waveguides are rather expensive when considered as counter-parts of optical fibers. For this reason, defect waveguides are used in such nanophotonic devices as miniaturized optoelectronic devices, especially in con-verters of optical signals to the electric ones and back [3]. They are also applied in optical sensors, which will be discussed later. However, it is instructive to compare photonic-crystal waveguides with optical fibers. Although the geometrical optics

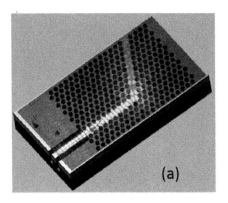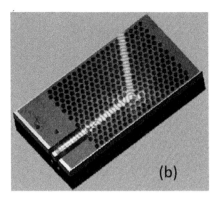

(a)   (b)

Figure 6.2 Defect photonic-crystal channels can guide modes without leakage and reflection around sharp bends. The bend angles in (a) and (b) are the same. The difference is in the optimization of the bend area. If the inclusions are regular everywhere, as it is in (a), the bend is not optimized and there is a mismatch. A small misplacement of inclusions in the bend area, as it is in (b), offers its matching with the defect waveguide, which is matched to free space. Reprinted from [5] with permission of the authors.

that governs conventional multi-mode fibers is not applicable to defect channels, the concept of the Brillouin waves is applicable [6], and the same phenomenon of total internal reflection (TIR) determines the wave propagation in fibers and in defect channels [6, 7]. Thus, in what concerns the common requirement of minimal decay and signal dispersion, optical fibers can be directly compared to defect waveguides. Multimode fiber waveguides possess higher dispersion, and the single-mode ones have a higher decay rate as compared to photonic-crystal waveguides [3]. However, conventional optical fibers are not usable in nanophotonic devices due to their large cross section and high bending losses.

Probably the most drastic advantages follow from two peculiarities of photonic-crystal waveguides. The first one is prohibition of propagation of light at the defect state frequency in the photonic crystal. The second one is a possibility to engineer the characteristic mode impedance, varying the shape of the inclusion cross section and the ratio inclusion thickness/lattice period. Both these advantages are illustrated by Figure 6.2. Mode leakage at bends is impossible because propagation at the mode frequency is prohibited in the photonic-crystal ambient. Next, a properly chosen geometry of inclusions allows one to engineer the mode characteristic impedance equal to that of free space. In this case, the guided mode matches to plane waves in free space and radiates from the surface of the photonic crystal without the need to taper the waveguide. In other words, specially designed antennas – e.g., horns – are not necessarily required for photonic-crystal radiating systems. The optimization of the bend area in this case is only needed in order to match the two straight defect waveguides. This matching represents a small misplacement of the inclusions in the bend area [5].

6.1.3 Photonic-Crystal Fibers

In photonic-crystal fibers, also called holey fibers, light propagates along a defect line. In other words, the defect channel is the fiber core. The defect

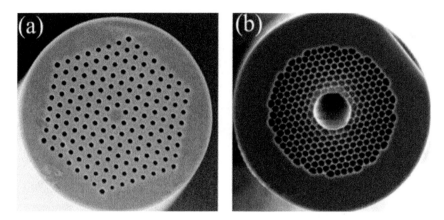

Figure 6.3   Photonic-crystal fibers with dielectric defect (a) and air-defect (b) longitudinal channels. Reprinted from [8] with permission of the editors.

can be either of the air type or the dielectric type, as shown in Figure 6.3. Propagation at the operation frequency across the photonic crystal is forbidden by the Bragg effect. In fact, if the frequency is at the center of the bandgap of a two-dimensional crystal, the propagation is also forbidden for the directions slightly tilted with respect to the transversal plane. Therefore, leakage to free space does not occur even for the mode whose angular spectrum covers all propagation directions. Really, if the wavevector is strongly tilted with respect to the cross section and forms a small angle with the fiber axis such that the wave experiences total internal reflection at the surface of the cladding. Due to the combination of TIR and bandgap effects, holey fibers turn out to be more robust to bending as compared to conventional fibers. They also have smaller cross sections. Practically, the wave guided in the core is fully decoupled with the ambient if the photonic crystal of air holes comprises 12–15 unit cells across the waveguide, and bending losses become significant when the radius of curvature approaches the fiber thickness [9].

Modern holey fibers are not expensive though they have negligibly small optical losses and dispersion in their operational band [10]. As a rule, these fibers are fabricated in the following way. First, one prepares a disk of fiberglass or elastomer and creates a regular array of millimeter-scale holes in it. Next, one quickly and strongly heats the sample to make it soft and stretches it in a special tower so that it elongates thousands of times. After stretching, the holes become submicron and the fiber thickness is maximally equal to a few dozens of microns. Finally, one inserts the fiber into a mm-thick shell or encapsulates it in a mm-thick laminate [9]. Fiberglass holey fibers are more expensive, but elastomer holey fibers are a bit more lossy.

Beyond telecommunications, holey fibers found an application in the THz technique, where they are used as endoscopes [11]. In this case, the holey fibers are manufactured from porous polymer matrices, and their manufacturing techniques are different. However, the techniques even more affordable and suitable for mass production (see, e.g., [10, 12]).

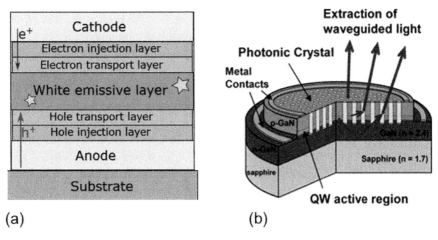

Figure 6.4 (a) Conventional LED is multilayered and the vertical emission is accompanied by high optical loss. The author's drawing is based on [13]. (b) In a LED comprising a two-dimensional photonic crystal of air holes, the waveguide modes cannot propagate, and the emission enhances. Reprinted from [14] with permission of the authors.

6.1.4 Photonic-Crystal Light-Emitting Diodes

Figure 6.4 illustrates another conventional application of two-dimensional photonic crystals the: enhancement of radiation from a *light-emitting diode* (LED). Physics of LEDs is based on an effect called *electroluminescence*, which is a physical effect inverse to the *photoelectric effect* that will be discussed in Chapter 10. For our purposes here, it is enough to know that in a LED, there is an active layer of a material that emits light when the multilayer composing the LED is biased by a DC voltage. This active layer may comprise a so-called *heterojunction*, or it may be a depletion region of a *p–n junction* in semiconductor or organic LEDs. Finally, it may be a so-called *quantum well* in the most advanced LEDs.[1] A LED generates either a non-coherent light or partially coherent (and very narrowband) light. Usually, this light, resulting from spontaneous emission, is not polarized. However, in the case of a quantum-well LED there is a polarization anisotropy. The most part of the emitted power in this case corresponds to the non-coherent and non-polarized light, but there is a noticeable quasi-monochromatic coherent component in the emission spectrum.

A typical LED structure emitting white light (see, e.g., in [13]) is shown in Figure 6.4(a). The substrate is backed by a back reflector (usually a polished metal). The central organic layer with heterojunctions emits a non-coherent broadband light. The extraction of this light can be vertical or lateral [15]. If the cathode is implemented from a polished metal the emitted light exists in the form of the waveguide mode of the multilayer structure. This guided mode leaves to free space from one of two lateral sides that is specially roughened so that to avoid total internal reflection on it (whereas the opposite side is mirrored). If the cathode is made from a transparent conductive material (such materials are, for example, oxides of several metals), the light is emitted from the top of the structure. In this case, the top surface is roughened.

The most common drawback of LEDs is their low efficiency due to high optical losses in the active layer [14, 15]. These losses accompany the propagation of the light in the horizontal plane which occurs even for LEDs with the emission though the top surface. This is so because a significant part of the emitted light is reflected from the cladding (electron and hole injection and transport layers and the electrodes) back to the emissive layer. Only a small part of the emitted light leaves the top surface of the LED, the power is mostly wasted heating the device. To prevent the losses one uses a photonic crystal of air holes piercing the active layer as depicted in Figure 6.4(b). It prohibits propagation of the light in the horizontal plane. The array of holes guides the photons to the top from the place where they were generated and the parasitic modes of the multilayer waveguide are not formed. To fully benefit from this property, it is desired that the photonic bandgap covers the maximal possible part of the emission spectrum. It is easier to achieve in quantum-well LEDs whose emission spectrum is comparatively narrow.

In this case, one uses the Bragg phenomenon to reduce the dissipative losses in the active layer. Another photonic-crystal solution explores the advantages of the localized defect state and offers control of the polarization and radiation pattern. Moreover, using defect states allows one to obtain a partially coherent radiation because the emission band drastically squeezes [15–18]. In Figure 6.5(a), we see a defect state formed by 19 inclusions of the dielectric type (18 reduced air holes and one absent at the center). In this case, the photonic crystal is an array of holes in the top transparent electrode, and the quantum-well LED is its substrate. Thus, the photonic crystal does not intersect with the active layer.

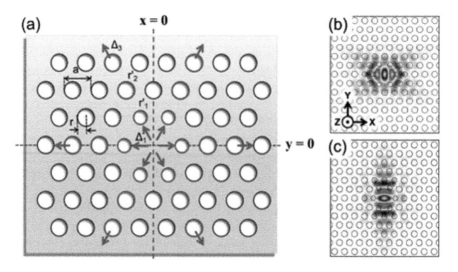

Figure 6.5   A photonic-crystal LED enhanced by a defect state. The top view (a) and the intensity maps of two coherent eigenmodes (b) and (c). Reprinted from [17] with permission. Copyright 2012 OSA.

Of course, some part of light emitted by the quantum well is polarized in the horizontal ($x - y$) plane and propagates along the z-axis, being radiated into free space from the top surface of the LED. For this component, the effective resonator is not formed in the defect area. However, this light is only a background of the enhanced radiation. The most part of the emission power is radiated from the effective aperture formed by the defect. And this emission is much stronger than the background one. First, there is anisotropy of the polarization state of the quantum well; second, light propagates mostly under substantial angles to the vertical direction, which correspond to TIR. Since, in the photonic crystal, light cannot propagate in the active layer, an effective channel is formed, directed toward the defect. The emitted light concentrates in the defect resonator, and its intensity increases nearly Q times compared to the structure without photonic-crystal layer. Here, Q is the resonator quality factor. In such defect resonators, it attains 500 at room temperatures [15–18]. Coupling of the emitting layer with the resonator drastically enhances the emission coherence. The relative bandwidth of emission fits the resonator's relative bandwidth that is equal to $1/Q$. In the case corresponding to the LED from work [17] that is depicted in Figure 6.5(a), in the operation band of the quantum-well emitter, there are two resonator modes. Color maps of these modes are shown in Figure 6.5(b) and (c). It is clear that we have an aperture-type radiation, and this aperture is especially engineered for directive patterns and favored polarizations. Photonic-crystal LEDs utilizing localized defect states are also more efficient than conventional LEDs. Though they are not as efficient as those utilizing the waveguide defects [18] and those offering radiation from the whole top surface [14], they have such advantages as directional radiation, high anisotropy of polarization, and high coherence. In their operation characteristics, these LEDs approach semiconductor lasers [15].

6.1.5 Photonic-Crystal Sensors

In fact, this is historically the earliest application of photonic crystals. The first works claiming dramatic influence of periodical composites and periodical composites with a defect to spontaneous emission of molecules were published well before the corresponding technologies appeared (see, e.g., in [19, 20]). When the microwave and millimeter-wave analogs of photonic crystals (electromagnetic crystals) appeared, it gave a new pulse to these studies (e.g., [21, 22]) and finally resulted in creation of two-dimensional and three-dimensional photonic crystals, discussed in the previous chapter.

There is a huge variety of photonic-crystal optical sensors exploiting defect states, including three-dimensional photonic crystals (see, e.g., in [16]). Here, we will briefly discuss the advantages of one type of these sensors exploiting a defect channel. In a conventional optical molecular sensor whose vertical cross section is schematically shown in Figure 6.6(a), the molecules emit light being excited by a surface wave accompanying the guided mode of the slab waveguide [23]. The spectroscopic analysis of this radiation allows one to identify these molecules because their emission occurs at frequencies slightly different from that of the pumping (laser source) light. The ratio of the emission peaks in the spectrum tells

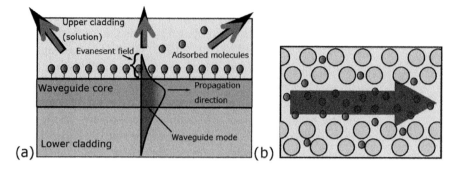

Figure 6.6 A conventional molecular optical sensor pumps the photoluminescence fluorescence or Raman radiation of molecules on the whole surface of the dielectric slab waveguide (a). The top view of the defect channel where the flowing analyte is concentrated and pumped by concentrated light propagating in the channel (b). Both (a) and (b) are the author's drawings, after [23].

about the ratio of the partial concentrations of different types of molecules. It is very important that the pumping light is spatially separated from the emission. The last one is very weak; its spectrum is continuous and partially overlaps with the spectral line of the laser source. Therefore, the spectroscopic analysis of the mixture of the source and emitted light would be very difficult. The main drawback of such a sensor is evident: emission is very weak because the pumping light energy is spread over a large area. Using a photonic crystal with a defect channel instead of a slab waveguide channel may squeeze both the flow of the liquid containing the analyte and the pumping light energy [16]. This technical solution is illustrated by Figure 6.6(b). Strong light concentration plays the main role, and the level of the emission increases by one to two orders of magnitude (see also Chapter 9).

6.1.6 Add-Drop Filters

Two parallel defect waveguides distanced from one another by a few photonic-crystal unit cells can be coupled via specially prepared defects. This coupling allows the photonic crystals to be applied for the enhancement and miniaturization of the so-called *optical add-drop filters* [24, 25].

In photonics, add-drop filters allow one to select one (or more) optical signal from a common waveguide carrying many signals and redirect it to a receiver coupled to a different waveguide. The waveguide in which all of the signals propagate is called the bus; the other one dedicated for the selected signal is called the drop. Add-drop filters are important components of optical informational systems [26]. Usually, add-drop filters use an optically large resonator system coupled with two waveguides. The resonator system employs its frequency selectivity and the interference effect in order to perform the function schematically illustrated in the top panel of Figure 6.7.

A usual resonator system is a ring resonator with mm sizes. It can be replaced by the structure shown in the bottom panel of Figure 6.7, which is as small as

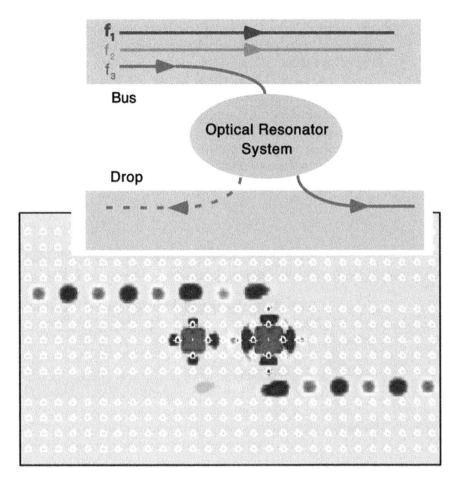

Figure 6.7 Add-drop filter sketch (top panel) and its implementation in a two-dimensional photonic crystal with two dielectric defect channels and two dielectric defects (electric field amplitude map). Reprinted from [24] with permission. Copyright 1998 OSA.

dozens of micrometers and operates with better efficiency. Here, the frequency of the redirected signal is equal to the resonant frequency of the defect state. The frequencies of all signals (in the example, there is only three, but there can be dozens) are within the bandgap. Whereas waves at frequencies f_1 and f_2 propagate in the first channel passing the two defects without perturbation, the wave at frequency f_3 due to the evanescent wave penetration to the photonic crystal excites the defect mode in the resonators, which in their turn excite the drop waveguide. Due to the presence of resonators, the photonic-crystal layer separating two defect channels in presence of two defects acts as a coupler for the signal at frequency f_3 but does not pass waves at two other frequencies.

However, this frequency-selective transmission is not all we need. The drop signal f_3 penetrating into the drop channel may, in principle, excite both backward and forward waves in it. Then, half of the signal energy will be lost and may even cause some parasitic interference. This is why one uses two or more defect resonators. The interference of signals transmitted into the drop guide via these resonators is destructive for the backward signal and constructive for the forward signal (see details in [24]). Alternatively, one may use a single resonator with degenerate modes. Degenerate modes are two different resonator modes with the same eigenfrequency that differ in the spatial distributions of the mode fields. They can be both excited and offer the needed interference for the dropped signal. The photonic-crystal add-drop filters are much smaller than the conventional ones and do not possess radiative loss inherent to conventional add-drop filters. Therefore, they are more efficient and their frequency selectivity is higher [25].

6.2 Emerging Applications of Photonic Crystals

6.2.1 Dense Wavelength Multiplexing and Demultiplexing

Before explaining the first emerging application of photonic crystals, so-called *photonic-crystal superprism* used in advanced demultiplexers and multiplexers, we have to discuss this application as such. Demultiplexers are passive devices that transfer signals of different frequencies from an optical waveguide transporting simultaneously many optical signals (bus guide) to one-signal optical waveguides terminated by receivers. Multiplexers perform the reciprocal function – collect signals from narrowband waveguides and send them into one bus waveguide.

The critical issue for both multiplexers and demultiplexers is cross talk due to near-field coupling of many narrowband waveguides. To avoid cross-talk, these waveguides must be spatially separated – distanced at least by several microns from one another. An alternative solution is to use time-division multiplexers where different signals are separated in time, which has obvious limitations. Since the number of carrier frequencies in optical information systems attains 320 and the operation band is the transparency band of pure silicon (in terms of wavelengths $\lambda = 1.47–1.55\,\mu m$), the adjacent carrier wavelengths have the relative difference of the order of 10^{-4}. Therefore, in optical information systems, wavelength-division multiplexers and demultiplexers are called dense wavelength-division multiplexers (DWDMs) and dense wavelength-division demultiplexers (DWDDMs). Both of these components are very important in both modern optoelectronics and prospective nanophotonics [27].

How do we spatially separate two waves with frequencies so close to one another? The frequency dispersion of the deviation angle of a prism of transparent natural materials is very modest. For spectroscopic purposes, when the adjacent wavelengths that must be spatially resolved differ from one another by several percentage points, dispersion offered by a glass prism depicted in Figure 6.8(a) is usually sufficiently strong. However, the deviation angle of a usual prism is very low. In order to operate as a frequency-dispersive component of a DWDM, such a prism should have a size exceeding 1 m.

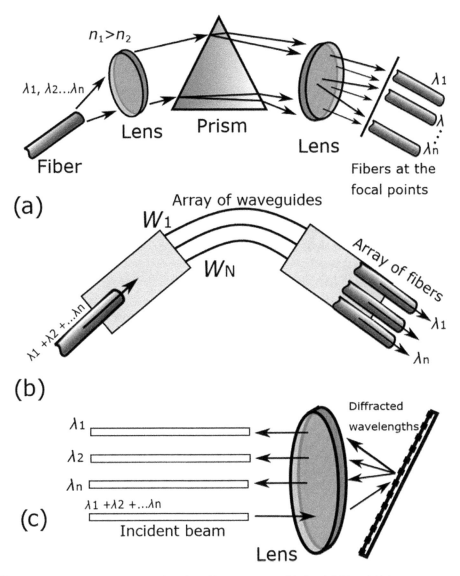

$n_1 > n_2$

$\lambda 1, \lambda 2 ... \lambda n$

Fiber

Lens   Prism   Lens

$\lambda 1$

λ

λn

Fibers at the focal points

(a)

Array of waveguides

W_1

W_N

$\lambda 1 + \lambda 2 + ... \lambda n$

Array of fibers

$\lambda 1$

λn

(b)

Diffracted wavelengths

$\lambda 1$

$\lambda 2$

λn

$\lambda 1 + \lambda 2 + ... \lambda n$

Incident beam

Lens

(c)

Figure 6.8 A prism for demultiplexing light signals in the visible spectrum is useful when $\Delta\lambda/\lambda \sim 0.1$ (a). A DWDDM based on an array of optical waveguides is a conventional but cumbersome solution (b). A DWDDM based on a reflecting-type phase diffraction grid grants a miniaturization but suffers of scattering losses (c). These three figures are drawn by the author, after [28].

The most practical solution for a DWDDM comprises an array of optical waveguides, as is schematically shown in Figure 6.8(b). These waveguides can be photonic-crystal waveguides or so-called integrated optical waveguides. At the input and output, there are multimode slab waveguides called free-propagation plates (FPPs). The first FPP is a matching device between the bus fiber and the waveguide array. In the second FPP, the waves with different frequencies coming from the waveguide array form different phase fronts. Though different, they are all convergent, and the wave beams with different frequencies converge to different points, finally coupling to different fibers. However, the size of the device is quite substantial – of the order of 20 c (see, e.g., in [27]).

A miniaturized DWDM or DWDDM exploits dispersion of a diffraction grid that is much higher than dispersion of a glass prism. Dispersion becomes stronger for higher diffraction orders. The size of a reflecting grid shown in Figure 6.8(c) can be as small as 1–2 cm and because it is a phase diffraction grid, the second-order or even the third-order diffraction lobe may dominate over the zero-order one (which does not exhibit dispersion). However, these devices have an inherent drawback – losses of the signal energy in them are high because the zeroth order of diffraction cannot be completely suppressed as well as other useless diffraction lobes. Scattering losses (20%–30%) cannot be avoided in such devices [28].

Advanced demultiplexers and multiplexers based on photonic crystals are smaller than their analogs based on the waveguide arrays and are more efficient, being practically lossless. In the following section, we will see that exploiting the giant dispersion of a photonic-crystal prism results in the size reduction.

6.2.2 Superprism

Dense wavelength-division multiplexers (DWDMs) and dense wavelength-division demultiplexers (DWDDMs) are conventionally based on arrayed waveguides or diffraction gratings and have substantial sizes; otherwise, it is impossible to avoid the cross talks. A photonic-crystal superprism (see, e.g., in [29–31]) performs the same demultiplexing and multiplexing functionalities, being much more compact and energy efficient.

Let us consider a DWDDM whose key component is a hollowed photonic crystal with the matrix permittivity $\varepsilon = 12$. We know that in the range $\omega a / 2\pi c = 0.185$–$0.225$, this lattice has superquadric isofrequencies. The photonic crystal is designed so this range of normalized frequencies nearly corresponds to the range of wavelengths $\lambda = 1.47$–1.55 μm where crystalline silicon has permittivity close to 12. Let the photonic crystal sample be a few mm large parallelepiped placed on top of a triangular prism performed of the same semiconductor, but without holes. The structure is shown in Figure 6.9(a). Usually, these two components are performed as a monolithic sample of trapezoidal cross section whose rectangular part is hollowed and triangular part is solid. The multi-frequency signal from the input optical waveguide is collimated by a lens and impinges the solid prism like a normally incident plane wave. In this case, the group velocities of all input signals $V_{g\,\text{input}}$ practically coincide. We want the output group velocities $V_{g1\,\text{out}}$ and $V_{g12,\text{out}}$ of two signals with slightly different frequencies ω_1 and ω_2 (the relative difference of these frequencies is of the order 0.0001–0.001) are directed so differently that the angle $\Delta\theta_{\text{out}}$ between them is sufficiently large (of the order of 1°) for signal separation.

How can we achieve this goal with such a small device without involving scattering mechanisms? It is achieved by exploiting huge frequency dispersion of the deviation angle granted by the superquadric shape of the isofrequency contours. In Figure 6.9(b), two wavevectors in the dielectric, with $\varepsilon = 12$

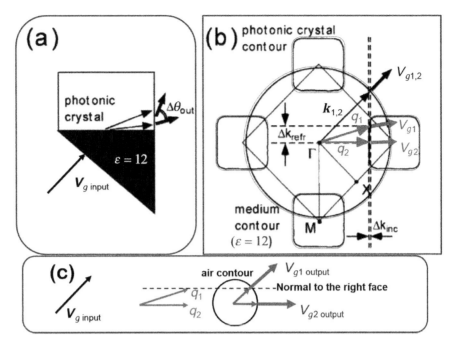

Figure 6.9 Photonic-crystal superprism: (a) sketch of the device, (b) illustration of the frequency separation inside the crystal, and (c) illustration of the frequency separation at the output. These three figures are the author's drawings, after [31].

corresponding to frequencies ω_1 and ω_2, are shown as black arrows $\mathbf{k}_{1,2}$ with a slight difference of lengths, $\Delta\mathbf{k}_{\mathrm{inc}}$. This notation stresses that both monochromatic waves are incident onto the interface of the photonic crystal (from a highly refractive dielectric). This incidence is oblique, and the incidence angle equals $45°$. In Figure 6.9(b), the normal to the interface is vertical, and the relevant part of the superquadric contour is almost vertical, too. This high slope of the isofrequency determines the very high difference between two Bloch wavevectors $\mathbf{q}_1$ and $\mathbf{q}_2$, shown as thin arrows in Figure 6.9(b). The difference in the refracted wavevectors $\mathbf{q}_2 - \mathbf{q}_1 = \Delta\mathbf{k}_{\mathrm{refr}}$ is incomparable larger than $\Delta\mathbf{k}_{\mathrm{inc}}$. Though the group velocities of these two waves, $\mathbf{V}_{g1}$ and $\mathbf{V}_{g2}$ (thick arrows), differ rather weakly, this difference is not relevant.

The output side of the photonic-crystal prism is the vertical side; namely, the vertical component of $\mathbf{q}_1$ and $\mathbf{q}_2$ is preserved in this refraction. A very small difference $\Delta\mathbf{k}_{\mathrm{inc}}$ results in a rather big difference Δk_{refr} in the vertical components of the wavevectors. This difference is preserved and results in a relatively large $\Delta\theta_{\mathrm{out}}$. It is shown in Figure 6.9(c) that the incident wave with the wavevector $\mathbf{q}_1$ transmitted to free space becomes more tilted to the horizontal axis, and its group velocity $V_{g1\,\mathrm{out}}$ is tilted at a large angle to the horizontal direction. Meanwhile, the incident wave with the wavevector $\mathbf{q}_2$ transmitted to free space keeps propagating horizontally.

As a result, the ray corresponding to two slightly different frequencies splits into two rays at the output side of the prism, and they diverge strongly. This is why such composite prism is called superprism. This design allows a compact realization of a practically lossless DWDDM: the total size of the whole device is as small as 1 cm [29]. Conventional demultiplexers based on arrays of waveguides with the same operational characteristics have the sizes about 10 cm.

It is worthwhile to note that in the superprism the reflection losses at the input and output interfaces are suppressed using antireflecting coatings. This is necessary; otherwise, the reflection loss would make it less efficient than the phase diffraction grating. The reflection loss at the internal interface – between the solid dielectric and the photonic crystal – is also practically removed using the antireflecting coating [30]. These three coatings are submicron dielectric layers (not shown in Figure 6.9 for simplicity).

Here we have considered a DWDDM based on a photonic crystal. Due to reciprocity, the same superprism can work in a DWDM and result in the similar compactness of the device. Why are superprism-based DWDDMs and DWDMs are not commercialized yet in spite of their advantages? It is a question of costs. Though two-dimensional photonic crystals are much cheaper than the three-dimensional ones, their conventional counterparts keep still cheaper. There is a hope that the development of nanotechnologies will reduce the difference in costs and the reduction of the size will open the door to the commercialization of these devices.

6.2.3 Pseudoscopic Frequency-Selective Imaging

In this section, we will consider also the so-called *photonic-crystal pseudolens*. Again, before explaining the operation of the photonic-crystal pseudolens we first discuss the application as such.

A *pseudolens* is a frequency-selective structure that creates the so-called *pseudoscopic image*, focusing only the wave beams generated by sources located close to the structure. Only strongly diverging wave beams converge behind the pseudolens. In this way, the objects under study located closely to the pseudolens are imaged, whereas the distant objects do not create parasitic images. Usual lenses and their stacks used in microscopes do not possess the pseudoscopic imaging functionality – they obviously focus the beams of parallel rays (plane waves).

Conventionally, a pseudoscopic imaging is achieved in an array of microlenses [32]. An advanced variant is the frequency-selective pseudoscopic array. *Frequency selectivity of the transmission* allows an advanced pseudolens to form the images only of those objects where are illuminated by a predefined light source. Focusing the light of the predefined source on the object under study, one wants to get rid of the parasitic images of other objects, perhaps also located in the proximity of the pseudolens. These objects can be illuminated by the sun or by the room lighting system and their image is white-light one. If the pseudolens is frequency selective, only a very small part of this parasitic light transmits through the pseudolens to the image area. Meanwhile, the narrowband light scattered by our specially illuminated object toward the pseudolens, transmits to the image area almost completely. Thus, the frequency

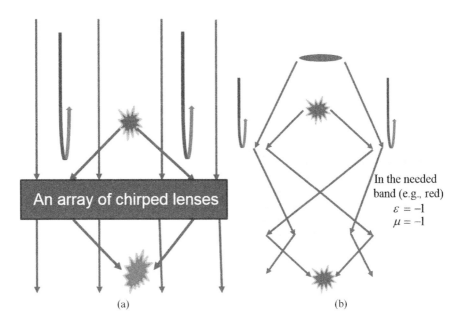

(a)　　　　　　　　　　　　　　　　(b)

Figure 6.10 (a) An array of chirped microlenses allows frequency-selective imaging of the objects of our interest located closely. Only the waves in the needed frequency range (here – red light) transmit. The waves produced by parasitic distant objects do not create parasitic images. (b) Veselago's pseudolens grants the absence of aberrations and better suppression of the parasitic images. Also, higher frequency selectivity is granted by a narrow frequency band where $\varepsilon \approx \mu \approx -1$.

selectivity of a pseudolens drastically increases the signal-to-noise ratio in the image. It is especially important in biomedical applications, when the images of the selected biological objects are recorded. As a rule, the frequency selectivity is achieved in arrays of so-called *chirped lenses* [33] – microscopic lenses with submicron patterning of their surfaces.

The emerging optical application called pseudoscopic *frequency-selective* imaging is illustrated by Figure 6.10(a). Usually, such a fine imaging system is dedicated for dynamic imaging of biological objects having sizes in the range 5–500 μm. Being located in a small water drop on an object glass they are imaged in the arrayed optoelectronic devices having also a microscopic size of a pixel [33, 34]. These devices are arrays (matrices) whose unit cell registers a pixel: charge-coupled device matrices, matrices of micro-photodiodes, or matrices of liquid crystal cells [35]. One unit cell of the matrix can be, nowadays, as small as 1.7 μm. Besides a so high spatial resolution, these devices detect the optical signals rather fast – with a millisecond or shorter delay. It enables the dynamic video of microscopic biological objects involved into Brown's movement.

However, microlens arrays, especially those of chirped lenses, are expensive devices. Moreover, they are not free of drawbacks, and their main drawback is aberrations, especially *coma aberration* and *astigmatism*. Both of them result from the loss of the wave beam homocentricity after the transmission through a lens.

Let us recall from the course of optics what is astigmatism. Every point of an object surface is a source and the radiation of this point source represents a divergent wave beam impinging the lens at all possible incidence angles from $\theta = 0$ to $\theta = \pm 90°$. However, only a part of this wave beam – that corresponding to small incidence angles – converges behind the lens to a point. Waves corresponding to substantial incidence angles converge to different points whose coordinate depends on the angle. Therefore, the point source is imaged as a finite-length line. This purely geometric optical effect results from the Snell's law and cannot be avoided for a solid glass object such as a lens. Next, let us recall what is coma aberration. It results from the different refraction of the rays belonging to the vertical and to the horizontal cross sections of the wave beam. Any two rays taken in the beam vertical section converge to one point; any two rays taken in the horizontal section converge to another point. Together with the astigmatism, the coma aberration results in the curved line that images a point source behind a lens.

Notice that the wave image of a point source even in absence of these aberrations would be not a point but a finite-area domain whose minimal size is restricted by the free-space diffraction of the wave beam. This spreading of the image is called the Abbe diffraction effect [36]. However, here we discuss only geometric optical effects and neglect the impact of the Abbe diffraction. The last one is not important because the size of the unit recording device is much larger than the wavelength. However, the aberrations make the size of the comma-like image of a point source significantly larger than $1.7\,\mu m$ and distort the shape of the finite object. For pseudoscopic imaging devices, it is the critical issue hindering the recognition of the objects. Significant efforts have been done to eliminate the aberration of pseudolenses. For example, recently one suggested replacing a chirped microlens array with a quite sophisticated metasurface [37].

However, already in 1968, V. G. Veselago, in work [38], pointed out another possibility to avoid aberrations in a pseudolens. In his technical solution, the pseudolens was a parallel plate. However, the material of this plate was very exotic for 1960s. It was a hypothetic continuous medium having the relative permittivity ε and permeability μ both equal to $\varepsilon = \mu = -1$. Veselago proves that the refractive index of this medium equals $n = -1$, and the wave impedance of the medium is equal to $\eta = \sqrt{\mu_0\mu/\varepsilon_0\varepsilon} = 120\,\pi\Omega$ – that of free space. Therefore, all rays diverging from a point object located at a sufficiently small distance from a parallel plate of such a material (this distance Δ should be smaller than the slab thickness D) intersect at one point inside the pseudolens, diverge toward the rear interface, and, behind it, again intersect in free space at one point. This second point is an *aberration-free image* of a point object granted by the *Veselago pseudolens*.

All the rays produced by more distant objects do not intersect at all neither inside nor outside the pseudolens, as one can understand from Figure 6.10(b). Only the objects located at the distances $\Delta \leq D$ from the slab are imaged as real ones. Also, the finite-size objects located at the distances smaller than $D/2$ are magnified, the objects located at the distance $\Delta = D/2$ is imaged with the same size, and the objects located at the distances $D/2 < \Delta < D$ are reduced in their images. Those objects that are located at $\Delta > D$ have the virtual image behind the Veselago pseudolens. However, in all cases, the imaging occurs without distortions of the object shape.

The layer of a hypothetic medium described by Veselago is not simply an elegant solution for an aberration-free pseudolens. It also does not require the antireflecting coating. Glass reflects a noticeable portion of light, whereas the *Veselago medium* does not reflect propagating waves at all. If the material parameters of the medium equal $\varepsilon = -1$ and $\mu = -1$, the slab interface is fully matched to free space (see detailed explanations in Chapter 3). Indeed, the regime $\varepsilon \approx -1$ and $\mu \approx -1$ is possible for any passive medium only in a narrow band of frequencies [39]. Since passive media with negative material parameters are necessarily frequency dispersive and the frequency dispersion is obviously accompanied by losses, it is reasonable to expect that the material parameters of the Veselago medium at frequencies beyond the operation band will be very different. For example, one of the parameters beyond the operational band changes the sign. This will make a substantially thick slab impenetrable because the regime $\mu < 0$ and $\varepsilon > 0$ as well as the regime $\mu > 0$ and $\varepsilon < 0$ corresponds to a fully reflecting medium.

Therefore, the search for and creation of the medium was considered by Veselago as a target deserving significant efforts. Really, an aberration-free and highly frequency-selective pseudoscopic imaging device looked extremely promising even in 1960s, especially for biomedical imaging. Veselago and other researchers tried to implement these material parameters in an artificial material. However, it turned out to be too difficult of a task, especially because continuous materials do not possess magnetic susceptibility at optical frequencies, where all natural media have $\mu \approx 1$. As to metamaterials, composed of inclusions exhibiting the resonant artificial magnetism, the issue of high magnetic losses turned out to be critical. Recall, that high losses is the same as substantial imaginary part of the material parameter. For the operation of the Veselago pseudolens, one needs not only $\text{Re}(\mu) \approx -1$ but also negligibly small $\text{Im}(\mu)$. There were many attempts to compensate the losses of a metamaterial with $\text{Re}(\mu) \approx \text{Re}(\varepsilon) \approx -1$ using the optical gain – i.e., injecting into the metamaterial active media (used in lasers) and pumping them by the visible or UV light. However, for the Veselago pseudolens operation, one practically needs both $|\text{Im}(\mu)| < 0.01$ and $|\text{Im}(\varepsilon)| < 0.01$ at the frequency where the real parts of these parameters approach (-1). Magnetic losses accompanied by the equivalent electric gain do not allow one to achieve the needed level of the impedance matching free space. Moreover, this regime corresponds to the generation of parasitic signals in the active insertions [40]. Most probably, the issue of magnetic losses in the Veselago medium cannot be resolved in principle.

Thus, it is not surprising that some authors tried to implement the regime of the Veselago pseudolens in structures that are not effectively continuous media. Though photonic crystals cannot be characterized by the effective permittivity and permeability, they may mimic continuous media in what concerns their response to propagating electromagnetic waves. In 1978, R. A. Silin published (in Russian) a theoretical paper [41] where the Veselago pseudolens operation was predicted for a layer of regular optical composite of a semiconductor with air holes. Nowadays, this composite is referred to as photonic crystal. Previously, in work [42] (also in Russian), the same author had suggested another, less adequate implementation of the Veselago pseudolens. It was not free of aberrations, but it was historically the first technical solution for a frequency-selective pseudolens. Therefore, we start the discussion of the photonic-crystal pseudolens from the historically first structure suggested by Silin.

6.2.4 Photonic-Crystal Pseudolens Operating at a Super-Quadric Isofrequency

Consider a periodic array of circular cylinders filled with air in a matrix with the relative permittivity $\varepsilon = 12$, as depicted in Figure 6.11. We have already considered refraction at an air interface with such a lattice, when we discussed the use of *isofrequencies* in Chapter 5. Here, it is relevant to plot isofrequencies in the plane $(k_x - k_y)$ outside the Brillouin zone, where they form closed contours periodically repeating with the period $2\pi/a$. In Figure 6.11 these isofrequency

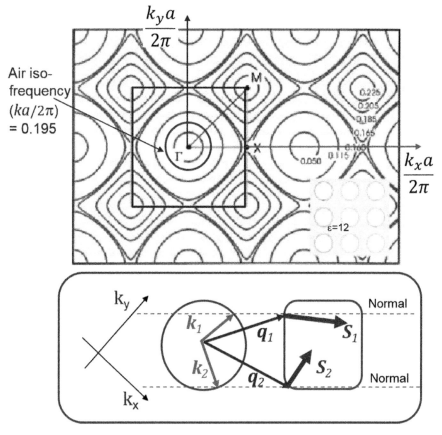

Figure 6.11 Top panel: Isofrequencies of a square array of air holes in a semiconductor matrix. Reprinted from [43]. with permission. Copyright 2002 American Physical Society. Bottom panel: Negative refraction corresponds to superquadric isofrequencies in the range $\omega a/2\pi c = 0.185 - 0.225$. Divergent incident wavevectors $\mathbf{k}_1$ and $\mathbf{k}_2$ result in convergent Poynting vectors $\mathbf{S}_1$ and $\mathbf{S}_2$, and a photonic crystal operates as a pseudolens.

contours correspond to the case when $\delta = 0.3a$ (a is the lattice period and δ is the hole diameter). The Brillouin zone where $k_x \equiv q_x$, $k_y \equiv q_y$ is outlined by blue color. When the normalized frequencies $\omega a/2\pi c$ are less than approximately 0.12, the isofrequencies are nearly circular, and the array of holes is effectively a continuous dielectric medium that is practically isotropic in the plane (x–y).

When the frequency grows, the isofrequency contours become noncircular, and the Bloch wavevector $\mathbf{q}$ has different length in different directions. This type of optical anisotropy is referred to as spatial dispersion because it cannot be explained within the model of a *continuous* dielectric medium. Continuous media obviously have either elliptic (see, e.g., in [36]) or hyperbolic (see, e.g., in [44]) isofrequency contours. In isotropic dielectrics isofrequency contours are circular (in three-dimensional, spherical surfaces). In the present case, the shape of isofrequencies becomes superquadric starting from $\omega a/2\pi c \approx 0.16$. Notice that the transition from closed isofrequencies surrounding the point Γ to isofrequencies surrounding the point M that occurs between $\omega a/2\pi c = 0.164$ and $\omega a/2\pi c = 0.165$ is sometimes called *topological transition* in the electrodynamics of metamaterials [45] (or *Lifshitz's transition* in the solid-state theory [46]). The isofrequency contour at the topological transition frequency is infinitely extended. This regime of the photonic-crystal operation is very interesting, but here we will concentrate on the isofrequency contours that correspond to the frequencies higher than the topological transition. Notice that negative refraction at interfaces with air holds at normalized frequencies $0.185 < \omega a/2\pi c < 0.25$.

At $\omega a/2\pi c > 0.185$, the isofrequency contours surrounding the point M become squeezing to this point until the normalized frequency attains the value $\omega a/2\pi c = 0.250$, where $\mathbf{q}$ corresponds to the Bragg regime ($q_x a = q_y a = \pi$). At the normalized frequencies $0.225 < \omega a/2\pi c < 0.250$, the isofrequency contours are nearly circular. However, here we inspect the interval of frequencies $0.185 < \omega a/2\pi c < 0.225$, where the contours are superquadric.

At the normalized frequency $\omega a/2\pi c = 0.195$, the size of the superquadric contour is equal to the corresponding isofrequency of free space. Therefore, negative refraction occurs for any incidence angle until the grazing incidence. In the bottom panel of Figure 6.11, this case is shown. Notice that this regime, initially suggested by Silin in work [42], was independently reproduced in [47], two decades later, because at that time, paper [42] was not yet translated from Russian.

We know that a pseudolens needs to be frequency selective, and its image should not suffer from reflections at the interfaces. Therefore, at the operation frequency, one needs impedance matching. Namely, if the wave impedance Z_0 of the incident wave referred to the interface of the slab is equal to the surface impedance, there is no reflection from the surface. The surface impedance of a substantially thick photonic-crystal slab is practically equal to the so-called Bloch impedance Z_B of the infinite photonic crystal. The *Bloch impedance* is defined in the theory of photonic crystals as the ratio of the tangential components of electric and magnetic fields of an eigenmode calculated in the middle, between two adjacent crystal planes. This concept is illustrated by Figure 6.12. In order to avoid reflections at the interfaces of the photonic-crystal slab, one

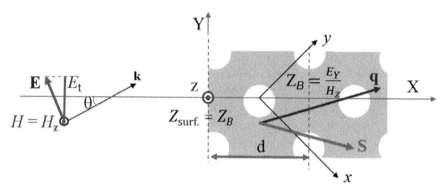

Figure 6.12 The Poynting vector **S** in the photonic crystal has a negative tilt to the normal axis X for all incidence angles θ. This all-angle negative refraction is accompanied by impedance matching of TM-polarized waves. The impedance Z_0^{TM} of free-space incident waves referred to the plane $X = 0$ equals to the surface impedance of the lattice $Z_{surf.}$ for any θ. The surface impedance equals to the Bloch impedance Z_B of an infinite lattice, defined as the ratio E_Y/H_Z calculated at the crystal planes parallel to the axis Y (shown by dashed lines). The equivalence of Z_B to Z_0^{TM} holds at $\omega a/2\pi c = 0.195$.

should perfectly match it to free space – i.e., equalize Z_B and Z_0 for any tilt angle θ of the incident wavevector **k**.

For TM-polarized waves, where $Z_0^{TM} = \eta \cos\theta$, this equivalence holds at $\omega a/2\pi c = 0.195$. At this frequency, all TM-polarized plane waves incident on our slab transmit into it without reflection. They experience the negative refraction at both interfaces of the photonic-crystal slab [43, 48]. If the incident TM-waves form a divergent wave beam behind the rear interface of the slab, the wave beam is convergent. Of course, this *parallel-plate focusing* holds in a very narrow frequency band. Both high-frequency selectivity and the reflection-free regime of the pseudoscopic imaging are advantageous features of this pseudolens.

However, this image is achieved only in TM-polarized light and, what is more important, suffers from the same aberrations as an array of microlenses. Only in a narrow angular range, the transmitted ray vectors are directed toward the same point. Rays with large tilts intersect at different points, as shown in Figure 6.13. This astigmatism makes point-to-point imaging impossible – the object spreads along the normal axis. The coma aberration is also present, and the object shape is distorted. Therefore, the regime based on the superquadric isofrequency, as it was suggested in works [42, 47], needed an improvement. To this end, in further theoretical works [41, 43, 48], the authors noticed the circular isofrequency contours in their photonic crystals and saw that a slab of these crystals operates like the Veselago medium. Notice that the results of the Russian paper [41] were reproduced in papers [43, 48]. The delay was the same as that of work [47] with respect to work [42] because the authors of [43, 48] were not familiar with [41].

6.2.5 Pseudolens for Aberration-Free Imaging

Look at Figure 6.11. In accordance with this figure, isofrequencies of the considered crystal between $\omega a/2\pi c = 0.225$ and $\omega a/2\pi c = 0.25$ are nearly circular – like

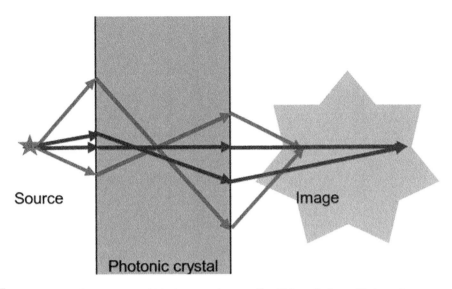

Figure 6.13   A pseudolens exploiting superquadric isofrequency does not offer a high-quality image. The image is distorted because rays with different incidence angles converge behind the slab at different points.

for a uniform isotropic medium. Of course, for a continuous isotropic medium, these circles surround point Γ, while in Figure 6.11, they are centered at point M. However, the interfaces of the photonic crystal are parallel neither to the lattice axis x nor to the lattice axis y. The tangential axis Y shown in Figure 6.12 is tilted by $45°$ with respect to the lattice axes. Considering the lattice in these new coordinates, as it is shown in Figure 6.12, we see that the new unit cell is enlarged to size d. Instead of the Bloch vector $\mathbf{q}$ of the generic lattice defined in the generic coordinate frame xy, one may consider the new Bloch wavevector $\mathbf{k}_{\mathrm{main}} = \mathbf{q} + (2\pi/d)\mathbf{X}_0$ – that of the lattice with the period d ($\mathbf{X}_0$ is the unit vector along x).

To better understand the physical meaning of this replacement of the Bloch wavevector, it is instructive to consider the isofrequencies of a slightly modified photonic crystal. The geometry of the new lattice is the same, but the relative permittivity of the matrix is now equal to 11 instead of 12. In accordance with [41], this modification is needed to equate the size of one of the circular isofrequencies of the lattice to the size of the corresponding isofrequency of free space. With the previous lattice ($\varepsilon = 12$), it is impossible (recall that the superquadric contour of the previous lattice at $\omega a/2\pi c = 0.195$ was equal to the air circle, and the circular contours of the previous lattice corresponded to higher frequencies and are smaller).

Two isofrequencies of the new lattice, $\omega a/2\pi c = 0.205$ and $\omega a/2\pi c = 0.210$, are depicted in Figure 6.14. The first one is still superquadric and is larger than the corresponding isofrequency of free space the (larger central circle). The operation frequency is that of the second contour – the smaller central circle. An incident wave with the wavevector $\mathbf{k}$ excites in the lattice the eigenmode with the Bloch wavevector $\mathbf{q}$. The length of this wavevector is large because its end is close to the corner of the Brillouin zone – point M. In other words, at

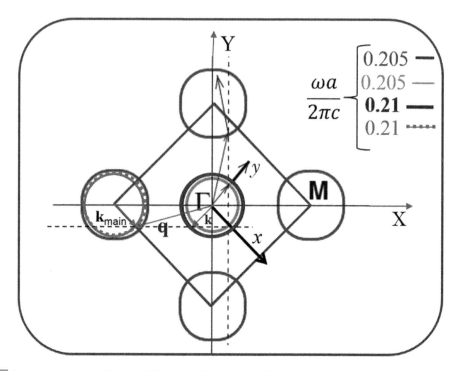

Figure 6.14 Two isofrequencies $\omega a/2\pi c = 0.205$ and $\omega a/2\pi c = 0.210$ of a photonic crystal with the relative permittivity of the matrix equal to 11. Two central circles are the corresponding isofrequencies of free space. At the frequency $\omega a/2\pi c = 0.21$, the slab of this photonic crystal, having the interfaces parallel to the axis Y, is a pseudolens operating in the Veselago-Silin regime.

the operation frequency, $q_x \approx \pi/a$ and $q_y \approx \pi/a$. Is it evident that the spatial harmonic corresponding to the Bloch wave of the generic lattice is the main spatial harmonic in the Bloch expansion of the eigenmode? No, it is not evident and, it is not so.

The incident wave impinges not the plane $x = 0$ but the tilted plane $X = 0$, having the period d and not a. Starting from the interface, it sees the lattice of air inclusions as an array with the period d. Then the true *Bloch wavevector* is $\mathbf{k}_{\mathrm{main}} = \mathbf{q} + (2\pi/d)\mathbf{X}_0$, as it is seen in Figure 6.15(a). Of course, the eigenmode can be also written as the *Bloch expansion* with the *Brillouin zone* of the size $2\pi/a$ (generic unit cell). However, the Bloch expansion with the Brillouin zone of the size $2\pi/d$ (enlarged unit cell) also makes sense and, since it is only another method for solving the same problem, must give the same result. This result was for the first time obtained by R. Silin in [50] and later reproduced in works [48, 49, 51]. The modified Bloch wavevector $\mathbf{k}_{\mathrm{main}} = \mathbf{q} + (2\pi/d)\mathbf{X}_0$ is the wavevector of the dominant spatial harmonic in the Bloch expansion for the generic unit cell. It is the fundamental spatial harmonic – the *Bloch wave* of the lattice with the enlarged unit cell. In this modified Bloch expansion, the amplitude of the Bloch wave exceeds those of the other spatial harmonics. One may tell that the enlarged unit cell is a more adequate choice for the lattice than the generic unit cell of the lattice because the lattice axes are tilted by $\pi/4$ with respect to the slab interfaces.

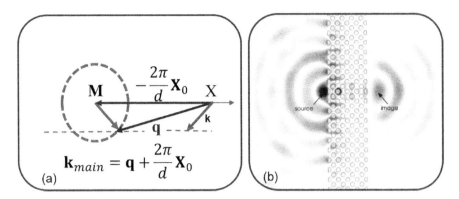

Figure 6.15 The modified Bloch wavevector $\mathbf{k}_{main} = \mathbf{q} + (2\pi/d)\mathbf{X}_0$ of the lattice considered in the frame XY is the wavevector of the dominant spatial harmonic in the Bloch expansion of the generic lattice. Therefore, the finite-thickness lattice mimics the Veselago medium slab when the interfaces are parallel to Y (a). Full-wave simulations (b) confirm point-to-point diffraction-limited imaging obtained in a photonic-crystal pseudolens exploiting the circular isofrequency. Reprinted from [43] with permission. Copyright 2002 American Physical Society.

From Figure 6.15(a), it is clear that this photonic crystal really mimics the Veselago medium. The absolute value of the true Bloch vector $\mathbf{k}_{main}$ is equal to that of the incident wavevector $\mathbf{k}$ and the tilt of $\mathbf{k}_{main}$ to the interface is exactly opposite to that of $\mathbf{k}$. This observation illustrated by Figure 6.15(a) refers to all incident plane waves. So all the incident rays experience at the slab interfaces the negative refraction corresponding to the refractive index equal to -1. The impedance matching condition is automatically satisfied for all incidence angles (for incident waves with TM-polarization, of course). The main spatial harmonic of all refracted waves is characterized by the refractive index of the Veselago medium – i.e., $n = -1$.

To the best of our knowledge, these ideas were published for the first time in papers [41, 42, 50] by R. A. Silin, whereas the works of other groups such as [43, 47–49] only confirmed them. Since all of these works were targeted to mimic the Veselago pseudolens, it is appropriate to call the parallel-plate aberration-free imaging in a photonic-crystal slab the *Veselago-Silin imaging*.

For lattices of dielectric free-standing cylinders, this regime holds if the incident waves are TE-polarized. In Figure 6.15(b), the color map of the electric field is depicted, obtained by full-wave numerical simulation of the lattice of cylinders having the thickness $\delta = 0.6a$ and permittivity $\varepsilon = 14$. The source of round shape with a strongly subwavelength diameter is imaged into a circular spot whose size is of the order of λ. There are practically no aberrations in the field intensity map. This is so because the circular isofrequency contour is exploited.

It is possible to design three-dimensional photonic crystals so that there is a spherical isofrequency (also centered around the corner of the generic lattice) equivalent to that of free space. Similar $45°$ tilt of the interfaces of such a photonic-crystal slab grants the opportunity to realize parallel-plate imaging without aberrations in both TE- and TM-polarizations of the incident light. It can even be obtained in unpolarized light if its band is narrow enough and centered at the needed frequency. This possibility was theoretically proven in [49].

Experimental studies confirmed both superquadric parallel-plate imaging, predicted in [42, 43, 47], and the Veselago-Silin imaging regime, predicted in [41, 43, 48, 50], for two-dimensional photonic crystals. These experiments were reported in two simultaneously published works [52] and [53]. Two years after that, an experiment performed by an international team of coauthors representing several scientific groups [54] demonstrated the Veselago-Silin parallel-plate imaging in three dimensions. Later, more specific experiments allowed one to study finer effects in such pseudolenses.

In the time of writing this text, photonic-crystal pseudolenses are not yet commercialized. This is probably because only three-dimensional imaging is interesting for biomedical applications, but fabrication of corresponding three-dimensional photonic crystals is still very expensive. In what concerns costs, such a pseudolens would probably yield even to chirped microlens arrays whose fabrication is based on rather affordable nanotechnologies. Extra costs for the absence of aberrations in a pseudoscopic dynamic image is nowadays too high. Hopefully, in the future, development of nanotechnologies will decrease this cost.

It is worth noticing that all these works have nothing to do with the idea of the *perfect imaging* predicted by J. B. Pendry for the Veselago pseudolens in work [55]. In accordance with this work, an effectively homogeneous slab of material with the refractive index equal to $n = -1$ theoretically allows perfect imaging of arbitrary small objects. Perfect imaging means the point-to-point imaging without an impact of the Abbe diffraction. To overcome the Abbe diffraction limit is the same as to shrink the image of a point-wise (strongly subwavelength) object to a strongly subwavelength spot. Pendry has shown it is theoretically possible due to resonant enhancement of evanescent waves growing across the Veselago pseudolens. We discussed this effect in Chapter 3. As to photonic-crystal pseudolenses, even those operating in the Veselago-Silin regime are diffraction-limited devices. They mimic a slab of the Veselago medium only for propagating waves. Therefore, the distortion of the object shape in the recorded image is absent, but the object subwavelength details are not imaged.

6.3 Conclusions

Photonic crystals represent the most practically and scientifically developed branch of nanophotonics. In the previous chapter, we discussed the basic physical principles underlying photonic crystals. In this chapter, we examined conventional (already commercialized) applications of photonic crystals, hoping to convey their practical importance and versatility. Two emerging applications – the aberration-free pseudoscopic imaging by a parallel plate of photonic crystals and photonic-crystal-based multiplexing/demultiplexing – form only a tip of the iceberg in what concerns the opportunities offered by photonic crystals to modern and future industry. The goal of these two chapters is to stimulate the imagination of young researchers in diverse fields of photonic crystals, which can be turned into reality. Finally, let us list some of prospective applications of photonic crystals that appear to be most exciting for us (ranked in accordance with our personal evaluation):

1. Obtaining the so-called frozen visible light (e.g., [56]);
2. Ultrafast and/or ultimately low-power signal switching (e.g., [57]);
3. Photonic-crystal lasers (e.g., [58]);
4. Nonlinear photonic crystals for new light sources – generators of second and third harmonics (e.g., [59]);
5. Aperiodic photonic crystals and quasicrystals for extremely narrowband optical filtering, optical sensing, etc. (e.g., [60]).

Control Questions

1. If the defect in a photonic crystal is formed by removal of a chain of inclusions, the defect state is called mini-band or mini-passband. Why it is called like this?
2. Is it possible for a mini-band of a defect channel to be so narrow that the dispersion of the defect state is practically absent and the defect state is at single frequency? If no, why? If yes, what can you say about the group velocity of the guided mode?
3. Can the group velocity of a guided mode in a defect channel be equal to the phase velocity? Explain your answer.
4. Compare the angular spectrum of guided modes in a photonic-crystal fiber with that of wave beams guided along a usual optical fiber due to total internal reflection. Which spectrum is broader and why?
5. Why are photonic-crystal LEDs more efficient than the conventional ones?
6. What is the main advantage and what is the main drawback of photonic-crystal optical sensors compared to those based on surface waves of a dielectric slab waveguide?
7. Why cannot aberrations be avoided in a lens (even in a spherical lens focusing all rays at one point)?
8. Why does the Veselago pseudolens promise no aberrations?
9. What is the main drawback of the photonic-crystal pseudolens exploiting the superquadric isofrequency?
10. Why is all-angle impedance matching in the two-dimensional photonic-crystal pseudolens possible for only a specific polarization (TM or TE)?
11. Why does the spatial harmonic with the reduced Bloch vector $\mathbf{k}_{main}$ and not that with the Bloch vector $\mathbf{q}$ of the generic lattice dominate in a photonic-crystal slab operating in the Veselago-Silin regime?

12. Why does the photonic-crystal superprism exploit the superquadric isofrequency?

13. Does the photonic-crystal superprism need all-angle impedance matching? Explain you answer.

Bibliography

[1] J. D. Joannopoulos, R. D. Mead, and J. N. Winn, *Photonic Crystals: Molding the Flow of Light*, Princeton University Press, 1995.

[2] K. Sakoda, *Optical Properties of Photonic Crystals*, Springer-Verlag, 2005.

[3] H. S. Dutta, A. K. Goyal, V. Srivastava, and S. Pal, "Coupling light in photonic-crystal waveguides: A review," *Photonics and Nanostructures - Fundamentals and Applications* **20**, 41–58 (2016).

[4] O. Painter and K. Srinivasan, "Localized defect states in two-dimensional photonic crystal slab waveguides: A simple model based upon symmetry analysis," *Physical Review* **68**, 035110 (2003).

[5] K. Rauscher, D. Erni, J. Smajic, Ch. Hafner, "Improved transmission for 60° photonic crystal waveguide bends," *Proceedings of Photonics and Electromagnetics Research Symposium*, Pisa, Italy, March 28–31, 2004.

[6] S. He, M. Popov, M. Qiu, and C. R. Simovski, "An explicit method for the analysis of guided waves in a line defect channel in a photonic crystal," *Microwave and Optical Technology Letters* **26**, 67–73 (2000).

[7] O. Painter, J. Vuckovic, and A. Scherer, "Defect modes of a two-dimensional photonic crystal in an optically thin dielectric slab," *Journal of the Optical Society of America B* **16**, 275–285 (1999).

[8] D. C. Zografopoulos, R. Asquini, E. E. Kriezis, A. d'Alessandro, and R. Beccherelli, "Guided-wave liquid-crystal photonics," *Lab on a Chip* **12**, 3598–3610 (2012).

[9] F. Poli, A. Cucinotta, and S. Selleri, *Photonic Crystal Fibers*, Springer, 2007.

[10] P. J. Roberts, F. Couny, H. Sabert, et al., "Ultimate low loss of hollow-core photonic crystal fibres," *Optics Express* **13**, 236–244 (2005).

[11] W. L. Chan, J. Deibel, D. M. Mittleman, W. L. Chan, J. Deibel, and D. M. Mittleman, "Imaging with terahertz radiation," *Reports on Progress in Physics* **70**, 1325–1379 (2007).

[12] A. Hassani, A. Dupuis, and M. Skorobogatiy, "Porous polymers for low-loss THz guiding," *Optics Express* **16**, 6340–6346 (2008).

[13] F. Zhao and D. Ma, "Approaches to high performance white organic light-emitting diodes for general lighting," *Material Chemistry Frontiers* **1**, 1933–1950 (2017).

[14] D. L. Barton and A. J. Fischer, "Photonic crystals improve LED efficiency," *SPIE Newsroom*, doi 10.1117/2.1200603.0160 (2006)

[15] G. F. Neumark, I. L. Krukovsky, and H. Jiang, eds., *Wide Bandgap Light Emitting Materials and Devices*, Wiley-VCH, 2007.

[16] B. Troia, A, Paolicelli, F. de Leonardis, and V. M. N. Passaro, "Photonic crystals for optical sensing: A review," Chapter 11 in *Advances in Photonic Crystals*, ed. V. M. N. Passaro, IntechOpen Ltd, 2013.

[17] H. Takagi, Y. Ota, N. Kumagai, S. Ishida, S. Iwamoto, and Y. Arakawa, "High Q H1 photonic crystal nanocavities with efficient vertical emission," *Optics Express* **20**, 28292–28300 (2012).

[18] X. X. Fu, B. Zhang, X. N. Kang, et al., "GaN-based light-emitting diodes with photonic crystals structures fabricated by porous anodic alumina template," *Optics Express* **19**, A1104–A1108 (2011).

[19] V. P. Bykov, "Spontaneous emission in a periodic structure," *Journal of Experimental and Theoretical Physics* **35**, 269–273 (1972).

[20] V. P. Bykov, "Spontaneous emission from a medium with a band spectrum," *Soviet Journal of Quantum Electronics* **4**, 861–871 (1975).

[21] E. Yablonovich, "Inhibited spontaneous emission in solid-state physics and electronics," *Physical Review Letters* **58**, 2059–2062 (1987).

[22] S. John, "Strong localization of photons in certain disordered dielectric superlattices," *Physical Review Letters* **58**, 2486–2489 (1987).

[23] A. Densmore, D. X. Xu, S. Janz, et al., "Sensitive label-free biomolecular detection using thin silicon waveguides," *Advances in Optical Technologies* **2008**, 725967 (2008).

[24] S. Fan, P. R. Villeneuve, J. D. Joannopoulos, and H. A. Haus, "Channel drop filters in photonic crystals," *Optics Express* **3**, 4–10 (1998).

[25] S. Fan, P. R. Villeneuve, J. D. Joannopoulos, M. J. Khan, C. Manolatou, and H. A. Haus, "Theoretical analysis of channel drop tunnelling processes," *Physical Review* **59**, 15882–15892 (1999).

[26] H. A. Haus and Y. Lai, "Narrow-band optical channel-dropping filter," *Journal of Lightwave Technology* **10**, 57–62 (1992).

[27] R. M. C. Siva Ram and M.Guruswamy, *WDM Optical Networks: Concepts, Designs, and Algorithms*, Prentice Hall, 2011)

[28] A. Anderson, "Optical wavelength-division multiplexing," Bachelor's work, University of Tartu, Faculty of Science and Technology, Computer Science Department, 2008. Available at http://kodu.ut.ee/eero/pdf/2008/ AndresAndersn-baka-pre.pdf.

[29] B. Momeni, J. Huang, M. Soltani, et al., "Compact wavelength demultiplexing using focusing negative index photonic crystal superprisms," *Optics Express* **14**, 2413–2422 (2006).

[30] A. Jugessur, L. Wu, A. Bakhtazad, A. Kirk, T. Krauss, and R. De La Rue, "Compact and integrated 2D photonic crystal super-prism filter device for wavelength demultiplexing applications," *Optics Express* **14**, 1632–1642 (2006).

[31] C. Luo, M. Soljacic, and J. D. Joannopoulos, "Superprism effect based on phase velocities," *Optics Letters* **29**, 745–748 (2004).

[32] J. Arai, H. Kawata, and F. Okano, "Microlens arrays for integral imaging system," *Applied Optics* **45**, 9066–9078 (2006).

[33] F. Wippermann, "Chirped refractive microlens arrays," Dr.-Eng. dissertation, Scientific Research Institute for Bioprocessing and Analytical Measurement Techniques, 1974.

[34] A. Katzir, A. C. Livanos, J. B. Shellan, and A. Yariv, "Chirped gratings in integrated optics," *IEEE Journal of Quantum Electronics* **QE-13**, 296–304 (1977).

[35] T. Yoshizawa, ed., *Handbook of Optical Metrology*, 2nd ed. CRC Press, 2017.

[36] M. Born and E. Wolf, *Principles of Optics*, 7th ed. Cambridge University Press, 2002.

[37] F. Aieta, P. Genevet, M. Kats, N. Yu, R. Blanchard, Z. Gaburro, and F. Capasso, "Aberration-free ultrathin flat lenses and axicons at telecom wavelengths based on plasmonic metasurfaces," *NanoLetters* **12**, 4932–4936 (2012).

[38] V. G. Veselago, "The electrodynamics of substances with simultaneously negative values of ε and μ," *Soviet Physics Uspekhi* **10**, 509–514 (1968).

[39] S. A. Tretyakov and S. I. Maslovski, "Veselago materials: What is possible and impossible about the dispersion of the constitutive parameters," *IEEE Antennas and Propagation Magazine* **49**, 37–43 (2007).

[40] S. Wuestner, A. Pusch, K. L. Tsakmakidis, J. M. Hamm, and O. Hess, "Gain and plasmon dynamics in active negative-index metamaterials," *Philosophical Transactions of the Royal Society A* **369**, 3525–3550 (2011).

[41] R. A. Silin, "Possibility of creating parallel-plate lenses without aberrations," *Optika i Spectroscopia* **44**, 109–110 (1978). In Russian.

[42] R. A. Silin, "Parallel-plate focusing of divergent wave beams," *Izvestia VUZov Radiofizika* **15**, 809–820 (1972). In Russian.

[43] C. Luo, S. G. Johnson, J. D. Joannopoulos, and J. B. Pendry, "All-angle negative refraction without negative effective index," *Physical Review* **65**, 201104 (2002).

[44] A. Poddubny, I. Iorsh, P. Belov, and Y. Kivshar, "Hyperbolic metamaterials," *Nature Photonics* **7**, 948–957 (2013).

[45] H. N. S. Krishnamoorthy, Z. Jacob, E. Narimanov, I. Kretzschmar, and V. M. Menon, "Topological transitions in metamaterials," *Science* **336**, 205–209 (2012).

[46] I. M. Lifshitz, "Anomalies of electron characteristics of a metal in the high-pressure region," *Soviet Physics (Journal of Experimental and Theoretical Physics)* **11**, 1130–1135 (1960).

[47] M. Notomi, "Negative refraction in photonic crystals," *Optical Quantum Electronics* **34**, 133–143 (2002).

[48] S. Foteinopoulou, E. N. Economou, and C. M. Soukoulis, "Refraction in media with a negative refractive index," *Physical Review Letters* **90**, 107402 (2003).

[49] C. Luo, S. G. Johnson, and J. D. Joannopoulos, "All-angle negative refraction in a three-dimensionally periodic photonic crystal," *Applied Physics Letters* **81**, 2352–2354 (2002).

[50] R. A. Silin, *Unusual Laws of Refraction and Reflection*, Fazis, 1999. In Russian.

[51] A. L. Efros and A. L. Pokrovsky, "Dielectric photonic crystal as medium with negative electric permittivity and magnetic permeability," *Solid State Communications* **129**, 643–647 (2004).

[52] E. Cubukcu, K. Aydin, E. Ozbay, S. Foteinopoulou, and C. M. Soukoulis, "Electromagnetic waves: Negative refraction by photonic crystals," *Nature* **423**, 604–605 (2003).

[53] P. V. Parimi, W. T. Lu, P. Vodo, and S. Sridhar, "Photonic crystals: Imaging by flat lens using negative refraction," *Nature* **426**, 404–405 (2003).

[54] Z. Lu, J. A. Murakowski, C. A. Schuetz, S. Shi, G. J. Schneider, and D. W. Prather, "Three-dimensional subwavelength imaging by a photonic-crystal flat lens using negative refraction at microwave frequencies," *Physical Review Letters* **95**, 153901 (2005).

[55] J. Pendry, "Negative refraction index makes perfect lens," *Physical Review Letters* **85**, 3966 (2000).

[56] Y. Yang, Y. Poo, R. Wu, Y. Gu, and P. Chen, "Experimental demonstration of one-way slow wave in waveguide involving gyromagnetic photonic crystals," *Applied Physics Letters* **102**, 231113 (2013).

[57] M. Inoue, A. Granovsky, O. Aktsipetrov, H. Uchida, and K. Nishimura, "Magnetophotonic crystals," in *Magnetic Nanostructures*, Springer Series in Material Science, vol. 94, ed. B. Aktas, Springer, pp. 29–43, 2007.

[58] O. Painter, R. K. Lee, et al., "Two-dimensional photonic band-gap defect mode laser," *Science* **284**, 1819–1821 (1999).

[59] V. Berger, "Nonlinear photonic crystals," *Physical Review Letters* **81**, 4136 (1998).

[60] Z. V. Vardeny, A. Nahata, and A. Agrawal, "Optics of photonic quasicrystals," *Nature Photonics* **7**, 177–187 (2013).

Note

1 A quantum well is a nanolayer of a low-bandgap semiconductor sandwiched between two plates of a high-bandgap semiconductor. It can be also a set of such sandwiches.

7 Plasmonics

7.1 Introduction

In order to control and process light at the subwavelength (nano) scale, we need to find ways to confine electromagnetic fields in subwavelength volumes, and we need to find waveguiding structures with subwavelength cross-section dimensions. It is obvious that such conventional approaches to light concentration as dielectric lenses or parabolic mirrors do not offer a possibility of achieving this goal, because their focusing ability is limited by diffraction. One needs other means, where propagating waves would excite oscillations that vary in space much faster than propagating waves do. In radio and microwave engineering, such devices are abundant. For example, an electrically small but resonant wire dipole antenna being excited by a propagating plane wave (whose field variations are on the scale of the wavelength λ_0) collects the incident wave power from the area of the order of λ_0^2 and delivers it into a tiny gap between the two antenna arms (which is very small compared to λ_0) where the load is connected.

One of the possible approaches toward solving these problems in optics is the use of noble metals. Electromagnetic waves cannot propagate inside bulk pieces of conductors, but resonant modes at interfaces between metals (negative permittivity) and dielectrics (positive permittivity) can have rather localized distributions of electromagnetic fields in space. One can think about an analogy with the radio-frequency circuits, where a resonator formed by a capacitor filled with a dielectric (negative reactance, positive permittivity, electric stored energy due to accumulated charge) and an inductor made of a conducting wire (positive reactance, "negative permittivity," magnetic stored energy due to conduction current) has dimensions very small compared to the wavelength.

Indeed, radio-frequency antennas made of metals can effectively confine electromagnetic fields, metal coaxial cables can guide waves, and LC-resonators can be very small, so can we use metal elements to do the same in the visible and infrared spectra? This question is addressed by plasmonics.

7.2 Surface Plasmon Polaritons

Surface plasmon polaritons are surface waves that can propagate along interfaces between dielectrics and metals. They are basically inhomogeneous plane waves whose "propagation constant" in the direction orthogonal to the interface is imaginary so that the fields exponentially decay away from the interface. If this decay factor is high, and the fields concentration at the surface is strong enough, such a surface-wave structure may possibly serve as a waveguide with a subwavelength cross-section size. If the propagation constant along the surface is large enough (much larger than the free-space constant), we also may be able to realize subwavelength field confinement along the surface.

7.2.1 Dispersion Equation

To understand the properties of these surface waves, we will need to introduce some preliminary notions regarding plane waves near planar interfaces. Here, we mean waves whose complex amplitudes depend on the position vector $\mathbf{r}$ as $\exp(-j\mathbf{k} \cdot \mathbf{r})$, where $\mathbf{r}$ can be a complex vector. In this meaning, evanescent waves are also plane waves, and their other name is *inhomogeneous plane waves*. Let us consider plane electromagnetic waves in an isotropic medium in the vicinity of a planar interface with a different material or with a planar boundary. In this situation, it is convenient to introduce a unit vector normal to the interface $\mathbf{n}$ and consider tangential to the surface (orthogonal to $\mathbf{n}$) components of the plane-wave fields, because these components are continuous across the interface.

This situation is illustrated in Figure 7.1. The wavevector $\mathbf{k}$ is split into its normal and tangential components: $\mathbf{k} = k_n\mathbf{n} + \mathbf{k}_t$, where $\mathbf{k}_t \cdot \mathbf{n} = 0$. Denoting the length of the wavevector $k = \omega\sqrt{\epsilon\mu}$, we write the dispersion equation for plane waves as

$$k_n^2 + k_t^2 = k^2. \tag{7.1}$$

Thus,

$$k_n = \sqrt{k^2 - k_t^2}. \tag{7.2}$$

Because both tangential field components are continuous across the interface, it is convenient to introduce wave impedance as the ratio of the *tangential to the*

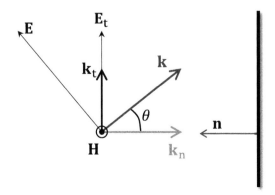

Figure 7.1   A transverse-magnetic (TM) polarized plane wave in a vicinity of a planar surface, shown by a vertical line.

interface components of the plane-wave fields. For this example of transverse-magnetic (TM) polarized waves, we have

$$\mathbf{E}_t = \sqrt{\frac{\mu}{\epsilon}} \cos\theta \, \mathbf{n} \times \mathbf{H} = \sqrt{\frac{\mu}{\epsilon}} \frac{k_n}{k} \mathbf{n} \times \mathbf{H} = \frac{k_n}{\omega\epsilon} \mathbf{n} \times \mathbf{H}. \tag{7.3}$$

Here, θ is the angle between vector $\mathbf{k}$ and the normal to the interface (the "incidence angle").

Now we are ready to find the dispersion equation for surface waves propagating along the interface and decaying in the normal direction. Equations (7.2) and (7.3) hold for plane-wave fields on both sides of the interface; we only need to substitute the corresponding material parameters of the two media and satisfy the boundary conditions, which tell us that the tangential fields must be continuous across the interface. Thus, we have

$$\frac{k_{n1}}{\omega\epsilon_1} = -\frac{k_{n2}}{\omega\epsilon_2} \tag{7.4}$$

(here we have taken into account that the unit vector $\mathbf{n}$ is directed oppositely on the other side of the interface).

Let us assume that the medium on one side is a lossless metal with the permittivity $\epsilon_1 < 0$, and the other half-space is filled with a lossless isotropic dielectric with the permittivity $\epsilon_2 > 0$. Since we want to find solutions that would exponentially decay away from the interface, the normal component of the wavevector of plane waves on both sides must be imaginary; that is, we are looking for solutions with $|k_t| > |k_{1,2}|$ (see (7.2)).

Because waves should decay in both the positive (medium 1) and the negative (medium 2) directions of the unit vector $\mathbf{n}$, the signs $k_{n1,2}$ should be opposite to each other. Thus, this equation can have nontrivial solutions for surface waves only if the signs of ϵ_1 and ϵ_2 are opposite, which is exactly the case of our interface between lossless dielectric (positive ϵ) and metal (negative ϵ). Substituting the values of $k_{n1,2}$ from (7.2) and solving for the tangential to the interface component of the propagation constant k_t, we find the dispersion equation

$$k_t = \omega\sqrt{\mu_0}\sqrt{\frac{\epsilon_1\epsilon_2}{\epsilon_1 + \epsilon_2}} = \omega\sqrt{\epsilon_0\mu_0}\sqrt{\frac{\epsilon_{r1}\epsilon_{r2}}{\epsilon_{r1} + \epsilon_{r2}}}. \tag{7.5}$$

The eigenvalue equation for the orthogonal (TE) polarization can be obtained upon replacing permittivities by the corresponding permeabilities in (7.4). For interfaces between metals and dielectrics, it has no solutions because both permeabilities are positive.

An example of dispersion curves for surface waves at interfaces between dielectrics and metals (neglecting losses in both media) is shown in Figure 7.2. Dispersion of the permittivity of the dielectric is neglected, and the dispersion

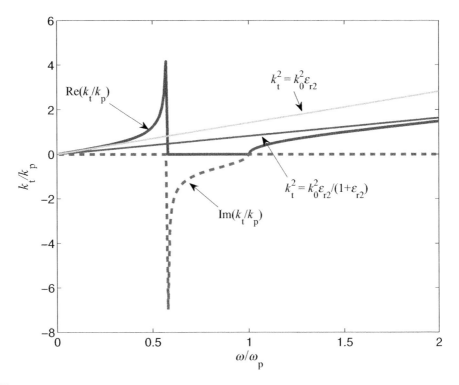

Figure 7.2 Example of dispersion curves for surface plasmon polaritons for the case of negligible absorption. The wavenumbers are normalized to the plasma wavenumber $k_p = \omega_p \sqrt{\epsilon_0 \mu_0}$. Plot parameters: $\epsilon_{r2} = 2$, the Drude model for lossless conductors.

of metal is modeled by the Drude model (2.18) with the loss factor $\nu = 0$. The resonant frequency (sometimes called the *surface plasmon frequency*)

$$\omega_{SPP} = \omega_p \sqrt{\frac{\epsilon_0}{\epsilon_0 + \epsilon_2}} \qquad (7.6)$$

corresponds to the point where the denominator of (7.5) tends to zero, so the propagation constant k_t tends to infinity. In the frequency range $\omega_{SPP} < \omega < \omega_p$, there is a stop band (no solutions with real values of k_t), and at $\omega > \omega_p$, the waves are not bounded to the surface because the condition $|k_t| > |k_{1,2}|$ is not satisfied.

For lossy metals and/or dielectrics, the propagation constant is complex at all frequencies, so there is some decay in the passbands and some propagation in the stop band; see a typical plot in Figure 7.3.

For applications in the visible and infrared ranges, silver is the usual metal of choice because of its relatively small dissipation losses as compared to other metals. Calculations based on the measured data for silver permittivity give the curves shown in Figure 7.4. The dielectric medium is the same as in the previous examples: a lossless dielectric $\epsilon_{r2} = 2$. We see that the surface plasmon polariton can propagate in the range from about 200 THz ($\lambda_0 = 1.5 \,\mu$m) to about 750 THz ($\lambda_0 = 0.4 \,\mu$m = 400 nm), covering the visible (about 390–700 nm) and infrared

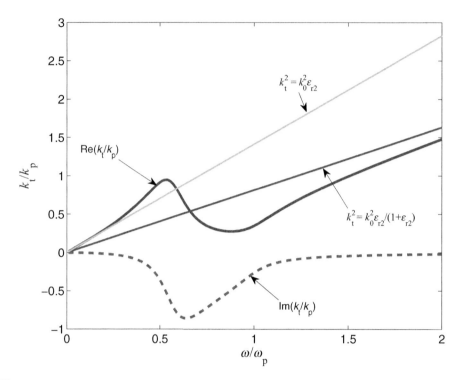

Example of dispersion curves for surface plasmon polaritons: absorptive metal. Plot parameters: $\epsilon_{r2} = 2$, Drude model for lossy conductors, $\nu/\omega_p = 0.2$.

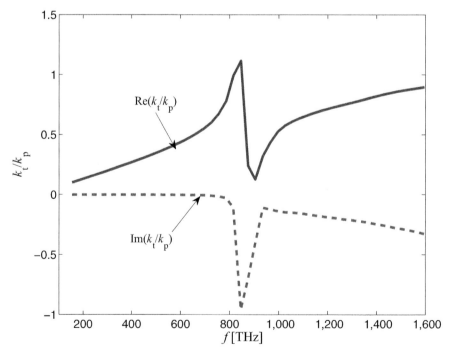

Dispersion curves for surface plasmon polaritons on an interface between silver and a dielectric ($\epsilon_{r2} = 2$). Experimental data for silver properties [1]. The propagation constant is normalized to the fitting value of the plasma wavenumber, corresponding to the plasma frequency $\omega_p/(2\pi) = 2175$ THz [2].

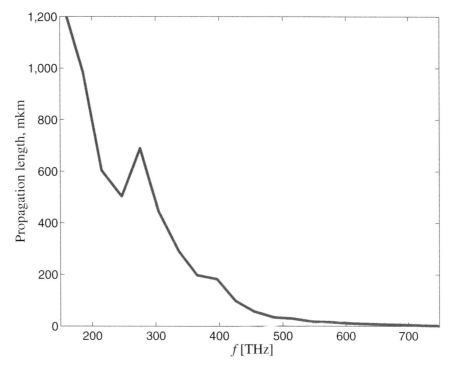

Figure 7.5 Propagation length of surface plasmon polaritons on an interface between silver and a dielectric ($\epsilon_2 = 2$). Experimental data for silver properties [1].

spectra. However, the properties of the wave are quite dissimilar at different frequencies. We note from the same plot that the decay rate, measured by the imaginary part of the propagation constant, grows with increasing frequency. It can be visualized better by plotting the so-called *propagation length*, which is equal to the inverse of the imaginary part of the propagation constant: $L = 1/\mathrm{Im}(k_t)$. This curve is shown in Figure 7.5. It is clear that at the high-frequency end of the visible spectrum, the propagation length is only a couple of microns, while in the infrared range, it can be two orders of magnitude larger than the wavelength.

Another important parameter is the confinement of the wave fields to the surface. After all, one of the main motivations for using plasmonic structures is a possibility of confining the field in subwavelength volumes. In this case, we look for a waveguiding structure whose cross-section size would be small compared to the wavelength. The field confinement is measured by the field decay rate in the directions normal to the surface – that is, by the inverse values of the normal component of the propagation constant (its imaginary part). The corresponding plots are shown in Figure 7.6. We see that while the fields decay rather quickly into the silver volume, fields extend quite far into the dielectric, especially at low frequencies. Actually, the "width" of the plasmonic waveguide is comparable to the free-space wavelength at the corresponding frequency.

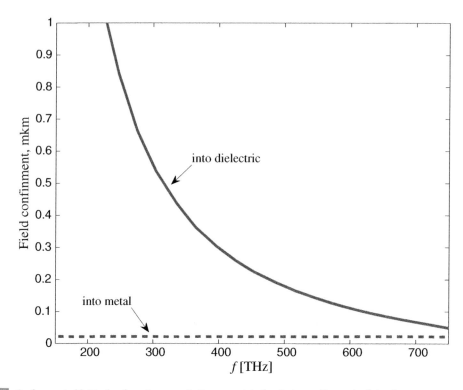

Figure 7.6 Confinement of fields of surface plasmon polaritons on an interface between silver and a dielectric ($\epsilon_{t2} = 2$). Experimental data for silver properties [1].

7.2.2 Slab Waveguides

Propagating surface plasmon polaritons have small propagation length due to losses in metal. In order to reduce losses, it is preferable to use waveguide topologies where it is possible to concentrate most of the fields in low-loss dielectrics or in the air. One possibility is to use the slab topology. Let us consider a three-layer structure, where a dielectric slab (permittivity ϵ_2) is bounded from both sides by semi-infinite metal spaces (permittivity ϵ_1). We would like to find a mode whose field maximum would be located inside the dielectric so that the fields at the interfaces with metal will be smaller than in the waveguide center. In case of plasmons on a single interface, which we considered in Section 7.2.1, the tangential fields at the metal surface are, in fact, at their maximum value (they exponentially decay in both directions away from the interface).

To get more physical insight, this layered waveguide structure can be viewed as a system of two coupled plasmons at the two metal-dielectric interfaces or as a metal parallel-plate waveguide with unideally conducting walls or as a planar dielectric waveguide with lossy cladding. The field solution can be found using the standard technique for analyzing open (in particular, planar) waveguides. Considering TM modes, the longitudinal component of the electric field in the two media $E_{z1,2}$ satisfies the Helmholtz equation

$$\nabla_t^2 E_{z1,2} + (k_{1,2}^2 - k_t^2)E_{z1,2} = 0. \tag{7.7}$$

The boundary conditions on the two interfaces read

$$\frac{\epsilon_1}{k_t^2 - k_1^2} \frac{\partial E_{z1}}{\partial n} = \frac{\epsilon_2}{k_t^2 - k_2^2} \frac{\partial E_{z2}}{\partial n}, \tag{7.8}$$

where **n** is the outward normal to the dielectric slab. These equations as well as the resulting dispersion equation are the same as for conventional dielectric-slab waveguides:

$$\tanh(k_{n2}d) = -\frac{k_{n1}\epsilon_2}{k_{n2}\epsilon_1} \qquad \text{odd modes} \tag{7.9}$$

$$\tanh(k_{n2}d) = -\frac{k_{n2}\epsilon_1}{k_{n1}\epsilon_2} \qquad \text{even modes,} \tag{7.10}$$

where d is the thickness of the dielectric slab. The solutions need to be found numerically, taking into account that the metal permittivity strongly depends on the frequency. The results show that, indeed, among even modes, there is a mode with a cosine-type distribution of the electric field in the dielectric slab such that the fields at the interfaces with metal are smaller than in the core center. Since, in this case, most of the power propagates in dielectric and not in metal, losses can be considerably smaller than in single-interface plasmonic waveguides. On the other hand, it is important to note that these relatively low-loss modes exist in the case when the thickness of the dielectric slab is comparable to or larger than the wavelength in the dielectric [3]. Thus, these waveguides do not offer a compact (subwavelength cross-section) solution. Compared to conventional planar optical waveguides, the advantages come from the fact that the waveguide is closed by metal walls, isolating the guided fields from the environment. Other modifications of plasmonic waveguides are metal strips, grooves, and hollow channels in bulk metal.

7.2.3 Surface Waves at Low Frequencies

As is obvious from Figure 7.2, if the operational frequency is much smaller than the plasma frequency of metal ($\omega \ll \omega_p$), the surface mode exists, but its propagation constant k_t is very close to that of plane waves in dielectric. The corresponding normal component of the propagation constant k_{n2} is close to zero. This conclusion is also seen directly from the dispersion equation (7.4): when the permittivity of metal ϵ_1 takes large negative values, the right-hand side is very small. Physically, this means that the wave is very weakly bound to the surface; in the limit of infinite conductivity of metal, it becomes simply a homogeneous plane wave propagating in the dielectric medium along the conducting surface. However, if the conductivity of the first medium is moderate, well-bounded surface waves can exist.

Considering again the equations for the normal to the interface components of the propagation constants in the two media $k_{n1,2}$, we can notice that the propagation constants are complex numbers if one of the contacting media is a lossy dielectric (complex permittivity). In this case, even if the real parts of both permittivities are positive, the field can be also bound to the surface and propagate along it. The propagation constant along the interface is also complex in this case, so the wave exponentially decays. A classical example is the radio-frequency surface wave over the Earth's surface. Average, soil or sea water are moderately conducting media. Surface waves can indeed propagate, and they are called Zenneck waves (1907). A comparative overview of the Zenneck wave and surface plasmons can be found in paper [4].

Surface waves can also propagate over metal surfaces at low frequencies if we structure the surface at the subwavelength scale. To understand these waves, let us revisit the derivation of the dispersion equation. Considering Figure 7.1, let us assume that to the right of the interface we have a dielectric material with the permittivity ϵ. The normal component of the propagation constant k_n should be a negative imaginary number so that the surface-wave fields would exponentially decay away from the interface. Thus, the corresponding wave impedance (7.3) is a negative imaginary quantity, a capacitive reactance. In order to satisfy the dispersion equation (7.4), we should bring it in contact with a surface that has a positive, inductive surface reactance. If we fill the left half-space with a metal and the frequency is well below the plasma frequency, that is impossible because the metal impedance is very close to zero. However, imagine that we corrugate a metal surface with electrically thin (and densely packed) grooves of depth d, orienting the grooves orthogonally to the tangential component of the electric field vector. At the bottom of the groove, the surface impedance is nearly zero (short circuit by the metal wall). But at the distance d from the groove bottom, we see the impedance of a short-circuited transmission line of the length d, which is proportional to $j\tan(kd)$. Here, k is the propagation constant of the material that fills the grooves (air, for example). Since the tangent function can take any value from $-\infty$ to $+\infty$, we can obviously choose the depth d so that the impedance at the surface of the corrugated metal plate takes the desired positive imaginary value.

Such structured surfaces (there are other possible realizations in addition to corrugations) are called *high-impedance surfaces* [5]. In microwave engineering, the use of structured surfaces to engineer dispersion of surface waves has been known for a long time. More recently, this possibility was also recognized by experts in optics, and the new name, "spoof surface plasmons," was proposed to describe surface waves over subwavelength structured surfaces.

Another possibility to adjust the surface impedance of a metal surface is to cover it with a dielectric slab. Using the simple transmission-line model, it can be seen that the nearly zero (short-circuit) impedance of metals at microwave frequencies is transformed into reactive (if dissipation is small) impedance at the surface of the dielectric slab. For some propagation constant of surface waves, it becomes possible to satisfy the surface-mode dispersion equation. Actually, the presence of surface modes on grounded dielectric layers creates problems in the design of printed-circuit components and microstrip antennas because these waves carry the electromagnetic energy away from the device components, increasing losses and creating undesired side lobes of microstrip antennas.

7.2.4 Excitation of Surface Plasmon Polaritons

Surface waves between two media characterized by wavenumbers $k_{1,2}$ are bound to the surface, meaning that the propagation constant along the interface k_t is larger than the wavenumbers in both media: $|k_t| > |k_{1,2}|$. It means that it is not possible to excite such waves by propagating plane waves incident on the interface: the exiting wave should be in phase synchronism with the surface wave, but the tangential component of the propagating wave is always *smaller* than the length of the wavevector $|k_{1,2}|$. Thus, we need a coupling device that would transfer propagating plane waves fields into fields whose variations in space would be fast enough.

One possibility is to use a prism in the full-internal-reflection regime. When a plane wave propagating inside the prism material is fully reflected from one of the prism faces, the field outside the prism is not zero (due to continuity of the tangential field components), but it decays exponentially away from the prism (because all the incident energy is reflected). Obviously, this exponentially decaying field is, in fact, a surface wave: the tangential component of the propagation constant is larger than the length of the wavevector in the medium outside the prism. Thus, if we bring such a prism close to an interface that supports surface plasmon polaritons, phase synchronism with the eigenwave of the interface mode becomes possible.

Another possibility is to use a periodical structure (grating). Let us consider again a dielectric/metal interface that supports surface plasmons. If we periodically perturb the dielectric surface – for example, cutting small parallel groves – the surface mode between the dielectric and metal will be not just a single inhomogeneous plane wave but an infinite series of Floquet harmonics with the wavenumbers

$$k_{tn} = k_t + n\frac{2\pi}{D}, \tag{7.11}$$

where $n = \pm1, 2, \ldots$, and D is the perturbation period. Properly selecting the period, it is possible to ensure that one of these harmonics is in phase sync with the incident plane wave; that is, k_{tn} is equal to the tangential component of the wavevector of the incident propagating plane wave.

Furthermore, coupling can be provided by positioning one or more small particles in the vicinity of the interface supporting surface plasmon polaritons. Either the particle should be considerably smaller than the wavelength in the surrounding space or it should have shape or material inhomogeneities on the subwavelength scale. In the near vicinity of subwavelength inhomogeneities, the spatial variations of fields are fast, and coupling to plasmon wave whose wavelength is smaller than that in propagating waves becomes possible. One can also understand it from the point of view of Fourier integral expansion of fields in the vicinity of a small inhomogeneity. In order to resolve the fine spatial structure of the fields, the Fourier spectrum must contain plane-wave components with large wavenumbers, which can come to sync with the surface mode. Typical examples of such couplers are metal or dielectric nanoparticles or tips of near-field scanning microscopes. Any inhomogeneity of the surface acts as a coupler between propagating waves and surface modes. Yet another example is the excitation of surface waves at surfaces of a finite size by illuminating the edge of the surface by propagating waves.

7.3 Plasmonic Nanoparticles

7.3.1 Polarizability

Let us consider an optically small (with respect to the wavelength in the surrounding space) metal sphere. Since the sphere is small, it is possible to model its response as that of an electric dipole, excited by external electric fields, which can be assumed to be approximately uniform over the particle volume. We denote the complex amplitude of the incident electric field at the position of the particle as $\mathbf{E}_{\text{inc}}$ and the complex amplitude of the incident-wave Poynting vector as $\mathbf{P}_{\text{inc}}$. We model the linear particle response by its polarizability α defined through

$$\mathbf{p} = \alpha\, \mathbf{E}_{\text{inc}}, \tag{7.12}$$

where $\mathbf{p}$ is the complex amplitude of the induced electric dipole moment.

The power extracted by a dipole inclusion (dipole moment $\mathbf{p}$) from a given incident field $\mathbf{E}_{\text{inc}}$ can be found from the classical formula

$$P_{\text{ext}} = \frac{1}{2}\text{Re}\int_V \mathbf{J}^* \cdot \mathbf{E}_{\text{inc}}\, dV = \frac{1}{2}\text{Re}(-j\omega p^* E_{\text{inc}})$$

$$= -\frac{\omega}{2}\text{Im}(\alpha)|E_{\text{inc}}|^2 = -\eta\omega\text{Im}(\alpha)P_{\text{inc}} \tag{7.13}$$

($\eta = \sqrt{\mu/\epsilon}$ is the wave impedance of the surrounding space and, as usual, the harmonic time dependence is in the form $\exp(j\omega t)$). Here, $\mathbf{J}$ is the volumetric electric current density inside the particle. This general formula is valid for any dispersive and lossy particle, assuming only that it is a small dipole particle and $\mathbf{E}_{\text{inc}}$ is uniform over its volume. The result in terms of the incident power density P_{inc} holds if the particle is excited by a propagating plane wave. The corresponding value of the incident time-averaged power flow density is

$$P_{\text{inc}} = \frac{1}{2\eta}|E_{\text{inc}}|^2. \tag{7.14}$$

The power that is scattered (reradiated) by the particle is the power radiated by the electric dipole $\mathbf{p}$:

$$P_{\text{sc}} = \frac{\mu_0\omega^4}{12\pi c}|p|^2 = \frac{\mu_0\omega^4|\alpha|^2}{12\pi c}|E_{\text{inc}}|^2 = \frac{k^4}{6\pi\epsilon^2}|\alpha|^2 P_{\text{inc}}, \tag{7.15}$$

where $k = \omega\sqrt{\epsilon\mu}$ is the wavenumbers in the surrounding space.

We can find the general expression for the scattering loss (radiation damping) factor in the dipole polarizability by equating the extracted and scattered powers for the case of a lossless particle (no absorption). This allows us to find an expression for the imaginary part of the inverse polarizability due to scattering loss (e.g., [6], eq. (4.82))

$$\frac{1}{\alpha} = \frac{\text{Re}(\alpha) - j\text{Im}(\alpha)}{|\alpha|^2} \tag{7.16}$$

as

$$\text{Im}\left(\frac{1}{\alpha}\right) = -\frac{\text{Im}(\alpha)}{|\alpha|^2} = \frac{\mu\omega^3}{6\pi c} = \frac{k^3}{6\pi\epsilon}. \tag{7.17}$$

This is a classical result that dates back to the work of M. Planck (1902).

We see that, in the general case, the inverse polarizability of a dipole particle has the form

$$\frac{1}{\alpha} = \xi + j\frac{k^3}{6\pi\epsilon}, \tag{7.18}$$

where the complex-valued parameter $\xi = \xi' + j\xi''$ depends on the particle size, shape, and material. The imaginary part ξ'' measures absorption in the particle, if it is lossy.

For electrically small particles, parameter ξ can be found using the quasi-static approximation, since the scattering properties have been already taken into account by the radiation damping term. For a small sphere made of an isotropic material with the permittivity ϵ_1, the quasi-static solution gives

$$\xi = \frac{1}{3\epsilon V}\frac{\epsilon_1 + 2\epsilon}{\epsilon_1 - \epsilon}, \tag{7.19}$$

where V is the volume of the sphere. This approximation of uniform internal field is adequate for Ag and Au nanospheres of radii smaller than 30 and 50 nm, respectively [7]. For a special case of a metal sphere modeled by the Drude permittivity

$$\epsilon_1 = \epsilon_0\left[1 - \frac{\omega_p^2}{\omega(\omega - j\nu)}\right], \tag{7.20}$$

we have

$$\xi' = \frac{\omega_p^2/3 - \omega^2}{\epsilon V \omega_p^2}, \qquad \xi'' = \frac{\omega\nu}{\epsilon V \omega_p^2}. \tag{7.21}$$

7.3.2 Localized Surface Plasmons

At the frequency where the real part of the permittivity of the sphere satisfies

$$\epsilon_1' = -2\epsilon' \tag{7.22}$$

(ϵ' is the real part of the host permittivity ϵ), the absolute value of parameter ξ (7.19) has a minimum, meaning that the amplitude of the particle polarizability (7.18) has a maximum: there is a resonance of the particle response. As seen from (7.21), for metal nanoparticles modeled by the Drude formula, the resonance frequency is $\omega_0 = \omega_p/3$. Condition (7.22) is called the Fröhlich resonance condition. This resonant effect is called the plasmonic resonance,

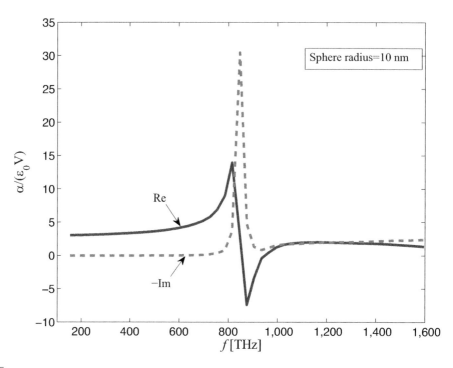

Figure 7.7 Normalized polarizabilities of spherical silver nanoparticles in a dielectric host $\epsilon = 2\epsilon_0$ for the sphere radius 10 nm. The radiation damping is negligible.

and the excited resonant mode is called the *localized surface plasmon.* The last name can be perhaps confusing, since the distribution of the polarization current (proportional to that of the internal electric field) is practically uniform, meaning that the internal field is not localized at the surface of the sphere. On the other hand, the fields in space around the particle are, in a certain sense, localized: they behave as the fields of a small electric dipole. In contrast to surface modes on planar interfaces, the fields do not decay exponentially away from the particle: reactive near fields decay as 1/distance^2 and 1/distance^3, while the radiated wave decays as 1/distance (assuming lossless background material).

Figures 7.7–7.9 show the normalized polarizability $\alpha/(\epsilon V)$ as a function of the frequency for silver nanoparticles in a dielectric host with the permittivity $\epsilon = 2\epsilon_0$. For moderate permittivities of the surrounding medium, the resonance is in the visible range, and the resonance frequency can be obviously tuned by varying the host permittivity. The dotted curves show the calculations based on the quasi-static model, neglecting the third-order scattering loss term in (7.18). In this approximation, the polarizability is proportional to the particle volume, and for the normalized polarizability, these curves are the same. We can clearly see that for very small spheres, the scattering loss is negligible as compared to the absorption in the particle (in the plot for the sphere of radius 10 nm, the curves cannot be distinguished). On the other hand, for larger spheres, the scattered power is comparable to or even dominates over absorption, as is clear from the plot for the radius equal to 50 nm.

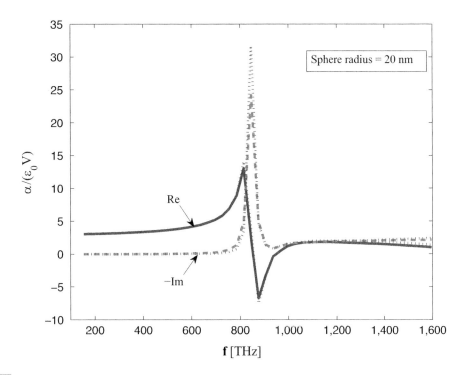

Figure 7.8 Normalized polarizabilities of spherical silver nanoparticles in a dielectric host $\epsilon = 2\epsilon_0$ for the sphere radius 20 nm. Dotted curves show the polarizability calculated neglecting the radiation damping term.

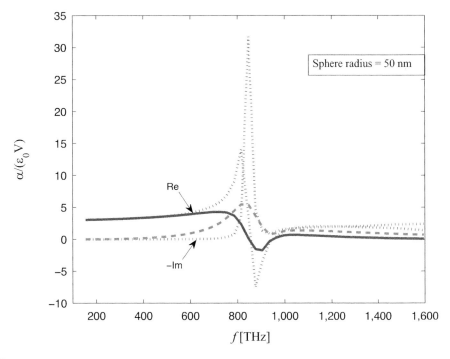

Figure 7.9 Normalized polarizabilities of spherical silver nanoparticles in a dielectric host $\epsilon = 2\epsilon_0$ for the sphere radius 50 nm. Dotted curves show the polarizability calculated neglecting the radiation damping term.

7.4 Chains of Plasmonic Nanoparticles

Surface plasmon polaritons can propagate along metal-dielectric interfaces, and to some extent, such structures can be used as waveguides with subwavelength cross sections. However, as we saw in Section 7.2, the field confinement is ensured only at the high-frequency end of the visible spectrum, and the propagation loss at these frequencies is rather high. Although various plasmon-polariton waveguide topologies, such as strip guides and grove waveguides, have been proposed and studied, considerable loss and poor field localization in the surrounding dielectric limit possible applications.

There is an alternative possibility to confine and guide light in a subwavelength waveguide: using chains of resonant nanoparticles – in particular, plasmonic nanoparticles. Each nanoparticle is an optically small resonantor, which can confine light in a subwavelength vicinity of the particle. Is it possible to adjust the particle interactions in a chain of plasmonic nanoparticles so that the excitation will propagate from particle to particle, as if it were a waveguide? This possibility was probably described for the first time in paper [8], where calculations for a finite-length chain were shown. Dispersion equation in the approximation of only near-field interactions was published in [9], and the analytical dynamic theory was developed in [10] for chains of arbitrary resonant dipoles. Full-wave analysis with numerical examples specifically for plasmonic nanoparticles can be found in [11].

Let us consider a periodical arrangement of small resonant particles along a certain line in free space or in an isotropic dielectric (along the axis z, as shown in Figure 7.10). For simplicity, we consider particles polarizable along the axis

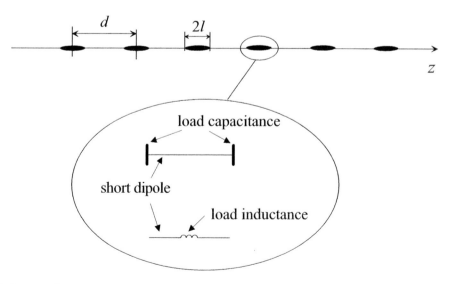

Figure 7.10 Geometry of a generic waveguiding structure formed by a periodical chain of small resonant electric-dipole particles. Reprinted from [10] by permission from Springer Nature Customer Service Centre GmbH, © 2000

direction so that all induced electric dipoles are oriented along this direction. For an arbitrarily chosen dipole on the line array (position $z = 0$), the electric dipole reads

$$p(0) = \alpha\, E_{\text{loc}}, \tag{7.23}$$

where E_{loc} is the z-component of the *local* field created by external sources and all the other particles. Assuming that there is no external field and looking for eigenwaves along the linear chain, we can make use of the Floquet theorem and write

$$p(nd) = e^{-jqnd} p(0), \tag{7.24}$$

where q is the propagation factor of waves traveling along the chain and d is the period. The local field is created by the other dipoles in the array:

$$E_{\text{loc}} = \sum_{n=-\infty,\, n\neq 0}^{\infty} \frac{1}{2\pi\epsilon_0} \left(\frac{1}{(|n|d)^3} + \frac{jk}{(nd)^2} \right) e^{-jk|n|d} e^{-jqnd} p(0), \tag{7.25}$$

where we have substituted the electric dipole fields (we use the usual notation $k = \omega\sqrt{\epsilon_0 \mu_0}$).

Let us denote by β the *interaction constant*

$$\beta = \sum_{n=-\infty,\, n\neq 0}^{\infty} \frac{1}{2\pi\epsilon_0} \left(\frac{1}{(|n|d)^3} + \frac{jk}{(nd)^2} \right) e^{-jk|n|d} e^{-jqnd}. \tag{7.26}$$

With this notation, at the position of the reference dipole $p(0)$, the local field is

$$E_{\text{loc}} = \beta\, p(0). \tag{7.27}$$

If there is no incident field, then $p(0) = \alpha E_{\text{loc}} = \alpha\beta\, p(0)$. Thus, we have the eigenvalue equation $\alpha\beta = 1$, or

$$\frac{1}{\alpha} = \sum_{n=-\infty,\, n\neq 0}^{\infty} \frac{1}{2\pi\epsilon_0} \left[\frac{1}{(|n|d)^3} + \frac{jk}{(nd)^2} \right] e^{-jk|n|d} e^{-jqnd}. \tag{7.28}$$

For lossless scatterers, the imaginary part of the inverse polarizability is known (7.17), and we get

$$\text{Re}\left\{ \frac{1}{\alpha} \right\} + j\frac{k^3}{6\pi\epsilon_0} = \sum_{n=-\infty,\, n\neq 0}^{\infty} \frac{1}{2\pi\epsilon_0} \left[\frac{1}{(|n|d)^3} + \frac{jk}{(nd)^2} \right] e^{-jk|n|d} e^{-jqnd}. \tag{7.29}$$

The imaginary part of this series can be calculated analytically, and the result reads

$$\text{Im}\{\beta\} = \frac{1}{\pi\epsilon_0 d^3} \begin{cases} \frac{(kd)^3}{6} + \frac{\pi}{4}[(qd)^2 - (kd)^2], & 0 < qd < kd \\[2mm] \frac{(kd)^3}{6}, & kd < qd < 2\pi - kd \\[2mm] \frac{(kd)^3}{6} + \frac{\pi}{4}[(2\pi - qd)^2 - (kd)^2], & 2\pi - kd < qd < 2\pi + kd \end{cases} \tag{7.30}$$

(this function is 2π-periodic with respect to qd) [10].

We see that if $q > k$ (more exactly, if $kd < qd < 2\pi - kd$), the imaginary parts of this equation cancel out, which means that propagating modes are possible if we tune the real part of the polarizability to be equal to the real part of the interaction constant β. Obviously, since $q > k$, the fields of the guided mode exponentially decay away from the axis z as $e^{-jk_\rho \rho}$ with the decay factor $k_\rho = \sqrt{k^2 - q^2}$. Numerically calculated examples with field distributions can be found, e.g., in [12].

Another interesting application of chains of plasmonic nanoparticles is in imaging with subwavelength resolution; see [13, 14].

7.5 Applications

Resonant phenomena in optically small metal nanoparticles can be used in a broad variety of applications, mainly exploiting possibilities to concentrate and resonantly enhance electromagnetic fields in optically small regions. Also, the ability to guide light along plasmonic structures is of practical interest. Let us list the main application areas:

- Optical circuits (metatronics); see Chapter 8
- Enhanced emission, fluorescence, and nonlinearities; see Chapter 9 and [15, chapter 9]
- Enhanced solar cells; see Chapter 10.
- Metamaterials, including superlenses; see Chapter 3 and papers [13, 14]
- Nanolasers
- Applications in biology and medicine (mainly in sensors); see Chapter 9 and also [15, 16, chapter 10];

The main practical problem: high dissipation losses in resonant metal nanoparticles or in resonant surface modes.

Problems and Control Questions

1. Surface plasmon polaritons can exist and propagate along flat metal-dielectric interfaces. What other structures can support surface plasmon polaritons? Discuss the necessary conditions for existence of surface modes.
2. Surface plasmon polaritons on infinite interfaces cannot be excited by propagating plane waves, but localized surface plasmons can be excited this way. Explain this difference.
3. Plot the dispersion curve for surface modes at a planar interface of a lossless metal (Drude dispersion, plasma frequency ω_p) with a lossless isotropic medium having Lorentz dispersion. Assume that the resonance frequency of the dielectric response is $\omega_0 = 1.5\omega_p$. Study the properties of waves as

functions of other parameters of the Lorentz material. Discuss the conditions for existence of surface plasmon polaritons.

4. The absorption cross section of a small particle is defined as the ratio of the absorbed power P_{abs} to the incident power density (the amplitude of the Poynting vector) P_{inc}:

$$\sigma_{abs} = \frac{P_{abs}}{P_{inc}}. \tag{7.31}$$

This value can be viewed as an effective area from which the particle "collects" energy, which is then dissipated into heat. Similarly, the total scattering cross section is defined as the ratio of the power scattered (reradiated) by the particle P_{sc} into space when it is illuminated by a plane wave with the incident power density P_{inc}:

$$\sigma_{sc} = \frac{P_{sc}}{P_{inc}}. \tag{7.32}$$

Finally, the extinction cross section is the sum $\sigma_{ext} = \sigma_{sc} + \sigma_{abs}$.

Consider an arbitrary linear electric dipole scatterer – for example, a plasmonic nanoparticle – and find if the values of these three cross sections have a fundamental upper bound. If yes, determine these bounds in general and, in particular, for metal nanoparticles modeled by the Drude formula. Discuss the physical meaning and implications of your results.

Hints

The inverse polarizability of any dipole scatterer can be written as

$$\frac{1}{\alpha} = \xi' + j\xi'' + j\frac{k^3}{6\pi\epsilon_0}. \tag{7.33}$$

The imaginary part ξ'' measures absorption *loss*, so it must be non-negative for any scatterer: $\xi'' \geq 0$. The real part ξ' can take any value – positive, negative, or zero.

Using the known formulas for electric dipoles, one can express the cross sections in terms of the polarizability. The power extracted from the field is

$$P_{ext} = -\frac{\omega}{2}\text{Im}(\alpha)|E_{inc}|^2 = -\eta_0\omega\text{Im}(\alpha)P_{inc}. \tag{7.34}$$

The scattered power reads

$$P_{sc} = \frac{\mu_0\omega^4|\alpha|^2}{12\pi c}|E_{inc}|^2 = \frac{k^4}{6\pi\epsilon_0^2}|\alpha|^2 P_{inc}. \tag{7.35}$$

The absorbed power is the difference between the power extracted by the particle from the incident field (7.34) and the power scattered by the same particle into the surrounding space (7.35):

$$P_{abs} = -\frac{\omega}{2}\text{Im}(\alpha)|E_{inc}|^2 - \frac{\mu_0\omega^4|\alpha|^2}{12\pi c}|E_{inc}|^2. \tag{7.36}$$

Next, we express all the cross sections in terms of the polarizability (7.33):

$$\sigma_{\text{abs}} = \frac{P_{\text{abs}}}{P_{\text{inc}}} = \frac{k}{\epsilon_0} \frac{\xi''}{\xi'^2 + \left(\xi'' + \frac{k^3}{6\pi\epsilon_0}\right)^2}, \tag{7.37}$$

$$\sigma_{\text{sc}} = \frac{P_{\text{sc}}}{P_{\text{inc}}} = \frac{k^4}{6\pi\epsilon_0^2} \frac{1}{\xi'^2 + \left(\xi''^2 + \frac{k^3}{6\pi\epsilon_0}\right)^2}, \tag{7.38}$$

$$\sigma_{\text{ext}} = \frac{P_{\text{ext}}}{P_{\text{inc}}} = \frac{k}{\epsilon_0} \frac{\xi'' + \frac{k^3}{6\pi\epsilon_0}}{\xi'^2 + \left(\xi'' + \frac{k^3}{6\pi\epsilon_0}\right)^2}. \tag{7.39}$$

Now you can study the bounds, remembering that $\xi'' \geq 0$. Answers can be found in [17].

Bibliography

[1] P. B. Johnson and R. W. Christy, "Optical constants of noble metals," *Physical Review B* **6**, 4370–4379 (1972).

[2] M. A. Ordal, R. J. Bell, R. W. Alexander, Jr, L. L. Long, and M. R. Querry, "Optical properties of fourteen metals in the infrared and far infrared: Al, Co, Cu, Au, Fe, Pb, Mo, Ni, Pd, Pt, Ag, Ti, V, and W," *Applied Optics* **24**, 4493–4499 (1985).

[3] I. P. Kaminow, W. L. Mammel, and H. P. Weber, "Metal-clad optical waveguides: Analytical and experimental study," *Applied Optics* **13**, 396–405 (1974).

[4] A. Sihvola, J. Qi, and I. V. Lindell, "Bridging the gap between plasmonics and Zenneck waves," *IEEE Antennas and Propagation Magazine* **52**, 124–136, (2010).

[5] D. Sievenpiper, L. Zhang, R. F. J. Broas, N. G. Alexopoulos, and E. Yablonovich, "High-impedance electromagnetic surfaces with a forbidden frequency band," *IEEE Transactions on Microwave Theory and Techniques* **47**, 2059–2074 (1999).

[6] S. Tretyakov, *Analytical Modeling in Applied Electromagnetics*, Artech House, 2003.

[7] V. Klimov, *Nanoplasmonics*, Pan Stanford, 2014, pp. 128–130.

[8] M. Quinten, A. Leitner, J. R. Krenn, and F. R. Aussenegg, "Electromagnetic energy transport via linear chains of silver nanoparticles," *Optics Letters* **23**, 1331–1333 (1998).

[9] M. L. Brongersma, J. W. Hartman, and H. A. Atwater, "Electromagnetic energy transfer and switching in nanoparticle chain arrays below the diffraction limit," *Physical Review B* **62**, 16356(R) (2000).

[10] S. A. Tretyakov, A. J. Viitanen, "Line of periodically arranged passive dipole scatterers," *Electrical Engineering* **82**, 353–361 (2000).

[11] W. H. Weber and G. W. Ford, "Propagation of optical excitations by dipolar interactions in metal nanoparticle chains," *Physical Review B* **70**, 125429 (2004).

[12] S. A. Maier, P. G. Kik, and H. A. Atwater, "Optical pulse propagation in metal nanoparticle chain waveguides," *Physical Review B* **67**, 205402 (2003).

[13] C. R. Simovski, A. J. Viitanen, and S. A. Tretyakov, "Resonator mode in chains of silver spheres and its possible application," *Physical Review E* **72**, 066606 (2005).

[14] S. Steshenko, F. Capolino, P. Alitalo, and S. Tretyakov, "Effective model and investigation of the near-field enhancement and subwavelength imaging properties of multilayer arrays of plasmonic nanospheres," *Physical Review E* **84**, 016607 (2011).

[15] S. Maier, *Plasmonics: Fundamentals and Applications*, Springer Science+ Business Media LLC, 2007.

[16] J. N. Anker, W. P. Hall, O. Lyandres, N. C. Shah, J. Zhao, and R. P. Van Duyne, "Biosensing with plasmonic nanosensors," *Nature Materials* **7**, 442–453 (2008).

[17] S. Tretyakov, "Maximizing absorption and scattering by dipole particles," *Plasmonics* **9**, 935–944 (2014).

8 Building Blocks for All-Optical Signal Processing: Metatronics

8.1 Introduction

8.1.1 Paradigm of All-Optical Signal Processing

In modern telecommunication systems, transmission of signals to far distances is done using optical techniques (fiber optics), but processing of signals is still done electronically. Thus, *optoelectronic devices* are needed to convert information from the optical band to the band of radio frequencies and back. An advanced optoelectronic decoder is shown in Figure 8.1. Optoelectronic convertors, even advanced ones like that of [1], form a bottleneck in the modern telecommunication systems because [1–3]

- Best-known optoelectronic convertors lose about 30% of the signal energy converting electronic energy into photons and back.
- They slow the transmission of messages. The delay restricts the frequency range of converted electronic signals by 60–100 GHz [2, 3].
- They are comparable with electronic processors in what concerns their size, weight, and power consumption [1, 2].

Modern electronic processors can be mass produced with high quality and at low cost, and their miniaturization in the past decades was very successful. Deeply submicron diodes, transistors, and other electronic devices make modern integrated electronic processors lightweight and compact. However, it is difficult to reduce power consumption of electronic devices. The main reason is the irreducible level of noise inherent in active devices and even in passive radio and microwave devices at room temperature (see Section 1.1). To ensure the sufficiently high signal-to-noise ratio, electronic processors as well as optoelectronic convertors have to consume significant electric power. Meanwhile, *thermal optical noise* at room temperature is negligibly small in the range $\lambda = 1260–1675$ nm. External optical noises can be screened. As a result, the electric power required to feed an operational amplifier based on an *all-optical transistor* is lower by three orders of magnitude than that required for its electronic analog with the same signal-to-noise ratio [4]. This estimation can be extended to an *all-optical processor* and even to an *all-optical computer*. Imagine that your computer consumes 100 mW from a small battery instead of the present 100 W taken from the electric power supply, and you will understand how important it is to develop optical analogs of electronic devices [4].

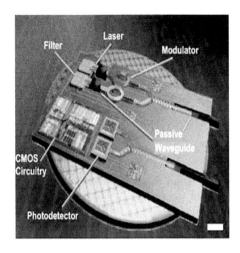

Figure 8.1 An advanced optoelectronic convertor (decoder). Reprinted from [1] with permission of the editors.

To reach this goal, one needs to create many new all-optical devices: all-optical switches, modulators, logic gates, memory cells, etc. This problem was formulated in 1980s (see, e.g., in [5]), and since 1980s, the main progress in this field has been related to a new scientific direction called *quantum computations* [3]. Generation, transmission, reception, and storage of quantum bits of information (called *qubits*) have been recently demonstrated experimentally [6]. Quantum computers will be super-compact. Most of them will be dedicated to work with optical signals, which implies very low power consumption. Such devices are called *linear optics quantum computers* [7, 8]. In this book, we discuss only one aspect of this wide field of research and development: so-called *metatronics*, which can potentially solve crucial problems for the creation of prospective *all-optical telecommunication systems*. The target of metatronics is all-optical signal processing before or after computations. The idea of metatronics is that all the electronic circuitry operating with alternating currents and voltages at radio frequencies can be emulated by optical circuitry operating with electric and magnetic fields of light. Metatronics offers an original way of tackling nano-optical data-processing tasks. In principle, metatronics intersects with optical quantum computing since it may also offer possibilities for nanoscale computations and data storage [9, 10]. However, presently, work is mainly concentrated on the processing of optical signals. Scientifically, metatronics merges circuit theory and design with nano-optics and uses the tools of metamaterials and plasmonics.

8.1.2 Metatronics Concept

The name metatronics was recently introduced by Nader Engheta with reference to the emulation of lumped circuit elements at optical wavelengths using building blocks of known metamaterials. In accordance with [10], metatronics is *metamaterial-inspired optical nano-circuitry*.

In microelectronics, the notion of a lumped circuit is a powerful concept in which a flow (current) of charges is related to another quantity (voltage) through the functions of lumped elements (resistor, inductor, capacitor, diode, transistor, etc.). The possibility of synthesizing circuits from lumped elements is the key difference between optics/photonics and radio engineering/electronics. Usual optical components operate with propagating waves and exploit either interference effects (lenses and phase diffraction grids) or require a long optical path for the signal (prism). One can say that the operation of optical device components is governed and mediated by wave propagation. For example, optical filters are either arrays of waveguides or stacks of dielectric and semiconductor layers. Any layer can be considered a piece of a transmission line for an incident plane wave. Therefore, an impedance matrix of any building block of any optical filter is that of a finite-length transmission line. Recalling these formulas, we may imagine how difficult it is to synthesize desired frequency dependence of the whole structure using such restrictive basic functions.

Meanwhile, in radio and microwave engineering, lumped circuit elements allow us to modularize a system. What is happening inside the element is not relevant to the connectivity and functionality of the element to the rest of the system. The notion of lumped elements is a powerful tool in discoveries of new functionalities, design, and innovations. The lumped L, C, R elements have one-element impedance matrices with basic frequency dependencies of their impedances – direct proportionality, inverse proportionality, and uniformity. Any frequency response to input signals can be approximated by polynomials engineered from these frequency dependencies. This is because the response of lumped elements is not restricted by the phases of waves. Why it is possible in electronics? Because electronics deals with conductivity currents and voltages. Voltage is an integral of the electric field over a small path. Conductivity current is a flow of electrons. Making the voltage relevant for the optical frequency range is possible – all we need is to make the element subwavelength. As to the conductivity currents, the inertia of electrons does not allow us to control them at optical frequencies in the same way as we do it at radio frequencies. However, if we recall Maxwell's equations for time-harmonic fields and currents,

$$\nabla \times \mathbf{H}(\mathbf{r}) = \mathbf{J} + j\omega \mathbf{D}(\mathbf{r}), \quad \nabla \times \mathbf{E}(\mathbf{r}) = -j\omega \mathbf{B}(\mathbf{r}), \tag{8.1}$$

we see (as Maxwell saw 150 years ago) that there is no principal difference between the density of the conductivity current $\mathbf{J}$ and that of the displacement current $\mathbf{J}_d = j\omega \mathbf{D} = j\omega \epsilon_0 \varepsilon_r \mathbf{E}$, where ε_r is the medium relative permittivity. The flow of photons is in this meaning similar to the flow of electrons.

Consider three types of subwavelength particles with absolute permittivity $\varepsilon = \epsilon_0 \varepsilon_p$, shown in Figure 8.2. The electric field inside an optically small particle determines the drop of voltage across it. The displacement current inside the particle, called polarization current, is equivalent to the conductivity current. Therefore, one may introduce local complex conductivity σ_{loc} inside the nanoparticle, writing $\mathbf{J}_d = \sigma_{\text{loc}} \mathbf{E}$:

$$\sigma_{\text{loc}} = j\omega \epsilon_0 \varepsilon_p. \tag{8.2}$$

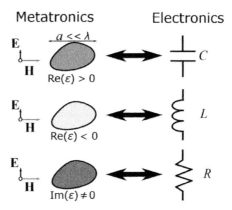

Metatronics   Electronics

Lumped optical elements emulating inductive, capacitive, and resistive electronic elements. Author's drawing, after [9].

Formula (8.2) establishes the point-wise relationship between the density of the polarization current inside the nanoparticle and the electric field inside it. Integrating the displacement current density over the cross section S of our particle, we may transit from $\mathbf{J}_d$ to the polarization current $\mathbf{I}_d$ in it and define the effective admittance and impedance of the particle. The current may be different in different cross sections, but we can calculate the averaged displacement current $\mathbf{I}_d^a$ integrating $\mathbf{I}_d$ along $\mathbf{E}$:

$$\mathbf{I}_d^a = j\omega\epsilon_0\varepsilon_p \frac{1}{l} \int_z \int_S \mathbf{E}\, dSdz. \tag{8.3}$$

Here, l is the particle length in the z-direction, and the axis z is parallel to $\mathbf{E}$. Since the drop of voltage V over the nanoparticle caused by field $\mathbf{E}$ is equal to

$$V = \int_z E_z dz, \tag{8.4}$$

we obtain from (8.3) an effective complex admittance Y_{int}, which relates the averaged polarization current I_d^a to V. This integral admittance of the nanoparticle, evidently depends on the shape of the particle. However, since the integral admittance is proportional to σ_{loc} (8.2), the following general observations can be made by just inspecting (8.2):

1. When $\varepsilon_p > 0$ and we can neglect the frequency dispersion of ε_p, the particle emulates a capacitor since $Y_{\mathrm{int}} \sim j\omega\epsilon_0\varepsilon_p$. This capacitance is physically related to the electric field created inside the particle by charges accumulated in every half-period near the extremities of the nanoparticle (for simplicity, here we neglect the fringe capacitance due to the electric fields outside of the particle volume).
2. When ε_p is close to an imaginary value $\varepsilon_p \approx -j|\varepsilon_p|$ (non-plasmonic metals and some semiconductors), the particle emulates a resistor with frequency-dependent resistance $R_{\mathrm{int}} \sim 1/(\omega\epsilon_0|\varepsilon_p|)$.
3. If the particle is made of a plasmonic metal, it can be considered an effective inductance. Really, its permittivity (in both the IR range and the long-wave

part of the visible range) is adequately described by the lossless Drude formula $\varepsilon_p \approx 1 - \omega_p^2/\omega^2$ (2.18). In the frequency range where $\omega \ll \omega_p$, we may approximately write $\varepsilon_p \approx -\omega_p^2/\omega^2$. After substituting this ε_p into (8.2), we find that the integral inductance is proportional to $L_{\text{int}} = 1/(\epsilon_0\omega_p^2)$ (times a shape-dependent factor). This inductance is determined by the magnetic flux inside the nanoparticle,[1] and in the optics of metals, it is called kinetic inductance (also, here, we neglect the inductance due to the magnetic flux outside of the particle volume).

The integration of the displacement current density results in additional geometrical factors for the integral values L_{int}, C_{int}, and R_{int}, but the frequency dependence for small particles is still determined by the local conductivity (8.2). The integral values of effective circuit parameters were derived for nanospheres, nanospheroids, nanodisks, and nanorods, in works [11, 12]. Thus, optical nanoparticles (as well as their dense clusters) successfully emulate L, C, and R electronic circuit elements.

8.1.3 The Problem of Optical Connectors in Metatronics

Submicron all-optical active devices not mediated by wave interference phenomena have been also suggested and studied (theoretically and experimentally). We can mention here subwavelength all-optical switches [13, 14], logical gates [15, 16], and diodes [17, 18], and even such advanced nanodevices as all-optical transistors [19, 20] and memory cells [21]. Nowadays, these lumped devices are under further development by several scientific groups. If we could have a set of efficient connectors for these ultimately small devices, we would realize all needed functionalities for all-optical information processing at subwavelength scale or at least at micron scale.

The key problem of metatronics is how to obtain connecting "wires" for photons [10]. Of course, these connectors can be pieces of optical waveguides – phase delays between components will not usually influence signal processing. However, optical waveguides that have sufficiently low dispersion and losses have substantial cross sections. Meanwhile, photonic "wires" should have subwavelength cross sections. Otherwise, the signal propagating in them will not see the subwavelength insertions in form of nanoparticles discussed earlier. Creation of subwavelength waveguides is a challenge for optics. Most known subwavelength optical waveguides are plasmonic ones. Plasmonic waveguides are narrowband, strongly dispersive, and lossy; see Chapter 7. At room temperature, optical signals in them decay over a few dozens of microns reducing below the noise level [22].

Since the key issue of metatronics is that of efficient optical connectors, in the following paragraphs, we concentrate specifically on it. Existing optical telecom operates in the band $\lambda = 1.4\text{--}1.7$ μm. In this band, silicon and fiber glass are practically lossless. This is the desirable frequency range for metatronics – in this case, a prospective all-optical processor based on metatronics will be compatible with the existing fiber optics (i.e., with Internet). However, in principle, metatronics can work also in the visible range or in the mid-IR range, where the low-loss atmospheric window enables the transfer of eye-safe laser signal in the form of diffraction-free beams [23]. In these cases, the frequency converters should be developed to connect processors and optical fibers. For the existing optical telecom, the waveguide cross section of the order of $\lambda/2$ and less is submicron one. Therefore, in the next section we call our subwavelength photonic "wires" *nanoguides*. The band of the efficient transmission of a nanoguide should cover the total band of optical signals under processing. Optical signals in the existing optical telecom systems require the relative bandwidth of the order of 7%–10%.

8.2 Epsilon-Near-Zero Metamaterial Platform for Metatronics

In electronics, ideal connecting wires that carry electrons from one circuit element to another are perfectly conducting wires. Electric current in wires is confined to wires simply because wires are in a nonconducting medium (free space or a dielectric). An ideal "connecting wire" for displacement current is free space, but what can be used as an isolator for displacement current? Just looking at the basic definition of displacement current, $\mathbf{J}_d = j\omega\epsilon_0\varepsilon_r\mathbf{E}$, we realize that in a medium with zero permittivity $\mathbf{J}_d = 0$; that is, such a medium is an ideal insulator for displacement current. Medium with zero permittivity grants $\mathbf{J}_d = 0$ and is an ideal insulator for the displacement current [24].

Another reason for considering the epsilon-near-zero platform is an interesting possibility of modifying the medium's effective permeability by immersing two-dimensional dielectric particles (infinite rods of arbitrary cross sections) in a two-dimensional epsilon-near-zero medium [25]. Such "doping" does not change the effective permittivity of the host, which remains near zero. The response of a large body can be tuned with a single impurity, including cases such as engineering *perfect magnetic conductor* and *epsilon-and-mu-near-zero* media with nonmagnetic constituents. Unfortunately, this concept is valid only for two-dimensional structures. Known microwave experiments [25, 26] use parallel-plate setups that emulate infinite uniform space by reflections in two parallel mirrors. Yet another interesting phenomenon associated with epsilon-near-zero media is wave tunneling through thin channels in metal bulk (at microwave frequencies, where metals are nearly perfect conductors) if the channels are filled by ENZ media [27, 28].

Media with zero permittivity do not exist in nature. However, metamaterials with a small relative permittivity can be, in principle, created even for the optical range. Both Lorentz and Drude types of dispersion allow the real part of the complex permittivity of a material to change the sign crossing zero at a certain frequency. Although at this frequency the imaginary part of ε_r is non-zero, it is not theoretically forbidden to have $|\text{Im}(\varepsilon_r)| \ll 1$ at the frequency where $\text{Re}(\varepsilon_r) = 0$, as well as to have both $|\text{Im}(\varepsilon_r)| \ll 1$ and $|\text{Re}(\varepsilon_r)| \ll 1$ in a certain frequency band. Papers [29–33] present results on realizations on ENZ media for optical applications, but more progress is needed to create low-loss composites needed for effective metatronics applications. Such prospective composite materials are called *epsilon-near-zero metamaterials* or *ENZ* metamaterials.

8.2.1 Hollow Waveguide with an Epsilon-Near-Zero Cladding

Let us consider a free-space gap between two half-spaces filled by an ENZ medium. For simplicity, let us study the ideal case when $\varepsilon_r = 0$ in the ENZ cladding. Any guided mode in this slab waveguide can be viewed as a superposition of two plane waves (Brillouin's waves), which are totally internally reflected from the two interfaces. In usual dielectric-slab waveguides, these waves have sufficiently large incidence angles to the interface. Meanwhile, a half-space of a medium with $\varepsilon_r = 0$ completely reflects propagating plane waves for any incidence angle. This fact immediately follows from Fresnel's reflection formulas for both TE- and TM-polarized waves if we nullify the refractive index of the reflecting medium. The only difference from a radio-frequency parallel-plate waveguide is the absence of the TEM mode in the ENZ waveguide. Working modes are either TE or TM modes.

In Figure 8.3(a), the distribution of the transverse components E_y and H_x of the lowest TM-mode is shown. The boundary conditions at the waveguide walls require $E_y|_{y=0} = E_y|_{y=b} = 0$. The axial electric field E_z and the transverse magnetic field H_x penetrate into the ENZ medium but strongly decay in it being the components of the TM-polarized surface waves excited at the waveguide interfaces. Formulas corresponding to the field distributions sketched in Figure 8.3(a) for the domain $0 < y < b$ are as follows:

$$E_y = Ae^{-j\beta} \sin \frac{\pi y}{b}, \quad H_x = \frac{\omega \epsilon_0 b}{\pi} Ae^{-j\beta} \cos \frac{\pi y}{b}. \tag{8.5}$$

Here, $\beta = \sqrt{k^2 - (\pi/b)^2}$ is the mode wavenumber. The cutoff wavelength is then $\lambda_c = 2b$. This waveguide is a single-mode one in the range $b < \lambda < 2b$, where the thickness of the waveguide is smaller than the wavelength in free space. A similar result holds for the TE-modes. If we replace the ambient with exactly zero permittivity $\varepsilon_r = 0$ with that with zero real part of ε_r and $|\mathrm{Im}(\varepsilon_r)| \ll 1$, the operation band of the waveguide does not change – in this case, $\lambda_c \approx 2b$. Such a waveguide can be really called a nanoguide because in the band of the optical telecom, $\lambda = 1.4–1.7\,\mu\mathrm{m}$, its thickness b is submicron.

For the main TM-mode in the epsilon-zero or ENZ waveguide, the magnetic field is antisymmetric with respect to the plane $y = b/2$, and we may identify the magnetic field averaged over the bottom half of the waveguide $0 < y < b/2$ with the effective surface current J_{eff} flowing through a given cross section $z = \mathrm{const}$ along the waveguide axis z. Respectively, the magnetic field averaged over the top half of the waveguide $b/2 < y < b$ is the effective surface current flowing though the same cross section z in the opposite direction. This way, we may treat the

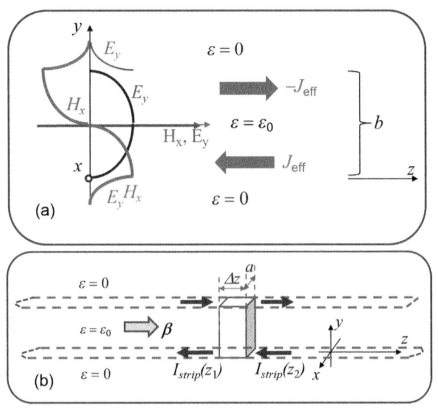

Figure 8.3 Two auxiliary problems: (a) where we deduce the subwavelength thickness b for a waveguide with the epsilon-zero cladding; and (b) where we prove the shunt admittance Y_{shunt} of a particle embedded into this waveguide to be proportional $j\omega\varepsilon_p$.

interfaces of this slab waveguide as effective surface current sheets similar to the case of perfectly conducting walls:

$$J_{\text{eff}} = \frac{2}{b}\int_0^{b/2} H_x \, dy. \tag{8.6}$$

The effective voltage between these two walls is evidently equal to

$$V_{\text{eff}} = \int_0^b E_y \, dy = 2\int_0^{b/2} E_y \, dy. \tag{8.7}$$

Now let us assume that a dielectric post with the relative permittivity ε_p and optically small thickness $\Delta z \ll \lambda_g = 2\pi/\beta$ and width $a \ll \lambda_g$ is introduced into such a waveguide, as it is shown in Figure 8.3(b). Let the z-coordinates of the front face 1 of the post be z_1, and for the rear face 2 of the post, we have $z_2 = z_1 + \Delta z$. Maxwell's equation gives us for the fields inside the post:

$$\frac{\partial H_x(y,z)}{\partial z} = j\omega\varepsilon_0\varepsilon_p E_y(y,z). \tag{8.8}$$

Averaging this equation over the post thickness Δz, we obtain

$$\frac{H_x(y, z_1 + \Delta z) - H_x(y, z_1)}{\Delta z} = j\omega\epsilon_0\varepsilon_p E_y^a(y), \tag{8.9}$$

where $E_y^a(y)$ is the electric field averaged over the post thickness at the given height y. Integrating (8.9) over y from 0 to $b/2$, we obtain in accordance with (8.6) and (8.7):

$$\frac{b}{2\Delta z}[J_{\text{eff}}(z_1 + \Delta z) - J_{\text{eff}}(z_1)] = j\omega\epsilon_0\varepsilon_p \frac{V}{2}. \tag{8.10}$$

Here, the voltage is averaged over the post thickness Δz and due to its smallness it can be referred to the post center $z = (z_1 + z_2)/2$.

Now let us consider the surface current flowing along the waveguide surface within an effective strip of width a stretched along z as shown in Figure 8.3(b). Integration of (8.10) over x within the strip (i.e., from $-a/2$ to $a/2$) gives the shunt admittance of the post:

$$I_{\text{strip}}(z_1 + \Delta z) - I_{\text{strip}}(z)(z_1) = Y_{\text{shunt}} V, \quad Y_{\text{shunt}} = j\omega\epsilon_0\varepsilon_p\xi, \tag{8.11}$$

where we have denoted

$$I_{\text{strip}} = \int_{-a/2}^{a/2} J_{\text{eff}}\, dx, \quad \xi = \frac{a\Delta z}{b}. \tag{8.12}$$

From (8.11), we can easily find the effective capacitance of the particle if it is performed of an epsilon-positive material: $C_{\text{eff}} = \epsilon_0\varepsilon_r\xi$. Epsilon-negative materials in the optical range are metals for which the Drude frequency dispersion holds. Neglecting the lossy factor and the relatively small term (-1) in the Drude permittivity formula (2.18), we may write $\varepsilon_r \approx -\omega_p^2/\omega^2$, which gives for the inductance length $L_{\text{eff}} = 1/\xi\epsilon_0\omega_p^2$. It is also easy to show that taking into account dielectric or metal losses makes the embedded particle equivalent to a parallel connection of either an inductor or a capacitor and a resistor.

Because the nanoparticle is effectively excited by fields created by currents flowing right at the position of the particle, it is not essential that the ENZ cladding plate has infinite extent. If the ENZ slab has a finite width w (which may be smaller than the wavelength and even close to a), the result (8.11) remains valid with a rather small correction to the value of ξ. Thus, we can state that an ENZ waveguide loaded by a small post in the optical range is equivalent to a two-wire line (as it is thought to be for low frequencies, where metal wires are close to perfect conductors) loaded by a shunt admittance: $Y_{\text{eff}} = j\omega\varepsilon_p\xi$. Here, the coefficient ξ is a positive value, depending on the post geometry and the waveguide width.

If we want to load an ENZ nanoguide by a resonant circuit, we can stack an epsilon-positive post with an epsilon-negative (plasmonic) one. The capacitance of the epsilon-positive part (dielectric) can resonate with the inductance of the plasmonic part of the dimer in the near-IR and even in the visible range if the permittivity of the dielectric is high enough. In Figure 8.4, there is an example of such, load experiencing resonance in the blue range, reproduced from paper [9],

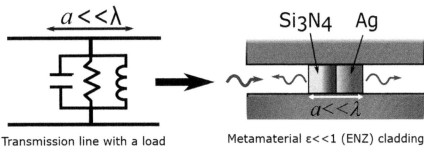

Figure 8.4 An ENZ nanoguide can be loaded so that it is equivalent to a two-wire transmission line shunted by a lumped parallel circuit. The corresponding post is a dimer formed by epsilon-positive (silicon nitride) and epsilon-negative (silver) particles. Author's drawing, after [9].

where the epsilon-positive material is silicon nitride (20 nm thick post) and epsilon-negative material is silver (also 20 nm). In the resonance band, the impact of the optical loss in silver is significant; therefore, the corresponding equivalent scheme comprises a resistor.

The series loading of an ENZ waveguide is also possible. For example, a small part of the epsilon-zero waveguide of length Δz can be made thinner than $\lambda/2$. Then, in this part, the waveguide is below the cutoff, and the main mode cannot propagate in it. However, if the length Δz is small enough ($\Delta z \ll \lambda_g$), the eigenmode can tunnel through this squeezed part, and it is possible to show that such waveguide patterning is equivalent to a capacitive load connected in series to an effective transmission line. Moreover, one can transform this series load into a resonant circuit filling the corresponding part of the waveguide with a plasmonic metal (negative-epsilon medium).

Although this approach is promising and opens up many unique possibilities for light processing at subwavelength scale, there are significant technological problems on the route toward practical implementations. Probably the most challenging issue is the creation of low-loss epsilon-near-zero claddings that would allow the propagation of waves along such ENZ waveguides to long distances [34–36]. For example, ENZ photonic wires for the use in mid-IR range described in paper [37] allow propagation only at distances of the order of the wavelength.

8.3 Free Space as a Possible Platform for Metatronics

The most straightforward solution for an optical connector is a wave beam propagating in free space. Gaussian beams in the visible range and near-IR ranges can be obtained with the effective width about 2−3 μm using conventional optics. If we drop the requirement of submicron sizes for all the components of metatronics, we can keep the main idea: optical signal processing at sub-wavelength scale without using optical interference phenomena. A few-micron collimation still grants to an *all-optical processor* a miniaturization compared to possible analogs utilizing conventional wave-interference approaches and

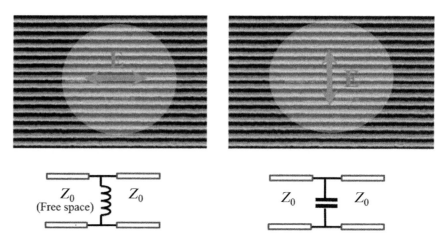

Figure 8.5 A metasurface of parallel nanowires may filter the signal with both polarizations of light, whereas the size of the illuminated spot restricts the minimal area of the metasurface. Left panel: the sheet impedance of the metasurface is inductive. Right panel: the sheet impedance of the same metasurface is capacitive.

existing optoelectronic convertors. Though this miniaturization will be not so drastic as that potentially granted by hypothetical lossless broadband ENZ media, it still offers compatibility with prospective all-optical devices that have area of a few square microns and a subwavelength thickness. Let us explain why and how this is possible.

First, consider the example where the device is a filter. A substantial width of the wave beam impinging this filter means that we cannot use nanoparticles for filtering. We need to find objects with corresponding transversal dimensions that are optically thin along the beam propagation direction (otherwise, the operation will be mediated by the wave interference) and can be described as lumped impedances: capacitances, inductances, and resistances. These objects are known – they are *metasurfaces*; see Chapter 4. Filtering can be performed by periodic metasurfaces such as grids of parallel metal or semiconductor nanobars/nanowires.

In Figure 8.5, we reproduce a typical atomic force microscope picture of such a metasurface illustrating two cases of the wave-beam transmission. In the left panel, the light of the normally incident wave beam is polarized in parallel to the nanowires and in the right panel the light is polarized transversely. The interaction of the illuminated part of the metasurface with a typical Gaussian wave beam is similar to that of the whole metasurface with a normally incident plane wave. The *sheet impedance* (also called grid impedance) of an optically dense grid of parallel metal wires is inductive when the electric field is polarized along the wires. When it is polarized across them, the sheet impedance is capacitive. Moreover, if the nanowires are made of a material experiencing resonant dispersion, such grid will be equivalent to a resonant circuit. If the electric field is polarized along the wires, this is a parallel circuit. Next, we analyze this case as an example.

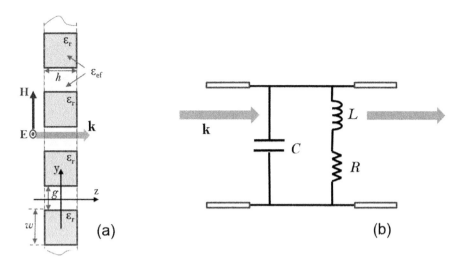

Figure 8.6 (a) Sectional view of the metasurface with resonant transmittance and reflectance. (b) Equivalent circuit of the metasurface.

8.3.1 Metasurface as a Parallel Resonant Circuit

Figure 8.6(a) shows a cross-section view of a metasurface operating as a stop-band filter at wavelengths $\lambda = 11\text{--}13\,\mu\text{m}$ [38]. In this example, the metasurface is a grid of parallel silicon nitride nanowires of thickness $h = 250$ nm and width $w = 225$ nm. The gap between adjacent nanowires is $g = 75$ nm. Let us show that its equivalent scheme is that presented in Figure 8.6(b) for the case when the electric field is polarized along nanowires. Here, the equivalent transmission line is free space that can be treated as a a TEM waveguide with the wave impedance $Z_0 = \eta$ and the wavenumber $k_0 = \omega/c$.

Consider the normal incidence of a plane wave on the grid depicted in Figure 8.6(a). Let the wave be polarized so that $\mathbf{H} = H\mathbf{y}_0$ is directed across the wires and $\mathbf{E} = E\mathbf{x}_0$ is directed along them. The wires are optically thin and have a rectangular cross section. The grid period $a = w + g$ is also optically small. Therefore, we may homogenize this grid, replacing it with a layer of the same thickness h and the averaged relative permittivity

$$\varepsilon_{\text{ef}} = \frac{\varepsilon_r \cdot w + 1 \cdot g}{a}, \tag{8.13}$$

where ε_r is the relative permittivity of silicon nitride. Then the jump of the magnetic fields across the grid can be found from Maxwell's equation for the effective dielectric layer with permittivity ε_{ef}:

$$\frac{\partial H}{\partial z} \approx \frac{H|_{z=+h/2} - H|_{z=-h/2}}{h} = -j\omega\epsilon_0\varepsilon_{\text{ef}}E. \tag{8.14}$$

Since h is optically small, the variation of E across the grid is negligible, and E in the right-hand side of (8.14) can be identified as the mean field at the central plane $z = 0$ of the effective layer. Meanwhile, the jump of the magnetic field across the grid in the left-hand side of (8.14) is substantial (due to large $|\varepsilon_r|$

in the right-hand side). In accordance with the Maxwell's boundary conditions, the jump of the tangential magnetic field across a thin sheet is equal to (with a negative sign) the surface current in this sheet. Here, we attribute this current to the central plane ($z = 0$) of the metasurface, and substituting (8.13) into (8.14), we obtain

$$J_g = Y_g E, \quad Y_g = j\omega\epsilon_0 h E \frac{\epsilon_r w + g}{a}. \tag{8.15}$$

The effective current J_g flows along the nanowires, and the scalar *sheet admittance* Y_g relates it to the electric field $\mathbf{E}$ having the same direction.[2] It is clear that the sheet admittance Y_g determined by (8.15) can be interpreted as $Y_g = Y_{nw} + Y_{ag}$ – i.e., a parallel connection of two admittances – that of a nanowire and that of the air gap between two adjacent nanowires:

$$Y_{nw} = \frac{j\omega\epsilon_0\epsilon_r wh}{a}, \quad Y_{ag} = \frac{j\omega\epsilon_0 gh}{a}. \tag{8.16}$$

Here, Y_{ag} is capacitive, and Y_{nw} is also capacitive if ϵ_r is non-resonant and positive.

However, silicon nitride is a semiconductor whose permittivity ϵ_r in the band $\lambda = 10$–$13\,\mu$m is essentially complex and experiences a resonance, approximated as follows [38]:

$$\epsilon_r = 1 + \frac{0.17}{1 - \left(\frac{\omega}{\omega_0}\right)^2 + j\gamma\frac{\omega}{\omega_0}}, \quad \frac{\omega_0}{2\pi} = 25 \text{ THz}, \quad \gamma = 0.14. \tag{8.17}$$

In the resonance band, unity can be neglected compared to the resonant term in the expression for ϵ_r. After the substitution of (8.17) into (8.16), we obtain the effective impedance of the nanowire $Z_{nw} = 1/Y_{nw}$ in the form

$$Z_{nw} = \frac{j\omega a}{\epsilon_0 wh\omega_0^2} + \frac{\gamma a}{\epsilon_0 wh\omega_0} = j\omega L + R, \tag{8.18}$$

which is a series connection of the effective inductance L and effective resistance R of the grid. It explains the equivalent circuit in Figure 8.6(b).

The amplitude transmission coefficient $T = 1 + R$ of the metasurface can be easily expressed through Y_g:

$$T = 1 + R = 1 + \frac{Y_0 - Y_{\text{input}}}{Y_0 + Y_{\text{input}}}, \quad Y_{\text{input}} \equiv Y_0 + Y_g, \tag{8.19}$$

where $Y_0 = 1/\eta_0 = \sqrt{\epsilon_0/\mu_0}$ is the admittance of free space and Y_{input} is the input admittance of the metasurface, which evidently results from a parallel connection of the sheet admittance and the admittance of free space. From (8.19), we obtain for the power transmittance of the metasurface:

$$\tau = |T|^2 = \left|\frac{1}{1 + Y_g\eta_0/2}\right|^2, \quad \eta_0 \equiv \sqrt{\frac{\mu_0}{\epsilon_0}}. \tag{8.20}$$

The minimum of $\tau_{\min} = 0.16$ is achieved when the value $|\epsilon_r|$ in Y_g attains its maximum – at the resonance of the circuit. Beyond the resonance band, $\tau \approx 0.9$.

Of course, the performance of this stopband filter is not stellar, but it is a good conceptual example that explains the equivalence of a resonant metasurface in free space to a resonant lumped load.

8.3.2 Metal-Dielectric Metasurface as a High-Order Filter

From Eq. (8.16), we see that the sheet impedance ($Z_g \equiv 1/Y_g$) of a solid ($g = 0$) dielectric ($\varepsilon_r > 1$) sheet is capacitive. If we substitute for ε_r the permittivity of a plasmonic metal adopting the lossless approximation ($1 - \omega_p^2/\omega^2$) and neglecting unity compared to ω_p^2/ω^2, we obtain an inductive grid impedance. Stacking a dielectric sheet and a plasmonic sheet results in a parallel connection of effective inductive and capacitive loads. Alternating dielectric and plasmonic strips in a single composite sheet, we obtain a series connection of L and C [40]. All these equivalent schemes are shown in Figure 8.7. Using this approach, one can engineer passband or stopband filters of the second or third orders – i.e., with practically acceptable slope of the amplitude-frequency characteristics at the band edges. An in-plane isotropic variant of a similar filter – with square plasmonic patches inserted into a high-permittivity sheet – was studied in [41]. Thus, the filtering functionality is really achievable with *filtering metasurfaces*, using free space as metatronics platform instead of nanoguides.

Finally, let us note that tightly focused wave beams (for optics, with the effective cross section about 2–3 μm), needed for metatronics devices based on the free-space platform, also can be formed by metasurfaces (see review [42] and Chapter 4). Using this approach, the need for conventional bulky lenses for beam focusing can be eliminated.

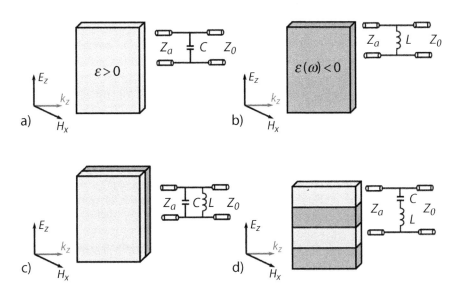

Figure 8.7 (a) The equivalent scheme for a dielectric sheet in free space. (b) The same scheme for a plasmonic sheet. (c) The same scheme for a stack of the dielectric and plasmonic strips. (d) The same scheme for a metasurface of alternating plasmonic and dielectric strips. Author's drawing, after [40].

8.4 Graphene Platform for Metatronics

A prospective all-optical processor utilizing wave beams in free space should comprise a lot of empty gaps inside it, and the level of miniaturization granted by such platform is not comparable with that promised by ENZ waveguides used as optical connectors. Since the progress in this field has been modest and the desire to engineer small all-optical chips is strong, the group of Engheta has theoretically (as of the time of writing) developed another platform for metatronics – that based on *graphene* – a stable two-dimensional crystal lattice (monolayer of carbon atoms) [43]. Due to its remarkable electronic, optical, and mechanical properties, many new applications have been suggested [44]. The use of graphene for metatronics was suggested in paper [45]. This application is based on the tunability of the surface conductivity of graphene in the terahertz and far-IR frequency regions. Thus, this type of metatronics (if or whenever it is implemented) will obviously demand frequency conversion of the near-IR signals carried by optical fibers.

Surface conductivity of graphene σ_g is a complex number defined by relation $J = \sigma_g E$, where E is the complex amplitude of the tangential electric field in the graphene sheet plane, and J is the surface current density. The term *surface conductivity* may be misleading because the surface admittance/impedance usually means the input admittance/impedance seen at the *surface* of a bulk body. In this case, however, surface conductivity models the response of a negligible-thickness sheet, relating the electric field at the sheet to the jump of the tangential magnetic field across it. It other words, this parameter is what we called earlier the *grid admittance* or *sheet admittance* of a metasurface. However, here we will adopt the terminology commonly used in the graphene research and call this parameter the surface conductivity.

For metatronics, it is important that σ_g is complex-valued and frequency dependent, and this dependence is controllable via the so-called *chemical potential* μ_c of graphene. Here we will not deviate into into solid-state physics and discuss the nature of the relationship between the chemical potential and the Fermi energy level E_F. For our purposes, it is enough to know that the chemical potential depends on the external static electric field (both normal and tangential components), which can be applied to graphene sheets using a DC voltage source. In Figure 8.8, the surface conductivity of graphene is depicted for several values of the Fermi level at zero absolute temperature [46]. These three values lie in the range of Fermi levels achievable due to practically applicable static electric fields. From Figure 8.8, we can expect that σ_g can be tuned to respond as an inductive or capacitive sheet by properly changing the bias voltage. For example, at 25 THz, the sheet is inductive and moderately lossy if $E_F = 0.06$ eV and capacitive and moderately lossy for $E_F = 0.08$ eV. Note, however, that the plot in Figure 8.8 is an idealization assuming zero temperature and approximate models of the graphene properties. In practice, it is usually easier to realize low-loss inductive response, and for realization of capacitive properties, metal-graphene metasurfaces may be preferable.

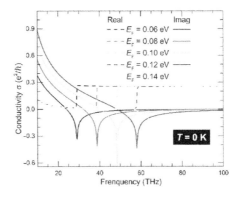

Figure 8.8 Imaginary and real parts of σ_g versus frequency for different values of the Fermi energy (i.e., different chemical potentials). Reprinted from [46] with permission of the authors.

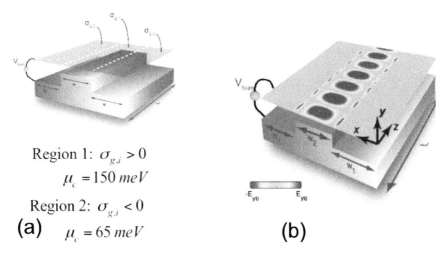

Region 1: $\sigma_{g,i} > 0$

$\mu_c = 150\ meV$

Region 2: $\sigma_{g,i} < 0$

(a) $\mu_c = 65\ meV$ **(b)**

Figure 8.9 (a) An effective wire of a graphene printed circuit board is an SPP waveguide. (b) Simulated color map of the electric field. Both figures are reprinted from [45] with permission of AAAS.

The potential use of nonhomogeneously biased graphene sheets for metatronics applications was discussed in paper [45]. The main idea is to bias different areas of a continuous graphene sheet with different bias voltages, creating areas with desired sheet impedances, either inductive or capacitive. Figure 8.9 shows a conceptual illustration of a graphene-sheet transmission line (a "wire"). Although the concept is rather attractive and promising, practical realizations, unfortunately, are very challenging. To the best of our knowledge, at this time, there are no experimental demonstrations of such devices. The main problems are in difficulties of realizing local biasing (the approach of varying substrate thickness, illustrated in Figure 8.9, requires excessively high bias voltages) and high level of resistive loss in graphene sheets. Note also that the initial simulations made in [45] assume cryogenic temperatures. Later studies have shown that it is possible to fabricate on a dielectric substrate so-called graphene *nanoribbons* operating at room temperature, whose inductive, capacitive, and resistive properties can be controlled via bias voltage [47–49]. In fact, these nanoribons are essentially nonlinear; however, their nonlinearity seems to be controllable [50, 51]. Experiments reported in works [50–52] promise that graphene-based metatronics devices may possibly work at room temperature. In the near future, we will see if it indeed is realistic.

8.5 Conclusions

Metatronics, which can be briefly defined as all-optical nano-circuitry, is a very exciting scientific direction. It promises us a novel class of ultrafast, low-power, broadband, and extremely small chips and processors. Its target is all-optical telecom without optoelectronic conversion and all-optical computers including those based on quantum effects. This target is still far from practical implementation. However, theoretically, it is achievable, and significant progress has been achieved on this path since the initial paper [9] was published. In this book, we have not reviewed nonlinear optical devices for metatronics – subwavelength optical diodes, transistors, modulators, etc. The essential point of metatronics is the possibility of properly connecting these devices. These connectors must also be compatible with all-optical filters. Thus, the *all-optical chip* must be compact – grant miniaturization as compared to conventional optical devices employing wave interference in dielectrics or in free space. We have seen that, currently, one develops three main platforms for connecting submicron all-optical devices:

- Nanoguides formed as free-space channels in ENZ background media. Filtering functionality is offered by nanoparticles inserted into these nanoguides.
- Free space where strongly collimated light beams propagate. Filtering functionality is offered by metasurfaces, and better connection to submicron devices is granted by focusing metasurfaces.
- Graphene sheets or graphene nanoribbons.

Among them, the most realistic platform is evidently the second one. However, it offers only limited miniaturization, and the use of conventional optics cannot be completely avoided. It also requires solving a lot of practical problems, especially concerning the compatibility of different devices in the presence of diffraction and other parasitic effects. This is probably the reason why this platform develops slowly. The third platform in what concerns the graphene nanoribbons develops fast. If this development is not blocked by any unexpected physical effect, one may possibly expect practical all-optical chips created in the next decade. As to the first platform, it seems for the instance more challenging. It is more difficult to believe in the commercialization of metatronics based on this platform in the future.

  In any case, at the present stage of these studies, we cannot reject any of the three possible technological platforms of metatronics. Perhaps, alternative, more practical or more advantageous platforms for metatronics will appear in the future. In our opinion, the target of metatronics – a practical, compact, and energy-efficient all-optical signal processor – definitely deserves more effort and resources than those invested in this science nowadays.

Problem and Control Questions

Problem

Using formula (8.15) for the grid admittance of an optically dense grid (metasurface) of parallel infinitely long dielectric nanobars of a material whose complex permittivity ε_r is described by formulas (8.17), design a stopband filter performed as a bilayer grid – two parallel identical metasurfaces separated by a certain gap d. The stopband is defined as the range of wavelengths in which the power transmittance does not exceed one-half of the maximal one. The explicit task is as follows:

1. Neglecting near-field interactions – i.e., assuming that in the space between the metasurfaces, there are only two plane waves (the fundamental Floquet harmonic), which is an adequate approximation if $d \geq h$ – derive the transmittance τ of a bilayer filter expressing it through Y_g and d.
2. Find analytically the optimal value for d that grants the minimal transmittance at λ_0, the wavelength at the center of the stopband.
3. Show numerically (e.g., in MATLAB) that the resonant ($\lambda_0 = 12\,\mu\text{m}$) transmittance for silicon nitride bilayer metasurface is lower than that for a single grid. Let the geometric parameters be equal $g = 75$ nm, $w = 225$ nm, $h = 250$ nm (see Figure 8.6).
4. Can the stopband of this bilayer filter be equal to that of the monolayer one? If yes, for which value of d?
5. Show numerically that the frequency dependence of the transmittance of the bilayer filter is closer to the ideal one (zero in the stopband and unity outside it) than that of the monolayer grid. What is the reason of this improvement? Hint – recall the resonance curve for two weakly coupled identical RLC-circuits.

Questions

1. Metatronics is an optical circuitry analogous to electronic circuits. What is an analog of an electron in metatronics: a light wave, a phonon, or a photon?
2. What is the main advantage of metatronics compared to the older idea of all-optical signal processing governed by wave processes?
3. Why does metatronics need ENZ nanoguides instead of already known plasmonic nanoguides? Why cannot a parallel-plate waveguides of silver plates operate in the optical range, similarly to its larger radio-frequency analog?
4. What are the advantages and disadvantages of graphene-based metatronics compared to metatronics based on wave beams in free space?

Bibliography

[1] V. J. Sorger, R. F. Oulton, R.-M. Maa, and X. Zhang, "Toward integrated plasmonic circuits," *MRS Bulletin* **37**, 728–738 (2012).

[2] D. D. Nolte, *Mind at Light Speed: A New Kind of Intelligence*, The Free Press (Simon and Schuster, Inc.) (2001).

[3] R. S. Tucker, "The role of optics in computing," *Nature Photonics* **4**, 405–410 (2010).

[4] E. Yablonovitch, "Why we need to replace the transistor, and what would be the newly required material properties?" *International Congress Lasers and Photonics'2016*, St. Petersburg, June 27–29, 2016, University ITMO, Book of Abstracts, p. 46.

[5] D. G. Feitelson, *Optical Computing: A Survey for Computer Scientists*, MIT Press (1988).

[6] J. J. Pla, K. Y. Tan, J. P. Dehollain, et al. "A single-atom electron spin qubit in silicon," *Nature* **489**, 541–545 (2012).

[7] P. Kok, W. J. Munro, K. Nemoto, et al. "Linear optical quantum computing with photonic qubits," *Reviews of Modern Physics* **79**, 135–174 (2007).

[8] D. Gevaux, "Optical quantum circuits: To the quantum level," *Nature Photonics* **2**, 337 (2008).

[9] N. Engheta, "Circuits with light at nanoscales: Optical nanocircuits inspired by metamaterials," *Science* **317**, 1698–1702 (2007).

[10] N. Engheta, "Taming light at the nanoscale," *Physics World* **23**, 31–34 (2010).

[11] N. Engheta, A. Salandrino, and A. Alú, "Circuit elements at optical frequencies: Nanoinductors, nanocapacitors, and nanoresistors," *Physical Review Letters* **95**, 095504 (2005).

[12] A. Polemi, A. Alú, and N. Engheta, "Nanocircuit loading of plasmonic waveguides," *IEEE Transactions on Antennas and Propagation* **60**, 4381–4390 (2012).

[13] I. Glesk, P. J. Bock, P. Cheben, et al."All-optical switching using nonlinear subwavelength Mach-Zehnder on silicon," *Optics Express* **19**, 14031–14039 (2011).

[14] B. Piccione, C. Cho, L. K. Van Vugt, and R. Agarwal, "All-optical active switching in individual semiconductor nanowires," *Nature Nanotechnology* **7**, 640–645 (2012).

[15] Z. Yang, Y. Fu, J. Yang, C. Hua, and J. Zhang, "Spin-encoded subwavelength all-optical logic gates based on single-element optical slot nanoantennas," *Nanoscale* **10**, 4523–4527 (2018).

[16] P. Singh, D. K. Tripathi, S. Jaiswal, and H. K. Dixit, "All-optical logic gates: designs, classification, and comparison," *Advanced in Optical Technologies* **14**, 275083 (2014).

[17] W. Li, B. Chen, C. Meng, et al. "Ultrafast all-optical graphene modulator," *Nano Letters* **14**, 955–958 (2014).

[18] Q. Yun-Ping, N. Xiang-Hong, B. Yu-Long, and W. Xiang-Xian, "All-optical diode of subwavelength single slit with multi-pair groove structure based on SPPs-CDEW hybrid model," *Acta Physica Sinica* **66**, 117102 (2017).

[19] X. Fang, K. F. MacDonald, and N. I. Zheludev, "Controlling light with light using coherent metadevices: All-optical transistor, summator and invertor," *Light: Science & Applications* **4**, 292 (2015).

[20] W. Chen, K. M. Beck, R. Bucker, et al., "All-optical switch and transistor gated by one stored photon," *Science* **341**, 768–770 (2013).

[21] E. Kuramochi and M. Notomi, "Optical memory: Phase-change memory," *Nature Photonics* **9**, 712–714 (2015).

[22] S. Bozhevolnyi, ed., *Plasmonic Nanoguides and Circuits*, Pan Stanford Publishing (2009).

[23] N. S. Prasad, "Optical communications in the mid-wave IR spectral band," *Journal of Optical and Fiber Communications Reports* **2**, 556–602 (2005)

[24] B. Edwards and N. Engheta, "Experimental verification of displacement-current conduits in metamaterials-inspired optical circuitry," *Physical Review Letters* **108**, 193902 (2012).

[25] I. Liberal, A. M. Mahmoud, Y. Li, B. Edwards, and N. Engheta, "Photonic doping of epsilon-near-zero media," *Science* **355**, 1058–1062 (2017).

[26] Z. Zhou, Y. Li, H. Li, W. Sun, I. Liberal, and N. Engheta, "Substrate-integrated photonic doping for near-zero-index devices," *Nature Communications* **10**, 4132 (2019).

[27] M. Silveirinha and N. Engheta, "Tunneling of electromagnetic energy through subwavelength channels and bends using ϵ-near-zero materials," *Physical Review Letters* **97**, 157403 (2006).

[28] R. Liu, Q. Cheng, T. Hand, et al., "Experimental demonstration of electromagnetic tunneling through an epsilon-near-zero metamaterial at microwave frequencies," *Physical Review Letters* **100**, 023903 (2008).

[29] A. V. Goncharenko and K.-R. Chen, "Strategy for designing epsilon-near-zero nanostructured metamaterials over a frequency range," *Journal of Nanophotonics* **4**, 041530 (2010).

[30] A. Monti, F. Bilotti, A. Toscano, and L. Vegni, "Possible implementation of epsilon-near-zero metamaterials working at optical frequencies," *Optics Communications* **285**, 3412–3418 (2012).

[31] P. Pinchuk and K. Jiang, "Broadband epsilon-near-zero metamaterials based on metal-polymer composite thin films," *Proceedings of SPIE* **9545**, 95450W (2015).

[32] R. Maas, J. Parsons, N. Engheta, and A. Polman, "Experimental realization of an epsilon-near-zero metamaterial at visible wavelengths," *Nature Photonics* **7**, 907–912 (2013).

[33] E. J. R. Vesseur, T. Coenen, H. Caglayan, N. Engheta, and A. Polman, "Experimental verification of $n = 0$ structures for visible light," *Nature Photonics* **7**, 907–912 (2013).

[34] M. H. Javani and M. I. Stockman, "Real and imaginary properties of epsilon-near-zero materials," *Physical Review Letters* **117**, 107404 (2016).

[35] A. J. Labelle, M. Bonifazi, Y. Tian, et al., "Broadband epsilon-near-zero reflectors enhance the quantum efficiency of thin solar cells at visible and infrared wavelengths," *ACS Applied Materials & Interfaces* **9**, 5556–5565 (2017).

[36] W. Lewandowski, M. Fruhnert, J. Mieczkowski, C. Rockstuhl, and E. Gorecka, "Dynamically self-assembled silver nanoparticles as a thermally tunable metamaterial," *Nature Communications* **6**, 6590 (2015).

[37] R. Liu, C. M. Roberts, Y. Zhong, V. A. Podolskiy, and D. Wasserman, "Epsilon-near-zero photonics wires," *ACS Photonics* **3**, 1045–1052 (2016).

[38] H. Caglayan, S.-H. Hong, B. Edwards, C. R. Kagan, and N. Engheta, "Near-infrared metatronic nanocircuits by design," *Physical Review Letters* **111**, 073904 (2013).

[39] Y. Sun, B. Edwards, A. Alù, and N. Engheta, "Experimental realization of optical lumped nanocircuits at infrared wavelengths," *Nature Materials* **11**, 208–2012 (2012).

[40] Y. Li, I. Liberal, and N. Engheta, "Engineering spectral dispersion with multi-ordered optical metasurfaces using insertion-loss method," *Proceedings of the CLEO International Conference on Quantum Electronics and Laser Science QELS'2016*, San Jose, CA, USA, June 5, 2016, paper FM2D.5(1–8).

[41] J. Cheng and H. Mossalaei, "Truly achromatic optical metasurfaces: A filter circuit theory-based design," *Journal of the Optical Society of America B* **32**, 2115–2120 (2015).

[42] S. B. Glybovski, S. A. Tretyakov, P. A. Belov, Y. S. Kivshar, and C. R. Simovski, "Metasurfaces: From microwaves to visible," *Physics Reports* **634**, 1–72 (2016).

[43] K. S. Novoselov, A. K. Geim, S. V. Morozov, D. Jiang, Y. Zhang, S. V. Dubonos, I. V. Grigorieva, and A. A. Firsov, "Electric field effect in atomically thin carbon films," *Science* **306**, 666–669 (2004).

[44] M. I. Katsnelson, *Graphene: Carbon in Two Dimensions*, Cambridge University Press (2012).

[45] A. Vakil and N. Engheta, "Transformation optics using graphene," *Science* **332**, 1291–1294 (2011).

[46] Y. Fan, F. Zhang, Q. Fu, and H. Li, "Harvesting plasmonic excitations in graphene for tunable terahertz/infrared metamaterials," in *Recent Advances in Graphene Research*, P.K. Nayak, ed., InTech Open (2016), doi:10.5772/61909.

[47] Q. Bao and K. P. Loh, "Graphene photonic, plasmonic, and broadband optoelectronic devices," *ACS Nano* **6**, 3677–3694 (2012).

[48] K. J. A. Ooi, H. S. Chu, L. K. Ang, and P. Bai, "Mid-infrared active graphene nanoribbon plasmonic waveguide devices," *Journal of the Optical Society of America B* **30**, 3111–3116 (2013).

[49] A. Tavousi, M. A. Mansouri-Birjandi, and M. Janfaza, "Optoelectronic application of graphene nanoribbon for mid-infrared bandpass filtering," *Applied Optics* **57**, 5800–5804 (2018).

[50] E. Carrasco, M. Tamagnone, J. R Mosig, T. Low, and J. Perruisseau-Carrier, "Gate-controlled mid-infrared light bending with aperiodic graphene nanoribbons array," *Nanotechnology* **26**, 134002 (2015).

[51] S. Zamani and R. Farghadan, "Graphene nanoribbon spin-photodetector," *Physical Review Applied* **10**, 034059 (2018).

[52] Q. Guo, R. Yu, C. Li, et al., "Efficient electrical detection of mid-infrared graphene plasmons at room temperature," *Nature Materials* **17**, 986–992 (2018).

Notes

1 See also Eq. (3.43).
2 The same result was obtained in [39] in a stricter and more difficult way.

Selected Topics of Optical Sensing

9.1 Introduction

9.1.1 Brillouin and Raman Radiations

Raman radiation, also called *Raman scattering*, was predicted by A. Smekal in 1923 in work [1] and experimentally revealed by C. Raman and his PhD student K. Krishnan [2] in 1928. This scattering effect represents one of the two known types of the so-called *combination* or *inelastic scattering*, which comprises also the so-called Brillouin radiation or *Brillouin scattering*. The last effect, predicted by L. Brillouin in 1922 in [3], was experimentally confirmed in 1926 by G. Landsberg and L. Mandelstam and thoroughly studied by them from 1926 to 1928 (see, e.g., in [4]). Common features of these two effects are new (absent in the incident wave spectrum) frequencies in the spectrum of the scattered electromagnetic field. If the source light is narrowband and can be considered as a monochromatic wave with frequency ω_L, inelastic scattering results in two new frequencies symmetrically located on the frequency axis with respect to ω_L with a relatively small shift Ω ($\Omega \ll \omega_L$).

Here, we give only a brief explanation of the Brillouin radiation as a macroscopic scattering phenomenon observed in solid dielectrics. Basically, it is parametric scattering by thermal oscillations of the crystal lattice. There are two types of these collective oscillations of the lattice nodes – *thermoacoustic waves* (also often called *thermal phonons*) and so-called *polarons*, which both result from thermal oscillations. Both of these groups of waves are stochastic fluctuations propagating in the crystal in all directions. Thermal phonons propagate with the speed of sound V_s, and polarons propagate with their specific speed V_p. These waves modulate the refractive index n of the crystal lattice in both time and space. The frequency range of these fluctuations is the same as that of thermal radiation – at room temperature, it is the far infrared range. When a light wave with frequency ω_L propagates in the crystal, it interacts with its space-time modulated lattice. This interaction is not important except for the case when the light wave and the wave of the modulated refractive index n are in phase synchronism. This spatial resonance shares out from the spatial spectrum of the lattice thermal fluctuations those waves that propagate under a certain angle θ with respect to the light wave. As a result of this parametric interaction, light waves at combination frequencies are born. This effect is called the Brillouin scattering.

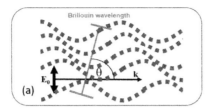 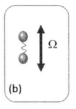 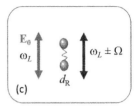

(a). Illustration of the Brillouin scattering. Scattering results from parametric modulation of the electromagnetic wave by a sound or polaron wave. The modulation creates two additional waves propagating obliquely to the source electromagnetic wavevector **k**. (b) and (c). The Raman scattering results from mixing the fundamental molecular vibration frequency Ω with the frequency of the electromagnetic wave ω_L.

Figure 9.1(a) illustrates the phase synchronism of a short-wavelength light wave and long-wavelength sound wave for some angle θ. Here, the picture plane is the cross section of the wave cone corresponding to the Brillouin scattering. One of these additional waves, that with the frequency $\omega_L + \Omega$, propagates under a sharp angle θ to the source wave. The other one, that with the frequency $\omega_L - \Omega$, propagates in the opposite direction. The frequency shift is given by formula $\Omega = 2\omega_L V_{s,p} \sin(\theta/2)/c$, where $V_{s,p}$ is the velocity of thermal-oscillation waves (phonons or polarons, respectively).

The target of our short explanation is to clarify that the Brillouin radiation is a *macroscopic type of inelastic scattering*. It is inelastic because two waves at two new frequencies, $\omega_L \pm \Omega$, are generated. These macroscopic effects may occur only in samples comprising a sufficient number of the lattice unit cells. Since this effect is macroscopic, it does not refer to nanophotonics, and we do not discuss it anymore.

In contrast to the Brillouin scattering, the Raman scattering is a *microscopic type of inelastic scattering*. It arises due to interaction of light with vibrational oscillations of individual molecules. The dipole moment induced in a molecule by the local field oscillating at the light source frequency ω_L also experiences oscillations at the frequency $\Omega \ll \omega_L$ of a vibrational eigenmode of a molecule. The molecule has many vibrational eigenmodes, and, depending on the molecule symmetry, one or more modes can interact with the incident light [1]. In practice, the main combination frequency (which corresponds to the smallest frequency shift) is usually the most important since it has the highest amplitude. In some models, this mixing is explained as a parametric modulation of the usual linear scattering; in some other models, it is considered as a nonlinear effect. Parametric mixing models assume that the properties of the scattering molecule vary in time due to mechanical vibrations of the molecule and that these vibrations exist independently of the presence of external light illumination (thermal oscillations). Nonlinear models assume that the scattering properties of the molecule are changed under light illumination.

To explain nanophotonic applications of Raman scattering, we use the parametric model and assume that the dipole moment of a molecule in the local field $\mathbf{E}_0 \sin \omega_L t$ can be expressed as

$$\mathbf{d}(t) = \left\{ \alpha_0(\omega_L) \sin \omega_L t + \alpha_S \sin[(\omega_L - \Omega)t + \Phi] + \alpha_{AS} \sin[(\omega_L + \Omega)t + \Psi] \right\} \mathbf{E}_0,$$

$$(9.1)$$

where α_0 is the usual polarizability of a molecule in the linear electromagnetic regime, whereas α_S and α_{AS} are called *Stokes molecular polarizability* and *anti-Stokes molecular polarizability*, respectively [5]. Both Stokes and anti-Stokes polarizabilities are also called *Raman polarizabilities*. They can be calculated for known molecules using different known quantum models of molecular vibrations in the field of an electromagnetic wave [6, 7]. If the impact of the environment is negligible, these models are equally and sufficiently accurate [8, 9].

The redshifted Raman frequency $\omega_R^- = \omega_L - \Omega$ is called the *Stokes frequency* and the blueshifted Raman frequency $\omega_R^+ = \omega_L + \Omega$ is called the *anti-Stokes frequency*. This terminology is related to the so-called *Stokes law*, postulated in 1852. In accordance with this obsolete law, the secondary radiation (scattering) in optics may occur only at frequencies either equal to or lower than the source light frequency ω_L. This law is correct for the linear (Rayleigh) scattering when the frequency of the scattered field is the same. It is also correct for the process of molecular secondary radiation called *phosphorescence* when the photon is absorbed by a molecule, and a part of its energy is spent on the emission of a secondary photon some time after the absorption.

A similar process called *fluorescence* is molecular reradiation, which occurs immediately after the absorption of the photon and therefore ends in one nanosecond or a few nanoseconds after the end of the source radiation. The Stokes law only holds for the linear fluorescence and does not hold for the nonlinear one. The nonlinear fluorescence as well as the Raman and Brilloun types of scattering, for which this law fails, were not known in 1852.

The regime when the amplitude of the Raman dipole moment of a molecule is directly proportional to the amplitude of the local field at the source frequency is called *spontaneous Raman scattering*. In this regime, the molecules emit the Raman radiation independently of one another with random phases Φ and Ψ in (9.1). These phases slowly (as compared to the light wave oscillations) vary in time so that the Raman signals are not coherent. This property is in harmony with the assumption of the parametric nature of the mixing process because, in this case, the phases of the combination frequency components are the sums or differences of the phase of the incident light and the initial phase of the vibrational oscillations (which is random and does not depend on the incident light). However, in many literature sources, both known types of the Raman scattering – spontaneous and stimulated – are considered as intrinsically nonlinear processes. In the following paragraphs, we dedicate a special discussion to this issue.

The Raman polarizabilities α_R that correspond to the dipole moment of a molecule $\mathbf{d}_R$ at one of the two main Raman frequencies (Stokes and anti-Stokes ones), in accordance with (9.1), do not depend on $\mathbf{E}_0$. Both *elastic scattering* (Rayleigh scattering) and inelastic Raman scattering of the incident light by a molecule are illustrated by the same drawing in Figure 9.1b. When α_R is a scalar,

$\mathbf{d}_R$ is directed along $\mathbf{E}_0$ as well as the main dipole moment (at the frequency ω_L). Therefore, in this case, the pattern of Raman scattering at both frequencies is identical and is the same as that of the Rayleigh scattering. It is another crucial difference between the Raman scattering and the Brillouin scattering, which is highly directive, and the waves at two combination frequencies propagate in opposite directions.

For arbitrary local field $\mathbf{E}_{\text{loc}}(\omega_L)$ acting on a molecule (not obviously that of a plane wave) we may write

$$\mathbf{d}_R(\omega_R) = \alpha_R \mathbf{E}_{\text{loc}}(\omega_L), \tag{9.2}$$

where α_R depends on neither $\mathbf{E}_{\text{loc}}$ nor ω_L if the source light frequency is in the visible range [10].

9.1.2 Spontaneous Raman Radiation: Is It a Nonlinear or a Parametric Effect?

Raman scattering used in surface-enhanced Raman scattering (SERS) (except some advanced schemes) is *spontaneous Raman scattering*, when the phases Φ and Ψ in (9.1) for a given molecule illuminated by a source wave (a flux of photons) randomly vary versus time. As we discussed in Section 9.1.1, this property is expected for parametric processes. The linear dependence of the dipole moment on the incident field magnitude is an evident argument against the nonlinear nature of the effect. However, in the dominant literature, starting with works [11–13], spontaneous Raman scattering has been commonly attributed to nonlinearity of molecular vibrations. For example, in popular books [7, 8, 14–18], this opinion, linking the spontaneous Raman scattering to *Kerr nonlinearity*, has been promoted. Authors of [19] explain how different nonlinearities of the molecular response turn out to be compensated in both Stokes and anti-Stokes polarizabilities, keeping both of them non-zero. First, the nonlinearity results in the generation of the second harmonic ($2\omega_L$). Second, the nonlinear mixing of this frequency with the anti-Stokes component results in the difference frequency $2\omega_L - (\omega_L + \Omega)$ – i.e., the Stokes frequency. The impact of these two nonlinearities cancels out, and the Stokes polarizability turns out to be independent of the intensity. A similar compensation of nonlinearities (involving the so-called *four-wave mixing*) is suggested in [19] for the anti-Stokes component.

In fact, there is also a classical model of the Raman radiation (see, e.g., in [20]). It considers a molecule as a classical oscillator, and in this model, the Raman radiation is purely a parametric effect. Its main drawback is the qualitatively incorrect prediction for the ratio between the intensities of the Stokes (rather strong) and anti-Stokes (very weak) components of the spontaneous Raman radiation. However, this classical model can be improved by using the classical molecular physics. In books [21, 22], the authors claim that an only deficient point of the classical model in work [20] is the deduction of (9.1) from the product $\alpha(t) \sin \omega_L t$, where $\alpha(t) \sim \sin(\Omega t + \phi)$ describes the linear polarizability of a molecule modulated by its harmonic vibration. It is not so simple because the molecule is not a single-state oscillator like a pendulum or an *LRC* electric circuit.

It is an oscillator governed by the *Maxwell–Boltzmann statistics*. This statistics determines the populations of the unit energetic intervals versus the oscillator energy. The energy of the anti-Stokes component is higher than the Stokes one, and, therefore, the corresponding energetic level is populated much weaker. This observation qualitatively explains why the ratio of the anti-Stokes component of the Raman signal to the Stokes one is much smaller than unity for ensembles of molecules. Thus, in accordance with classical models [21, 22], the parametric insight of the spontaneous Raman scattering is fully adequate.

It is important to note that there is also a coherent type of Raman scattering called *stimulated Raman scattering*, which is an essentially nonlinear process. It occurs when the local field is very high and strongly monochromatic. Parameter α_{AS} in (9.1) comprises a part proportional to the Kerr (third-order) nonlinearity. In this case, the four-wave mixing process really arises in a molecule that probes the Kerr nonlinearity and results in synchronization of the anti-Stokes oscillations with the local field oscillations. When the local field is high enough, all molecules of the ensemble synchronize and oscillate in phase.

In other words, the phase Ψ in (9.1) is stable in time – i.e., the anti-Stokes component of Raman radiation is coherent. Coherence allows this radiation component to grow significantly and become comparable to (or even higher than) the Stokes one. This effect is widely used for the enhancement of the operation characteristics of the so-called *vibrational atomic force microscopes* [23–25]. Moreover, when molecules emit coherent radiation, they stimulate the Raman radiation of one another, which may result in the increase of the Raman radiation. This increase implies that the induced radiation prevails over the spontaneous radiation. The growth rate is restricted by the nonlinearity of the medium. The energy of the source radiation transfers into the energy of the anti-Stokes Raman radiation. The structure where this process occurs is called a *Raman laser*. Notice that this regime is possible only at the anti-Stokes frequency; the nonlinearity of α_S is insufficient. The Stokes component remains spontaneous and incoherent.

As to spontaneous Raman scattering, there is no unique point of view in the scientific community. The classical model of [21, 22] still does not give an adequate prediction for α_{AS} in (9.1), and the anti-Stokes component of Raman radiation can be correctly predicted only by the quantum models. These models (though different in different works) are equivalent in what concerns the result. Most of them is based on the concept of so-called *virtual states*, also called vibronic or Herzberg–Teller states. This concept, implying both electromagnetic and mechanic nonlinearity of the molecular vibrations, was introduced into the theory of spontaneous Raman scattering by Shorygin [11]. However, a quantum model of spontaneous Raman radiation suggested in later work [6] describes it as a parametric process without involving virtual states, and the authors of [26] support this description. A unique point of view was elaborated by the scientific community only for stimulated Raman scattering – nonlinear in both quantum and classical models.

9.1.3 Surface-Enhanced Raman Scattering for Molecular Sensing

Up to the year 1974, the Raman scattering was considered as an important tool for optical analysis, applicable only to statistically large molecular ensembles called *analytes*. Analytes were studied (usually) in liquids and (rarely) in solid specimens. Spectra of Raman radiation in ensembles of different molecules represent sets of clearly outlined and very narrow spectral lines whose pairs are fingerprints of specific molecules. The molecular content of the analyte and concentrations of specific molecules in it are easily retrieved from the spectroscopic analysis of the intensity of Raman scattering. Specialists in optical sensing tried to enhance this important radiation to increase sensitivity and analyze smaller analyte volumes. However, the goal to detect and locate separate molecules was not topical at that time.

In 1974, Martin Fleischmann [27, 28][1] managed to drastically enhance the Raman radiation of pyridine molecules using a properly prepared substrate. Fleischmann's idea was to increase the Raman signal, increasing the amount of excited molecules. Since organic molecules in ambient water attach to the surface of silver, Fleischmann and his students poured a water solution of pyridine on a silver substrate. Earlier, flat silver substrates had already been studied, and it was understood that for molecules attached to the interface silver/air the Raman radiation enhances nearly twice for grazing incidence of the source light, decreases for small incidence angles and vanishes for the normal incidence. Thus, a flat metal surface, even in the optical range, behaves quite similarly to a perfectly conducting plane. In the experiment reported in [29], the expected mechanism of further enhancement of Raman radiation was the increase of the amount of molecules attached to the silver surface. For this purpose, the surface was corrugated using chemical etching to maximally increase the surface area. It was expected that corrugations resulting from chemical etching should transform the flat surface into a densely packed array of nearly hemispherical protrusions with the characteristic sizes 10–100 nm. Such corrugations should have multiplied the area of the flat interface silver/air by nearly 2.5. Meanwhile, it was expected that some molecules would be shadowed by the protrusions, which could decrease the gain compared to 2.5. However, this shadow effect could not be estimated easily because the protrusions were smaller than the wavelengths of the visible light, and the geometric optics was not applicable. Thus, only an experiment could show the actual gain in the Raman radiation compared to that of the flat surface.

However, the result of the experiment reported in work [29] was striking: this gain turned out to be tenfold. The unexpected increase of Raman radiation was attributed in [29] to presumable formation of many spherical nanoparticles that could separate from the silver substrate due to roughening and could form a dense multilayer array of nanospheres, increasing the effective surface of silver by nearly 10 times. The result obtained in [29] manifested the start of the era of *surface-enhanced Raman scattering* (SERS) in the optical sensing.[2]

Further studies of SERS performed since 1974 have shown that its explanation in [29] was incorrect. The amount of separate nanoparticles obtained in the first experiments was minor, and they did not form densely packed multilayer arrays. Moreover, the tenfold increase of the Raman radiation in [29] was only an integral gain averaged over the whole surface. The majority of the pyridine molecules were either located in the flat gaps between the plasmonic protrusions of size 10–100 nm or attached to very small protrusions, smaller than 10 nm, which were also present. It was understood that at least one of the mechanisms of the enhancement is plasmon resonance experienced by substantial (10–20 nm and larger) protrusions and separate nanoparticles, whereas protrusions smaller than 10–20 nm do not experience plasmon resonance. For molecules located beyond substantial protrusions, their Raman radiation should have increased weakly or not at all. From the integral gain, which in further experiments was increased drastically compared to the initial tenfold enhancement (because the amount of plasmonic protrusions was increased), one managed to estimate the order of magnitude for the gain in the intensity of Raman radiation for molecules located on the protrusions experiencing plasmon resonance. It turned out to be the value of the order 10^5–10^6. This estimate retrieved from the experimental results was in line with the theoretical models developed in 1979–1982 by several groups of researchers. Then it was understood that SERS is potentially capable of detecting and locating separate molecules in real time. Unlike an electron microscope requiring long scanning procedures, this sensing enabled a simultaneous tracking of many molecules located at special points, for which a huge enhancement of Raman radiation is expected [30]. For example, a molecule located on top of a plasmonic nanopyramide, as depicted in Figure 9.2(a), produces Raman radiation with the intensity nearly 10^7 higher than the same molecule does in free space [31]. Such signals can be detected for a quite small ensemble of molecules located on the nanopyramides. Presently, several advanced SERS substrates are known that grant a possibility of detecting separate molecules. At the end of this chapter, we discuss these substrates. In the next section, we concentrate on the enhancement of the local field granted by the plasmon resonance. It is illustrated by Figure 9.2(b).

In early 1980s, two competing theories explaining the huge gain granted by silver-based and gold-based SERS were developed and have resulted in two bodies of mutually controversial literature. One theory, supported mainly by physicists (see, e.g., in [5, 32–35]), presented the gain in the Raman radiation granted by SERS as basically an electromagnetic effect. Though the Raman radiation can be slightly modified by chemical processes, the order of magnitude for the *Raman gain* in SERS is completely explained by the electromagnetic model. Another party, mainly electrochemists, promoted an alternative theory where the Raman gain in SERS equally resulted from the electromagnetic and chemical factors of the same magnitude. The chemical enhancement factor in this model originated from the effect of so-called *charge-transfer complexes* (see, e.g., in [36–41]). In the next section, we concentrate on the electromagnetic model and briefly discuss the charge-transfer complexes in the chapter appendix.

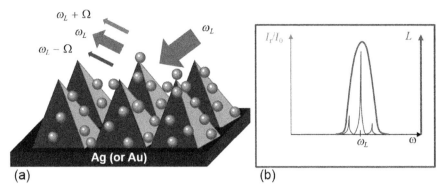

Figure 9.2 Concept of SERS as a tool for sensing molecules. (a) The sketch showing molecules as spheres on pyramidal plasmonic protrusions. (b) A typical intensity spectrum of the signal reflected from the SERS substrate shown together with the local field enhancement. The light source frequency ω_L coincides with the plasmon resonance frequency of the protrusions, and the plasmon resonance band corresponding to the enhancement of the local field covers both Raman frequencies.

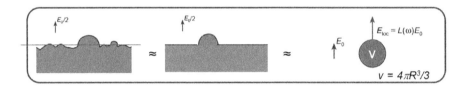

Figure 9.3 A hemispherical plasmonic protrusion that arises on a flat silver or gold surface due to roughening when illuminated by light with vertical polarization is nearly equivalent to a plasmonic sphere in the external field $\mathbf{E}_0(\omega_L)$. A molecule located on this protrusion is excited by local field enhanced $L(\omega_L)$ times compared to $E_0(\omega_L)$.

9.2 Electromagnetic Model of SERS: Local Field Enhancement

In order to understand the nature of surface-enhanced Raman scattering (SERS), let us consider a chemically roughened surface of silver with protrusions whose size is varying from 1–2 nm to 50–100 nm. Corrugations smaller than 10 nm exhibit no localized plasmonic resonances, and the part of the surface covered by such small protrusions can be approximately considered as a flat surface. These flat areas surround isolated protrusions with radii sufficient for plasmonic resonance ($5 < R < 50$ nm). Consider such an isolated protrusion of hemispherical shape on a flat surface of metal, as shown in Figure 9.3. The diameter of the protrusion smaller than 100 nm allows us to neglect the high-order multipoles in the description of the protrusion polarization response. Thus, in the conventional SERS schemes, the plasmonic protrusions are behaving as dipole scatterers.

Let the incident electric field $\mathbf{E}^i$ be polarized vertically as it is in the case of grazing incidence. Although in practice the incidence of illuminating light wave is not grazing, the horizontal component of the incident field is not important in this type of SERS. Though silver is not a perfect conductor, the skin depth is small and the mirror image principle is approximately applicable to the flat surface on which the silver hemisphere is located. Thus, we may replace the hemispherical protrusion on the flat silver by an entire silver sphere illuminated by the external field $E_0 = 2E^i$. The enhancement of the Raman radiation by the roughened surface is calculated with respect to the reference flat surface. Therefore, we treat E_0 as the amplitude of the source wave.

For the dipole moment of the sphere d_{sphere} of radius R and relative permittivity ε_{met}, we have (see Chapter 7):

$$d_{\text{sphere}} = 4\pi\epsilon_0 R^3 \frac{\varepsilon_{\text{met}} - 1}{\varepsilon_{\text{met}} + 2} E_0. \tag{9.3}$$

Vectorial notations are not relevant here since all the vectors are vertical. Recall that the bulk polarization density P_{met} of any medium described by permittivity is proportional to the electric field inside the medium; i.e., $P_{\text{met}} = \epsilon_0(\varepsilon_{\text{met}} - 1) E_{\text{inside}}$ (this relation is a rewritten definition of permittivity through the relation $\mathbf{D} = \epsilon_0 \varepsilon_{\text{met}} \mathbf{E} = \epsilon_0 \mathbf{E} + P_{\text{met}}$). Since the dipole moment of a sphere is the product of its volume v and its bulk polarization P_{met}, we have

$$d_{\text{sphere}} = v P_{\text{met}} = \frac{4}{3}\pi R^3 \epsilon_0(\varepsilon_{\text{met}} - 1) E_{\text{inside}}. \tag{9.4}$$

Comparing (9.3) and (9.4) allows us to express the electric field inside the sphere through the external field E_0:

$$E_{\text{inside}} = \frac{3E_0}{\varepsilon_{\text{met}} + 2} \equiv A(\omega)E_0, \tag{9.5}$$

where we have denoted $A(\omega) = 3/(\varepsilon_{\text{met}} + 2)$.

A molecule located on top of the sphere is excited by the local field E_{loc} created by the whole environment. To express E_{loc} through the internal field E_{inside}, we can notice that on the top of our sphere, both vectors $\mathbf{E}_{\text{loc}}$ and $\mathbf{E}_{\text{inside}}$ are directed normally to the metal–air interface. Then E_{loc} and E_{inside} are related via the continuity of the normal component of the electric displacement vector $\mathbf{D}$. In other words, we have the relation

$$\epsilon_0 E_{\text{loc}} = \epsilon_0 E_{\text{inside}} \varepsilon_{\text{met}}. \tag{9.6}$$

Thus, in the local field, the external field of amplitude E_0 and frequency ω_L is enhanced $L(\omega_L)$ times:

$$E_{\text{loc}} = A(\omega_L)E_0 \varepsilon_{\text{met}}(\omega_L) \equiv L(\omega_L)E_0, \quad L(\omega_L) = \frac{3\varepsilon_{\text{met}}(\omega_L)}{\varepsilon_{\text{met}}(\omega_L) + 2}. \tag{9.7}$$

Frequency dispersion of ε_{met} in the visible frequency range is discussed in Chapter 7. There is a frequency of the localized surface plasmon ω_{res} at which $L(\omega)$ is maximal. At this frequency, $\text{Re}(\varepsilon_{\text{met}}) \equiv \varepsilon'_{\text{met}} = -2$, whereas the imaginary part of the complex permittivity $\varepsilon''_{\text{met}}$ is quite small. In the band of the plasmon, resonance of a sphere L for silver attains nearly 60.

In plasmonics, factor $L(\omega)$ is called the *local field enhancement* (LFE) factor. Usually, instead of the complex value $L = E_{\text{loc}}(\omega_L)/E_0$ one considers the absolute value of this ratio, also denoted as L. LFE shows how much the electric field

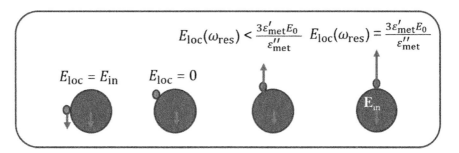

Figure 9.4 Illustration of the averaging of the locality factor over all possible positions of a molecule on the surface of the spherical protrusion.

acting on the molecule is enhanced due to the presence of the protrusion. Formula (9.7) gives the LFE only for the special case when the molecule is located on top of a protrusion. In this case, the local field is maximal. A more detailed study allows one to derive $L(\omega)$ for any position of the molecule. Inspecting other possible positions, we notice that the local field decreases when the molecule goes down from the top of the protrusion to its equatorial plane (that corresponds to the flat interface). There is even a point at which L is close to zero. This is so because the real part of the permittivity of metal is negative. Therefore, as it follows from (9.6), the local field $\mathbf{E}_{\text{loc}}$ on top of the sphere is directed oppositely to $\mathbf{E}_{\text{inside}}$. However, for the equatorial location of the molecule on the sphere both $\mathbf{E}_{\text{inside}}$ and $\mathbf{E}_{\text{loc}}$ are tangential to the surface of the *plasmonic nanoparticle* (PNP) and the continuity of the tangential components of the electric field gives for this case $E_{\text{inside}} = E_{\text{loc}}$.

This means that $\mathbf{E}_{\text{loc}}$ changes sign if the molecule at the sphere surface passes from the top to the equatorial plane. Therefore, there is a point when the local field vanishes and the position-averaged L must be obviously smaller than the maximal L corresponding to the location of the molecule on the top of the sphere. These speculations are illustrated by Figure 9.4. Omitting a simple but rather long derivation, the result for the position-averaged field enhancement factor can be written in the following form:

$$L(\omega) = \left| \frac{\varepsilon_{\text{met}}}{\varepsilon_{\text{met}} + 2} \right|, \quad L(\omega_{\text{res}}) = \frac{|\varepsilon_{\text{met}}(\omega_{\text{res}})|}{\varepsilon''_{\text{met}}(\omega_{\text{res}})}. \tag{9.8}$$

In other words, the position-averaged LFE is triply smaller than LFE on top of the protrusion. For silver, the position-averaged LFE at the plasmon resonance is close to $60/3 = 20$.

In accordance with (9.2) (where E_0 is replaced by E_{loc}), factor $L(\omega_{\text{res}})$ shows the gain in the Raman dipole moment if $\omega_L = \omega_{\text{res}}$:

$$d_R(\omega_R) = \alpha_R E_{\text{loc}}(\omega_L) = \alpha_R L(\omega_{\text{res}}) E_0. \tag{9.9}$$

The intensity of Raman radiation is proportional to the square power of the radiating dipole moment. If we admit that we measure in SERS the radiation produced by the molecules, the Raman gain should be equal to $G_R = |d_R/d_{R0}|^2$, where d_{R0} is the Raman dipole moment in the absence of the protrusion.

A hemispherical protrusion on the flat surface increases the Raman dipole moment of a molecule located at an arbitrary point of its surface $|L(\omega_{res})|$ times. Then, for a silver substrate completely covered by hemispherical protrusions, we would expect the impact of protrusions to be equal to $G_R = |L(\omega_{res})|^2 \approx 400$. In fact, this value should be lower, because even the maximally dense arrangement of hemispherical protrusions leaves some flat areas between them. In these flat areas, the dipole moments of the molecules are not enhanced.

Here, we have assumed that the protrusion is hemispherical. In fact, the chemical etching – the technique used in early schemes of SERS – results in arrays of nearly ellipsoidal protrusions with significant deviations of their eccentricity. Approximately one-half of them are oblate semi-ellipsoids, and one-half are prolate semi-ellipsoids. Their plasmon resonances hold at slightly different frequencies. However, their plasmon resonance bands overlap, and LFE for a prolate ellipsoid is higher than that for an oblate ellipsoid. Therefore, the damage due to the presence of oblate protrusions is compensated by the presence of prolate ones. As a result, the mean values of the Raman dipole moment turns out to be the same as if the roughened surface was composed by hemispheres. In early SERS techniques, one tried to pack these protrusions as densely as possible. However, it was difficult to achieve using chemical etching, and the amount of protrusions that were too small (and could not have pronounced plasmonic properties) was significant. Nowadays, several affordable technologies are available for SERS. For example, a SERS substrate can represent a random array of solid silver or gold nanospheres on a dielectric substrate. These particles transit to the substrate (and attach to it) from a colloidal suspension [22]. For such SERS substrates, the theoretical model presented here is applicable without involving the mirror image principle.

As we have already mentioned, for chemically prepared substrates of silver or gold the Raman gain G_R turned out to be much higher than $L(\omega_L)^2$. For silver SERS substrates, one practically obtains $G_R \sim 10^4$–10^6 for high density of protrusions depending on their shapes. This experimental fact indicates the presence of another factor entering G_R besides of L^2. In the literature on SERS, one may find two dominating hypotheses on this second factor. The first model is purely electromagnetic. This model was suggested in works [5, 32] and explains that the measured Raman radiation is basically produced by the same protrusion that enhanced the local field. This enhancement holds on the first stage of the process and corresponds to the enhanced molecular vibration. In the second stage, the energy of the molecular vibration is not radiated into free space but transferred to the plasmonic protrusion in which the dipole moment $\mathbf{d}_2$ is induced at the Raman frequency. So, besides the dipole moment $\mathbf{d}_{sphere}$ excited in the protrusion by the source field at frequency ω_L, another dipole $\mathbf{d}_2$ is excited by the molecule at ω_R. Obviously, large protrusions radiate much more effectively than small molecules.

For a long time, the electrochemical community disputed the point that the protrusion/nanoparticle may radiate more energy than the enhanced dipole moment $\mathbf{d}_R$ of molecule. The argument was simple: the molecular vibration is the source of the Raman radiation, and the energy of this vibration is enhanced only L^2 times. Therefore, protrusion cannot extract more energy. However, for a radio engineer familiar with antenna theory (in particular, the use of resonant parasitic antenna elements, such as in Yagi-Uda antenna arrays), there is nothing surprising in the fact that $d_2 > d_R$ and that the radiated energy is enhanced more than L^2 times. In fact, the action of the protrusion is twofold. First, it allows a molecule to extract more energy from the source wave. Second, it works as a transmitting antenna, enhancing the extracted energy. This way, radiation of energy of the enhanced molecular vibration is also enhanced.

In fact, the dimer of an emitter and a scatterer is a unique radiating system in which the plasmonic protrusion is the main radiating part [42]. The molecule is an element that generates the Raman frequencies and efficiently couples the protrusion to the source wave. For instance, in the Yagi-Uda antenna array, a small active dipole is the element efficiently coupling the radiating system of big passive dipoles to the voltage generator.

The enhancement of the total dipole moment of the dimer at the Raman frequency compared to the primary dipole moment d_R of the molecule grants the enhancement of the radiated power that is called the radiative *Purcell effect*. In order to properly introduce the *radiative Purcell factor* and non-radiative (dissipative) Purcell factor, we have to temporary divert from the topic of SERS and delve into another domain of nanophotonics, discussing Purcell's effect.

The role of the Purcell effect in SERS is still discussed and not commonly accepted by the experts in this field or, generally, by the experts in Raman radiation. The field where the critical role of the Purcell effect was fully recognized by the optical community is *plasmon-enhanced fluorescence* (PEF) [9]. For this reason, we start the discussion of the Purcell effect and Purcell factor namely for the case of PEF, where these concepts are commonly adopted. After considering the Purcell effect in PEF, we will return to SERS and explain that, in SERS, we encounter the same effect and deal with the same enhancement factor as in PEF.

9.3 Fluorescence

In order to explain the Purcell effect in PEF, let us first understand what fluorescence is. We have already defined fluorescence as emission of light at certain frequencies by a substance that has previously absorbed light of other frequencies. Linear fluorescence obeys the Stokes law (see Section 9.1); i.e., the absorbed light has higher frequencies than the emitted one. The most important case of fluorescence corresponds to the primary radiation in the ultraviolet (UV) range. While the source light is invisible, the emitted light can be in the visible region. Then the fluorescent substance becomes a clearly distinct color source. Unlike phosphorescent materials (which continue to emit light for a long time after the radiation source has been turned off), fluorescent materials cease to glow after a small delay in the range of ns. This slight delay is called *fluorescence lifetime*. Notice that even one ns is a quite long lifetime for an optical transition (OT). Therefore, the excited state from which the fluorescence photon is emitted

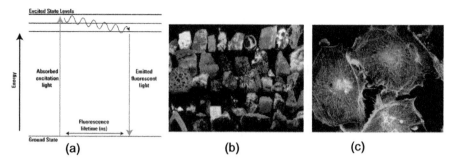

Figure 9.5 (a) Fluorescence of a molecule excited by a photon with a higher frequency than that of the emitted photon. (b) Fluorescent inhomogeneities in a mineral. Available in https://en.wikipedia.org/wiki/ Fluorescence. Reprint permitted by the author (Hannes Grobe/AWI Own work 26.04.2006). (c) Eukaryotic cells fluorescent in a biological medium. Available at https://warwick.ac.uk/services/ris/ impactinnovation/impact/analyticalguide/fluorescence. Reprint permitted by the illustrator (Dr. Ian Hancox).

is called *metastable optical state*. In the linear fluorescence, this is the bottom state in the excitation energetic zone of a molecule separated from the ground state by a gap. This gap, evidently, is equal to the energy of the emitted photon.

First, a molecule almost instantaneously absorbs the pumping UV radiation and is excited; a transition occurs from the ground state to the state above the metastable one. Then it relaxes to the metastable state. The relaxation process is very short (less than 1 ns after the absorption of the UV photon). Since further relaxation is impossible, the emission of a visible-range photon occurs afterward. Of course, not all molecules of the fluorescent substance relax to the fluorescence state and emit visible light. Most of them experience the backward OT from the upper state directly to the ground state. This process, in accordance with the Einstein–Smoluchowski theory, is nothing but the usual linear scattering of molecules – elastic (Rayleigh) scattering. This is not our subject, and we concentrate on OTs illustrated by Figure 9.5(a).

In order to exhibit measurable fluorescence, the fluorescence substance needs a noticeable percentage of fluorescent molecules and metastable states with long lifetimes. These states exist in long multi-atomic molecules, where the orbital movement of electrons in atoms hybridizes with molecular vibrations. The corresponding hybrid states are called *electron vibrational states*. The highest electron-vibrational states are excited by UV radiation. The lowest of them is metastable, and the backward OT from it to the ground state corresponds to visible photon emission. The higher the electron-vibrational state is, the shorter is the lifetime of this state. This property refers to any quantum system (see, e.g., in Chapter 10), and for molecular electron vibrations, the contrast of lifetimes for different states is especially high.

Fluorescence is an important tool for the detection and localization of *living cells* containing many long molecules – especially *eukaryotic cells, prokaryotic cells, fungal cells, egg and stem cells* of animals and humans, and some bacteria. It is also used in the *defectoscopy of minerals* – those minerals whose crystal lattices

9.3.1 Fluorescence for Spectroscopy and Imaging of Small Objects

Fluorescence of a movable tag is a very important optical tool of biological optical sensing. A *fluorescent tag*, schematically shown in Figure 9.6(a) as a bulb, is a unique tool for tracking living cells to which fluorescent molecules attach by *Van der Waals forces*: stem cells, cancer cells, leucocytes, cyanobacteria, animal cells, and prokaryotic cells. A conventional fluorescent tag is a set of fluorescent molecules attached to a dielectric nanoparticle (of sizes 5–100 nm). There are also single-molecule fluorescent tags. They can penetrate into a living cell, attach to its nucleus, and illuminate the interior of the cell that may become visible from the cell exterior. Usually, these fluorescent tags are proteins. The discovery of their fluorescence (which earned the Nobel Prize in Chemistry in 2008) resulted in a breakthrough in both cancer diagnostics and genetic engineering. More details on single-molecule fluorescent tags can be found in [43–45].

Conventional fluorescent tags used for other applications are based on so-called *rhodamine* and *rhodanide* molecules. Rhodamine (chemical formula NCS) is a rather long (1.6 nm) and very thin chain of atoms. Rhodanide (also called fluorescent isothiocyanate – FITC) looks like a tree of atoms and has a length of nearly 1 nm. In a conventional (multi-molecular) tag the ends of molecules attach to a dielectric surface (due to the same Van der Waals forces that cause the attachment of their free ends to the cell). In a colloidal suspension of dielectric nanospheres, these molecules assemble on the spheres, and this self-assembly results in arrays of molecules that can be thought of as radially arranged around a dielectric core, as it is shown in Figure 9.6(b). When this tag meets a living

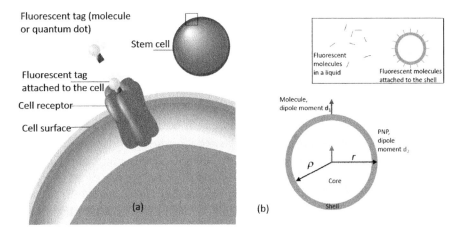

Figure 9.6   (a) A conventional fluorescent tag on a stem cell. Author's drawing's after the picture available on the National Institute of Health's website, https://stemcells.nih.gov/info/2001report/appendixE.htm. (b) An advanced fluorescent tag formed by a core–shell PNP and a set of fluorescent molecules attached to its dielectric shell (inset). Here, the dipole $\mathbf{d}_1$ of a fluorescent molecule induces the dipole $\mathbf{d}_2$ in the PNP.

cell, the free ends of these molecules attach to those places of the cell where the Van der Waals forces are maximal. For stem cells sketched in Figure 9.6(a), these places are the cell receptors.

To excite the tag, the UV radiation (projection signal) is injected into the domain under study. Using a millimeter-thick optical endoscope, it is possible to do it even with a human or animal body. On large cells with several receptors, like the stem cells shown in Figure 9.6(a), many fluorescent tags are attached after a certain time. Then a strong fluorescence shares the stem cells from those cells to which these tags cannot attach and even from those to which only one to two tags attach. Fluorescence allows one to distinguish the stem cells and separate them from all other cells in a biological liquid. It is only one of multiple biomedical applications of fluorescent tags. Among them, the most striking is probably the visualization of the cell processes in a conventionally dark areas, e.g., inside a living body. This visualization is called *dynamic tracking of cell movement* [46]. A simple illumination of a dark area by visible light through an endoscope does not allow one to visualize cells because scattering by cells is negligibly small and the light absorbed by the cells is spent heating them. Meanwhile, the UV signal emitted from the endoscope converts into visible light on the surface of a stem cell enhanced by the array of tags, and the cell emits an optical signal that is captured by the so-called *EMCCD camera* located on the endoscope. This camera is a matrix of microscopic light detectors called *electron-multiplying charge-coupled devices* (more details can be found in [47]).

However, the fluorescent image obtained in the described way is still quite poor and strongly distorted by optical noises. An improvement of the spatial resolution of these images has been achieved by applying statistical analysis to spectroscopic measurements [48, 49]. This approach requires a long observation time and/or multiple repetitions of the same experiment. Another way to improve the image is to make the optical image brighter so that the optical signal would exceed optical noises [50–54]. It is a promising approach to strong reduction of the total observation time and to improvement of the dynamic characteristics of tracking. Strong enhancement of fluorescence allows one even to record a video of cells [55].

9.3.2 Plasmon-Enhanced Fluorescence in Biosensing

How can this strong enhancement be achieved? It is achieved replacing a solid dielectric sphere by a *core–shell structure*, as it is shown in Figure 9.6(b). This plasmon-enhanced tag produces more intensive and more narrowband light than a similar tag without the plasmonic core. In work [50], the presence of silver nanocore in the sphere of diameter 100 nm increased the fluorescence intensity up to 50 times compared to the fluorescence of a similar tag with a solid dielectric core. The corresponding structure is shown in Figure 9.6(b).

Core–shell nanoparticles with a gold or silver core and thin latex or silica nanoshell can be fabricated using rather cheap chemical technologies that are commercially available in colloidal solutions. Optical properties of PNPs are determined by the plasmonic core diameter and the shell thickness (i.e., the distance between the plasmonic metal and the fluorescent molecules). Varying the thickness, one may tune the central frequency of the plasmon resonance band ω_{res} to the central frequency of the fluorescence spectral line ω_0.

Now, following works [51, 52], let us understand why this increase happens. Unlike SERS, there is no local enhancement of the source electric field. Really, the source wave is the UV light wave with frequencies very distant from ω_0. However, each fluorescent molecule (in the array of molecules sitting on the tag surface) excites a dipole moment in the PNP, and this secondary dipole is of higher amplitude than the primary dipole, and the radiating polarization fills a relatively large volume of the particle. This is the reason for the enhancement of fluorescent radiation.

Since the fluorescence of different molecules is not mutually coherent, there is no need to have in mind the array effects in the fluorescent array sitting on the PNP. One may consider an only radially arranged satellite molecule (1) with the primary dipole moment $\mathbf{d}_1$, corresponding to its optical transition to the ground state, accompanied by the emission of the photon $\hbar\omega_0$, and the induced dipole moment $\mathbf{d}_2$, induced by this photon at the same frequency ω_0 in the PNP (2). This situation is depicted in Figure 9.6. The key point of the process is that the absolute value of the dipole moment $\mathbf{d}_2$ turns out to be higher than that of $\mathbf{d}_1$.

For an experienced physicist, it is quite easy to understand why it may be so. The PNP is a much more substantial resonant radiator, comprising millions of lattice ions, whereas a molecule comprises a dozen of atoms or even less. Therefore, the radiation capacity of a PNP exceeds that of a molecule by orders of magnitude. For a radio engineer, the analogy with a two-element Uda-Yagi antenna array is evident. In this array, a small, weakly radiating coupling element excites a large passive resonant element, which radiates effectively. This way the radiated power dramatically increases.

Since every molecule emits a visible-light photon faster than when the plasmonic core is absent, it becomes ready for absorbing the next UV photon earlier. Therefore, the absorption of the source photons per unit time in the presence of the plasmonic core grows, and the fluorescence is enhanced accordingly. Plasmonic enhancement of the fluorescent tag in the regime of continuous pumping is a particular (but very important) case of a fundamental physical effect. It is the aforementioned Purcell effect.

9.4 Purcell Effect

The Purcell effect is, by definition, the enhancement of emission of any quantum emitter granted by its near-field coupling with a classical open resonator [9]. This enhancement was theoretically revealed by Edward Purcell in work [57]. This work, dating back to 1946, was awarded the Nobel Prize in Physics in 1952, after the experimental confirmation of Purcell's predictions. Historically, the Purcell effect was predicted and experimentally confirmed for quantum emitters of radio-frequency radiation from magnetized barium nuclei injected into an open cavity implemented as a vessel of distilled water. Due to huge permittivity of water at radio frequencies (about 80), this cavity experiences the so-called *Mie resonance* at the wavelength λ, which is much larger than its size D. If one applies a static magnetic field to a set of particles with non-zero spins (such as atoms of barium), these particles experience the so-called *nuclear magnetic resonance*: resonant precession of magnetic moment around the static magnetic field direction. It holds at the so-called *Larmour frequency* (proportional to the magnetic field). Due to this rotation, the magnetic moments become capable of emitting radio-frequency photons. Nuclear magnetic resonance can be juxtaposed to the Mie resonance of the water-filled cavity tuning the static magnetic field. Excited by a source oscillating at the Mie resonance frequency, the cavity effectively becomes a resonant magnetic dipole and efficiently reradiates the energy borrowed from the quantum emitters (precessing barium atoms) into free space.

Since the energy of a radio-frequency photon is fixed (at the Larmour frequency), the enhancement of the radiated power can be considered as a reduction of the precession time. In the quantum language, the precession time is the lifetime of the excited state of barium [57]. Thus, enhancing emission is the same as reducing the emission lifetime compared to that of the same emitter in free space [57]. The factor describing the enhancement of the decay rate (reduction of the emission lifetime) is called the *Purcell factor*. Let us understand now how this factor qualitatively depends on the resonator size in terms of the resonance wavelength.

The larger the resonant cavity compared to the wavelength in free space, the stronger the field confinement inside it is, and the weaker is its coupling to free space. Really, the high-order resonant modes of a dielectric cavity are formed by several periods of standing waves of polarization. These standing waves of the cavity are called *whispering gallery* waves. At these high-frequency resonances, the cavity radiates very weakly (in different parts of the cavity, the field has opposite phases, and the radiation from these parts cancels out). In the lossless case of a spherical cavity, the electromagnetic field of the whispering gallery mode is completely confined inside. Meanwhile, at low-order resonance frequencies – those of the magnetic and electric Mie resonance of the dipole type – the same cavity radiates efficiently. Respectively, the Purcell factor F_P in the whispering gallery regime (when the ratio D/λ is large) is low, whereas the Purcell factor in the regime of Mie resonances (when the ratio D/λ is small) is high. Fixing the wavelength, the cavity size reduction is favorable for F_P. This effect is three-dimensional, and it is reasonable to assume that F_P is proportional to $(\lambda/D)^3$.

However, efficient radiation from a cavity means high radiative losses, and the resonator quality factor Q is smaller when the cavity size reduces. Definitely, Q of a Mie resonator is lower than Q of a whispering gallery resonator. Low resonator quality means lower amplitude of the excited field for the same field source. To maximize the radiation from the cavity, we have to increase Q. A dilemma arises between the resonator volume and the resonator quality. The dilemma was resolved by Purcell in terms of the so-called *resonance mode volume V*. It is an effective volume in which the field of the resonator mode is concentrated.

In accordance with [57], the Purcell factor at the resonance frequency (coinciding with the emission frequency) is as follows:

$$F_P = \frac{3}{4\pi^2} \frac{\lambda_{\text{eff}}^3 Q}{V}, \qquad (9.10)$$

where λ_{eff} is the resonator mode wavelength in the material of the cavity. For a spherical cavity, V is close to its physical volume.

Formula (9.10) implies that the quantum emitter is embedded into an open resonator. However, the Purcell effect also holds for emitters located outside of the resonator if coupled with it by near fields. Therefore, the concept of the Purcell effect extends to other types of scattering resonators such as a PNP or another *plasmonic nanoantenna* with which a fluorescent emitter is coupled by near fields [9, 58–62]. The extension of the Purcell effect to plasmonics and the difficulty with the calculation of the effective volume V for a nanoantenna urged the researchers to deduce alternative formulas for the Purcell factor. For the general case of a reciprocal resonant environment it was expressed through the imaginary part of the so-called *Green function of the emitter environment* calculated at the position of the emitter [9, 58]. Also, quantum models of the Purcell effect were developed (see, e.g., in [59, 60]). These models are difficult to use in practice, but, as explained earlier (see also in [61]), Purcell's effect is a classical effect, and for calculations of the Purcell factor F_P, the quantum nature of the emitter does not matter.

Unlike the initial experiments confirming the Purcell effect for the barium ions in almost lossless water, most scattering resonators used in nanophotonics are plasmonic – i.e., essentially lossy objects. Therefore, the reduction of the emission lifetime is not obviously the same as the enhancement of radiation because some part of emitted energy is dissipated in the resonator. In order to take this effect into account, we introduce the aforementioned radiative and non-radiative Purcell's factors, whose sum gives the total Purcell factor. Designing a nanostructure for PEF, the goal is to increase the radiative Purcell factor and to decrease the non-radiative one. In a properly designed nanostructure targeted for maximal PEF, the non-radiative Purcell factor is negligible.

As it was noticed in [61], the highest values of the Purcell factor can be achieved by using resonant dipole scatterers. This observation fits the initial formula (9.10), since Q can be made high enough for a low-loss dipole, whereas volume V of a dipole resonator is obviously subwavelength. If the scatterer is a dipole collinear or parallel to the dipole of the quantum emitter, the vector $\mathbf{d}_1$ of the emitter dipole moment sums up with the vector $\mathbf{d}_2$ of the scatterer dipole moment. In the classical scheme of PEF, the scatterer is a PNP, as shown in Figure 9.6(b). In this case, the complex amplitude of the total dipole moment calculated at the emission frequency ω_0 equals $d_1 + d_2$, and we may write, for the lifetime of spontaneous emission in the absence (τ_1) and in the presence (τ_2) of scatterer 2,

$$\tau_1 = \frac{\hbar\omega_0}{\omega_0^4 |d_1|^2 / 3\pi\epsilon_0 c^3}, \qquad \tau_2 = \frac{\hbar\omega_0}{\omega_0^4 |d_1 + d_2|^2 / 3\pi\varepsilon_0 c^3}. \qquad (9.11)$$

Really, when a quantum emitter is isolated, it radiates the power ($\omega_0^4 |d_1|^2/3 \pi \varepsilon_0 c^3$) during the emission of a photon with the energy $\hbar \omega_0$ that corresponds to τ_1 in (9.11). In the presence of a PNP, the molecule induces dipole moment d_2 in the PNP, and the total dipole moment equals $d_1 + d_2$. The radiated power becomes equal to ($\omega_0^4 |d_1 + d_2|^2/3\pi\varepsilon_0 c^3$), and we obtain τ_2 in (9.11).

In general, for the case of the dipole resonator, we have for the radiative Purcell factor

$$F_P = \left| 1 + \frac{d_2}{d_1} \right|^2 . \tag{9.12}$$

In work [61], it is strictly shown that formula (9.12) is equivalent to the formula that expresses the radiative Purcell factor via the Green function of the dipole resonator. The equivalence of the last one to the initial Eq. (9.10) in the purely lossless case was earlier proven in [58]. In [61], other equivalent representations for the Purcell factor were obtained.

A fluorescent molecule during emission can be considered a *two-level system*. The upper energetic level of the two-level system has a finite width called either the *spectral line width* or the optical transition band. The width of the spectral line is evidently determined by the lifetime of the excited state. Thus, the Purcell factor is not a constant for a given resonator; it depends on the frequency as a resonant function. For the purpose of PEF, one has to maximize $F_P(\omega_0)$, which is the resonant value of F_P. This maximization demands the coincidence of the fluorescence frequency ω_0 and the frequency of the plasmon resonance ω_{res}.

In work [50], the resonant Purcell factor measured for a tag of rhodamine molecules operating in human's blood attained 50, which corresponded to nearly sevenfold dipole moment $d_1 + d_2$ compared to d_1. However, we will see in the following section that $F_P = 50$ is far from the maximally achievable gain offered by an optimally designed PNP to a fluorescent molecule. In paper [51] (theoretically) and in [53] (experimentally), the authors obtained $F_P \approx 100$–150. However, even $F_P = 150$ is not a limit for PEF. Now, let us understand how it is possible to further maximize F_P once the equivalence $\omega_0 = \omega_{\text{res}}$ already holds and what is the achievable limit of the Purcell factor in PEF. To understand it, we need to study in more detail the near-field interaction of two dipoles, where one of them is a quantum emitter (1) and another one is a PNP (2); see Figure 9.6(b).

9.4.1 Main Physical Effects in Plasmon-Enhanced Fluorescence and Their Circuit Model

Recent extensive studies of near-field coupling of a quantum emitter and a PNP have revealed the following physical effects inherent to PEF:

- Purcell effect
- Lamb shift
- Fano resonance
- Rabi spectral splitting
- Fluorescence quenching

Here, we will concentrate on the first effect, corresponding to a relatively weak coupling of the quantum emitter and the PNP. A brief explanation of four other effects will be given in the appendix to this chapter.

Besides the lifetime of the excited state (linked to the radiated power by formulas [9.11]), another important characteristic of fluorescence is the emission spectrum. In the absence of a PNP, and in the case, the absence of the so-called *weak coupling*, this spectrum is Lorentzian. This property can be understood from the quantum theory of a fluorescent emitter, which delivers the following expression for the spectrum S_1 of a fluorescent emitter 1 (see, e.g., in section 8.5 of [9]):

$$d_1(\omega) = d_{10}S_1(\omega), \quad S_1(\omega) = \frac{1}{1 - \frac{\omega^2}{\omega_0^2} + j\frac{\omega\gamma_1}{\omega_0^2}}. \tag{9.13}$$

Here, d_{10} is a scalar parameter with the dimensionality $A{\cdot}s{\cdot}m$ called in quantum mechanics the *matrix element of the dipole optical transition* . For fluorescent molecules or quantum dots, d_{10} is their fingerprint, similarly to the Raman frequency shift for usual molecules. The value γ_1 with the dimensionality $1/s$ is called the *decay rate of the excited state*. For practically usable fluorescent molecules and for quantum dots, $\gamma_1 \approx 1/\tau_1$; i.e., the decay rate can be measured directly via the emission time. Obviously, $\gamma_1 \ll \omega_0$. This ratio has the order of 10^{-3} or even smaller.

In the case of weak near-field coupling, the spectrum of the quantum emitter 1 does not change in the presence of a classical scatterer 2, and the dipole moment d_1 in the range of the fluorescence does not change. At the resonance, it is equal to $d_1(\omega_0) = d_{10}\omega_0/j\gamma_1$. Thus, emitter 1 induces in the scatterer (a PNP) a dipole moment $\mathbf{d}_2$, whose spectrum is also nearly Lorentzian, especially if the resonance frequencies exactly coincide: $d_2(\omega) = d_{20}S_2(\omega)$. At the resonance, $S_2(\omega_0) = -j\omega_0/\gamma_2$, and $d_2(\omega_0) = d_{20}\omega_0/j\gamma_2$. We see that, in the case of weak coupling, both quantum emitter 1 and PNP 2 behave as two classical coupled dipole antennas: one active and one passive.

9.4.2 Purcell Factor of a Plasmonic Nanoparticle

As we have already mentioned, weak near-field interaction of a quantum emitter 1 and a classical scatterer 2 corresponds to the Purcell effect. Let us study this effect in the case of PEF when the scatterer is a spherical PNP. For our purposes, it is enough to find the magnitude d_{20} and the decay rate γ_2 of the PNP. Then, knowing d_{10} and γ_1, we can find the resonant value of the radiative Purcell factor from (9.12):

$$F_P(\omega_0) = \left| 1 + \frac{d_{20}\gamma_1}{d_{10}\gamma_2} \right|^2. \tag{9.14}$$

Values d_{20} and γ_2 can be expressed via the polarization parameters of the PNP. The polarizability α_2 of the core–shell particle depicted in Figure 9.3 is well known (see, e.g., formula (6.82) of book [62]). This polarizability (as well as the polarizability of any other dipole scatterer including the quantum emitter itself)

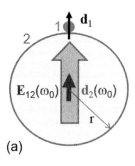 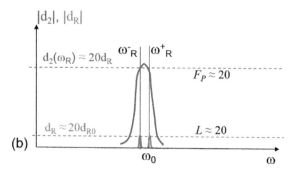

(a)   (b)

Figure 9.7 (a) Mechanism of the plasmon-enhanced emission for a spherical PNP: dipole $\mathbf{d}_1$ radiating within a narrow spectral line creates the near field $\mathbf{E}_{12}$ inducing the dipole $\mathbf{d}_2$, and the plasmon resonance frequency ω_0 $d_2 \gg d_1$. (b) Illustration of the approximate formula $G_R \approx L^4$; if both Raman frequencies $\omega_\pm$ are within the plasmon resonance band, the same mechanism as in (a) holds, and the Purcell factor $F_P \approx L$. Here, local field enhancement L offers the L-fold increase of the Raman dipole moment d_R due to the presence of the PNP.

can be related to the dipole effective length l, radiation resistance R_{rad}, dissipation resistance R_{dis}, and reactance X.

By definition, we have $\alpha_2 = d_2/E_{\mathrm{loc}}$, where E_{loc} is the field acting on the scatterer. Vector notations are removed because the spherical particle is isotropic, and its dipole is directed along $\mathbf{E}_{\mathrm{loc}}$. In the case of PEF, $E_{\mathrm{loc}} = E_{21}$, where E_{21} is the electric field produced by quantum emitter 1 at the center of PNP 2, as shown in Figure 9.7(a). Here, we take into account that the pumping radiation weakly interacts with the PNP having the plasmon resonance frequency far from its spectrum. Dipole moment d_2 is related to the effective *polarization current* I_2 as $d_2 = I_2 l_2/j\omega$. Recall that the relation $\mathbf{J} = j\omega\mathbf{P}$ between bulk polarization current density and bulk polarization density of any polarizable medium makes the conduction and polarization currents equivalent. Therefore, nothing changes if we replace the core–shell PNP with an equivalent wire of *effective length l_2* and attribute effective *impedance Z_2* to its center. Current I_2 flows in this wire and, being multiplied by Z_2, results in the *electromotive force* (EMF) $\mathcal{E}_2$, induced in the scatterer (PNP) by the quantum emitter. This EMF is equal to the product of the external field E_{21} and the effective length l_2 of the scatterer: $\mathcal{E}_2 = E_{21}l_2$. Thus, we can find its polarizability:

$$\alpha_2 = \frac{I_2 l_2}{j\omega E_{21}} = \frac{E_{21} l_2}{Z_2} \frac{l_2}{j\omega E_{21}} = \frac{l_2^2}{j\omega Z_2}. \tag{9.15}$$

Here, $Z_2 = R_2 + jX_2$, where $R_2 = R_{2\mathrm{rad}} + R_{2\mathrm{dis}}$ is the *effective resistance* of the scatterer and X_2 is its *effective reactance* (see Chapter 7):

$$X_2 = \omega L_2 - \frac{1}{\omega C_2}. \tag{9.16}$$

Substituting (9.16) into (9.15) allows us to express α_2 in the Lorenztian form:

$$\alpha_2(\omega) = \frac{l_2^2 C_2}{1 - \frac{\omega^2}{\omega_0^2} + j\frac{\omega\gamma_2}{\omega_0^2}}, \quad \gamma_2 = R_2 C_2 \omega_0^2. \tag{9.17}$$

Comparing (9.17) to formula (6.82) of [62], we easily specify the values $R_2 C_2$ and $l_2^2 C_2$ of the core–shell scatterer in terms of the permittivities of the metal core and dielectric shell, the core radius ρ, and the outer radius r of the shell. For our purposes, these long formulas are not needed; it is only instructive to see one more time that the circuit (LRC) parameters of a plasmonic dipole can be found through Lorentzian dispersion parameters.

What is important for us is the fact that the effective resistance of the PNP and its effective capacitance resulting from the comparison of (9.17) with the electromagnetic model of the core–shell PNP turn out to be much higher than those of fluorescent molecules (they can be found, e.g., in [9]). Thus, γ_2 is incomparably higher than γ_1. Therefore, γ_2 can be identified with the decay rate of the whole dimer. This means that the radiation in the case of strong Purcell effect is practically created not by the quantum emitter but by the classical scatterer. The quantum emitter only accepts energy from the pumping source and transfers it to the scatterer mainly by near fields. It is a non-radiative transfer.

In Eq. (9.14), d_{20} is the product of the interaction field E_{21} and the polarizability α_2 taken in the static limit – i.e., $\alpha_2(\omega = 0) = l_2^2 C_2$. Since E_{21} is created by the quantum emitter, it is proportional to its dipole moment d_1 with a certain coefficient A_{ee} called the *interaction factor*. At the resonance frequency, we have $E_{21}(\omega_0) = A_{ee} d_1(\omega_0) = A_{ee} d_{10} \omega_0 / j\gamma_1$. Thus, from (9.17), we obtain

$$d_2(\omega_0) = \frac{d_{20}\omega_0}{j\gamma_2} = \alpha_2(\omega_0) A_{ee} \frac{d_{10}\omega_0}{j\gamma_1} \qquad (9.18)$$

or

$$\frac{d_{20}\gamma_1}{d_{10}\gamma_2} = -j\frac{A_{ee} l_2^2}{\omega_0 R_2(\omega_0)} \qquad (9.19)$$

because $\gamma_2 = R_2 C_2 \omega_0^2$ and C_2 cancel out. Substituting this result into (9.14), we find the resonant Purcell factor that does not depend on the parameters of the quantum emitter at all:

$$F_P = \left| 1 - j\frac{A_{ee} l_2^2}{\omega_0 R_2(\omega_0)} \right|^2 . \qquad (9.20)$$

Radio engineers know that any dipole radiator of effective length l_2 has the radiation resistance

$$R_{2\mathrm{rad}} = \eta \frac{k^2 l_2^2}{6\pi}, \qquad (9.21)$$

where $k = \omega/c$ is the ambient (free-space) wavenumber and $\eta = \sqrt{\mu_0/\epsilon_0}$ is the ambient wave impedance. If the core–shell PNP is properly designed, its dissipative resistance is much smaller than the radiation resistance, and we have $R_2 \approx R_{2\mathrm{rad}}$. Then, $F_P \gg 1$, and formula (9.20) at $k_0 = \omega_0/c$, after the substitution of (9.21) and neglecting the unity, takes the form

$$F_P \approx \left(\frac{6\pi c^2 A_{ee}}{\omega_0^3 \eta} \right)^2 . \qquad (9.22)$$

This formula describes the radiative Purcell factor when the fluorescence frequency is equal to the resonance frequency of the plasmonic core in the shell.

Now, let us find A_{ee} an external electric field created at the center of the core-shell PNP 2 by a unit radially polarized dipole 1 located on its surface. Since A_{ee} is the external field, it is calculated in the absence of the PNP. The impact of the PNP in our model is fully described by its polarizability α_2 that takes into account both the core and shell. Therefore, if we adopt the assumption that the two scatterers interact as two electric dipoles referred to their centers, and the dipole of the PNP (2) is distanced from the dipole of our molecule (1) by the outer radius r of the PNP, the quasi-static approximation for finding A_{ee} can be used:

$$A_{ee}(\omega_0) \approx \frac{1}{2\pi\epsilon_0\varepsilon_a r^3}, \tag{9.23}$$

where ε_a is the ambient relative permittivity. For simplicity, we further assume that $\varepsilon_a = 1$ (free space). The use of this approximation is reasonable because the radius r is much smaller than the wavelength.

Then, substituting (9.21) and (9.23) into (9.20), we obtain an approximate relation for the resonant Purcell factor of a low-loss core–shell PNP excited by a radially oriented molecule located at the shell surface:

$$F_P(\omega_0) \approx \frac{9}{(k_0 r)^6}. \tag{9.24}$$

This $F_P(\omega_0)$ does not depend on the core radius ρ and the shell thickness δ because, for the given $r = \rho + \delta$, these two parameters are both determined by the resonance frequency $\omega_0 = k_0 c$.

The Purcell effect is very strong if $(k_0 r) \ll 1$, and, in accordance with (9.24), in order to increase the radiation enhancement, one needs to reduce r. However, this reduction is restricted by two factors. The first factor is the minimal allowed radius ρ of the plasmonic core. In order to experience the plasmon resonance, the core of gold or silver should have a sufficient radius (of the order of 5–10 nm). Otherwise, it is not a piece of a bulk silver or gold but a nanocrystal of natural atoms. Its plasmon resonance is suppressed by two effects: bound charges at the metal surface and spatial dispersion of the nanocrystal lattice (see, e.g., in [62]). These effects drastically decrease the magnitude of the plasmon resonance for PNPs with $\rho < 5$ nm, and the key assumption of low losses in the PNP ($R_2 \approx R_{2\text{rad}}$) becomes invalid.

The second factor restricting the minimum of r is the minimal allowed thickness δ of the dielectric shell. When the shell is too thin, our second key assumption (a sufficiently weak interaction with the molecule) is not valid anymore. The strong interaction results in the substitution of the Purcell effect by the other aforementioned effects, which are parasitic for PEF. Among them, the worst one is fluorescence quenching, when $|d_1 + d_2| < |d_1|$ and $F_P < 1$ [63]. This phenomenon occurs when the thickness δ critically decreases (practically, when it becomes smaller than 4–5 nm). If we reduce δ to 1–3 nm, the absolute value of vector $\mathbf{d}_1$, due to the feedback of the PNP, increases and approaches d_2. However, simultaneously, the phase of the dipole moment $\mathbf{d}_2$ at the resonance frequency turns out to be opposite. Dipole moments $\mathbf{d}_1$ and $\mathbf{d}_2$ compensate one another and form a weakly radiating quadrupole [64].

A shell of thickness $\delta = 5$–10 nm is sufficient to make the interaction of the core of 10 nm $< \rho < 60$ nm with a molecule located on the shell surface weak enough so that it is possible to neglect the backward action of $\mathbf{d}_2$ onto $\mathbf{d}_1$. Then the dominating effect is the Purcell effect, and our approximations seem to be suitable for qualitative estimations of F_P.

However, both silica and polystyrene – materials used in the known PEF techniques for the shells of PNPs – are nanoporous. In silica, these pores are of Angström thickness and quite long (dozens of nm). Thus, shells of the optimal thickness $\delta = 5$–10 nm allow the molecular diffusion from the gaseous or liquid ambient to the core. Only gold stands the presence of these molecules, whereas silver interacts with sulphur and oxygen. As to silver, to preserve it from sulphurization and oxidation, one would need a shell of thickness 40–50 nm. Therefore, the experimenters prefer to use gold.

For PEF based on the PNPs with a gold core, the optimal internal and external radii are equal to $\rho = 20$–25 nm and $r = 30$–40 nm, respectively. However, the plasmon resonance in this case holds in the range where gold is quite lossy. In this case, the approximation $R_2 \approx R_{2\text{rad}}$, resulting in our estimate (9.24), is not applicable. Instead, dissipative losses make $F_P \approx 50$ for such PNPs [65]. The highest achievable values of F_P correspond to such core–shell PNPs in which the core is dielectric and the shell is plasmonic. This is so because for such core–shell particles, the plasmon resonance turns out to be in the near-infrared range, where the dissipation is low and the approximation $R_2 \approx R_{2\text{rad}}$ becomes adequate. In this case, the required regime of weak interaction can be ensured if we cover the PNP with an additional dielectric shell. However, in practice, one uses for this purpose long organic molecules called ligands, which attach to the metal surface and form an effective outer dielectric shell with the relative permittivity close to unity.

In work [51], the resonant Purcell factor $F_P(\omega_0) = 144$ was predicted for this structure by complex numerical simulations. A radially polarized molecule was distanced by 3.8 nm (the size of the ligand) from a gold nanoshell of thickness $\delta = 3.3$ nm centered by a silica core of radius $\rho = 64$ nm. In this case, $r = 71.1$ nm, and the plasmon resonance of the PNP holds at 353 THz ($\lambda_0 = 850$ nm). Analytical formula (9.24) predicts for this case the value $F_P(\omega_0) \approx 428$. We see that our simplistic estimate (9.24) cannot reach quantitative accuracy in this case because it fully neglects the dissipative loss.

In fact, even in this low-loss case, the decay rate enhancement in presence of the PNP holds, due to not only the increased radiation but also the dissipative loss arising in the PNP. This effect is described by the aforementioned non-radiative Purcell factor. This value is not related to the dipole moment of the system. The non-radiative Purcell factor was derived using the circuit model of PEF in work [61]. At the resonance, this value has the same order of magnitude as the radiative Purcell factor. This means that the input resistance of the PNP cannot be identified with the radiative resistance, as we did for simplicity in deriving formula (9.24). This formula correctly predicts only the order of magnitude for the radiative Purcell factor, but the predicted value is 2–3 times higher than the correct radiative Purcell factor F_P. The highest value for F_P practically achievable for a radially polarized molecule located on the surface of an optimized core–shell PNP was obtained in [54] and equals $F_P \approx 200$.

Further enhancement of molecular fluorescence can be achieved by replacing the *conventional nanoantenna* implemented as a core–shell PNP by more advanced ones, e.g., the so-called *bow-tie nanoantenna*. The quantum emitter should be centered in the antenna gap. In such advanced nanostructures, the radiative Purcell factor F_P may attain 500–600 [52, 55].

9.5 Purcell Effect in SERS

Now we return to SERS, understanding that SERS refers to the case of weak coupling of molecules and resonant scatterers. For the special case of a molecule attached to a hemispherical protrusion or a spherical PNP, it is weak coupling between the Raman dipole d_R of the molecule and the plasmonic nanosphere. It is not strange that coupling remains electromagnetically weak even in the case when the distance between the molecule and the metal surface is as small as the radius of the molecule. First of all, the fundamental molecular vibration is robust and cannot be perturbed so easily as the excited state of the fluorescent molecule does. Second, the dynamic polarizability of the molecule at two Raman frequencies is very small. The circuit model of the polarizability of a QE implies smallness of the effective length l_1. As a result, the coupling coefficient κ of the dimer given by (9.46) is small even if A_{ee} is large.

Therefore, in SERS, one does not observe Rabi splitting of the Raman spectral line as well as Lamb shifts and Fano resonances. In SERS, backward action of the PNP to the spectrum of the Raman emission is negligible. Only the Purcell effect – the increase of the total dipole moment $d_1 + d_2$ of the dimer compared to $d_1 \equiv d_R$ – is observed at the Raman frequencies.

In accordance with (9.9), we have $d_R = \alpha_R L(\omega_L)E_0$, where $L(\omega_L) \approx 20$ for a silver nanosphere. The dipole moment $d_1 = d_R$ induces in nanosphere 2 the dipole moment d_2 at frequency ω_R:

$$d_2 = \alpha_2(\omega_R)E_{12}(\omega_R) = \alpha_2(\omega_R)A_{ee}d_1, \tag{9.25}$$

where the polarizability α_2 of the metal nanosphere of radius R is equal, in accordance with (9.3), to

$$\alpha_2 = 4\pi\epsilon_0 R^3 \frac{\varepsilon_{\text{met}} - 1}{\varepsilon_{\text{met}} + 2}. \tag{9.26}$$

Under the condition that $\omega_R = \omega_L \pm \Omega \approx \omega_L$ – i.e., if the Raman frequency is within the band of the plasmon resonance, as depicted in Figure 9.2(b) – we have

$$\alpha_2(\omega_R) \approx \alpha_2(\omega_L) \approx \frac{12\pi\epsilon_0 R^3}{j\varepsilon''_{\text{met}}(\omega_L)}. \tag{9.27}$$

For a top-located molecule, the dipole moments $\mathbf{d}_1$ and $\mathbf{d}_2$ are collinear like in the case of PEF. The interaction between them in the case of SERS is the same as described earlier, and to find the dipole field E_{12}, we can use formula (9.23) that in the present notations takes the form

$$A_{ee} \approx \frac{1}{2\pi\epsilon_0 R^3}.$$ (9.28)

From (9.25), (9.27), and (9.28), we obtain $d_2(\omega_R) \approx 6d_1(\omega_R)/j\varepsilon''_{met}(\omega_L)$. Taking into account that $|\varepsilon'_{met}(\omega_L)| = 2-$, i.e., $3|\varepsilon'_{met}(\omega_L)| = 6 -$ we come to the following absolute value of d_2 at the Raman frequency:

$$|d_2(\omega_R)| \approx 3 \left| \frac{d_1(\omega_R)\varepsilon'_{met}(\omega_L)}{\varepsilon''_{met}(\omega_L)} \right|.$$ (9.29)

Recall that for the top-located molecules we had the same coefficient 3 in formula (9.7). However, this coefficient 3 transformed into unity after the position averaging in formula (9.8). The same holds for the relation between d_2 and d_1. Coefficient 3 in (9.29) yields to unity after averaging over all possible positions of the vertically polarized molecule on the sphere. For this position-averaged Purcell factor, we obtain

$$F_P = \left| \frac{d_1 + d_2}{d_1} \right|^2 \approx \left| \frac{d_2}{d_1} \right|^2 \approx \left| \frac{\varepsilon'_{met}(\omega_L)}{\varepsilon''_{met}(\omega_L)} \right|^2.$$ (9.30)

In accordance with (9.30), the Purcell factor turns out to be equal to the square power of the position-averaged local field enhancement at the same frequency! Thus, the Raman gain

$$G_R = \left| \frac{d_1 + d_2}{\alpha_R E_0} \right|^2,$$ (9.31)

which is, evidently, the product of the local field intensity enhancement

$$L^2 = \left| \frac{d_1}{\alpha_R E_0} \right|$$ (9.32)

and the radiative Purcell factor

$$F_P = \left| \frac{d_1 + d_2}{d_1} \right|^2,$$ (9.33)

is described by the famous approximate formula $G_R \approx L^4$ because

$$G_R = \left| \frac{d_1 + d_2}{d_1} \right|^2 \left| \frac{d_1}{\alpha_R E_0} \right|^2 = [L(\omega_L)]^2 [L(\omega_R)]^2.$$ (9.34)

In the standard situation illustrated by Figure 9.7(b) – when the Raman frequency ω_R is close to the plasmon resonance frequency ω_0 and $\omega_L = \omega_0 -$ we have $G_R = [L(\omega_L)]^2 [L(\omega_L)]^2 \approx [L(\omega_L)]^4 \approx [L(\omega_R)]^4$. If the Raman shift is noticeable and $L(\omega_R) \neq L(\omega_L)$, we have, in accordance with (9.34), $G_R = [L(\omega_0)^2][L(\omega_R)]^2$.

Result (9.34), though proven here for a hemispherical plasmonic protrusion, holds for any SERS substrates because the local field intensity enhancement L^2 is equal to the Purcell factor. This equivalence was proven for spheroidal protrusions in [66] and for arbitrary substrates in [67]. This explains why SERS based on a properly roughened silver plate offers the gain in Raman radiation of the order of 10^5 for molecules located on the protrusions, and the averaged gain for all molecules is of the order of 10^4.

Again, it is worth while to note that the presented explanation of SERS in terms of the Purcell factor is not commonly adopted. It appears that the majority of researchers working in the field of SERS consider Purcell's effect as something alien to SERS. Only recently has this insight of the electromagnetic Raman gain as a product of the local intensity enhancement and the Purcell factor (which turn out to be equivalent to one another) been supported; see e.g., work [68]. At present, few researchers share the point of view of SERS as well as the insight of the classical nature of linear effects observed in PEF.

9.6 Advanced SERS

As it was already mentioned, in advanced SERS substrates prepared of practically identical protrusions, the Raman gain of top-located molecules is as high as $G_R \sim 10^7$–10^8. In regular plasmonic structures, the electromagnetic interaction between plasmonic inclusions is coherent and can be constructive, additionally enhancing the microscopic electric field at the nanopatterned surface as compared to the single inclusion. As to the Purcell factor, it is worth while to repeat that the position-averaged locality factor for electric intensity L^2 is always equal to the position-averaged Purcell factor. This property can be deduced from the reciprocity theorem involving the apparatus of the Green function through which both the Purcell factor and local field intensity enhancement can be approximately expressed [69]. We do not present here this proof; we only note that the result $F_P = L^2(\omega_R)$ remains valid, which implies the validity of the result (9.34) also for advanced SERS substrates if they do not include nonreciprocal structural elements.

Amazing Raman gains $G_R \sim 10^{12}$–10^{14} for molecules located in nanometer plasmonic crevices were numerically simulated in [70] and experimentally revealed even earlier in [31]. However, in practice, it is rather difficult to prepare regular arrays of hemispherical protrusions with such dense packaging. For advanced SERS, one often fabricates regular arrays of plasmonic nanorods [71] or nanopyramides [72] in which the top-located molecules show the Raman gain up to $G_R \sim 10^{12}$–10^{14} and the position-averaged Raman gain $G_R \sim 10^5$–10^6. The SEM image of one of such structures is shown in Figure 9.8(a). Notice that the simulations of extraordinarily high values of the Raman gain performed in [70] for such structures were not confirmed experimentally in [72] and were later disclaimed. These classical simulations did not take into account an important quantum effect: the tunneling of electrons through the nanometer gap between two closely located plasmonic surfaces. Taking into account this tunneling, one still obtains a rather high value, $G_R \sim 10^8$, and this order of magnitude for the Raman gain in such SERS schemes has been experimentally confirmed in the arrays like that shown in Figure 9.8(a).

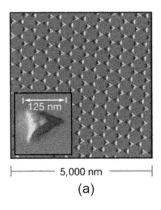
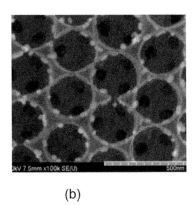

(a)   (b)

Figure 9.8 (a) Regular plasmonic nanopyramides (SEM image) for advanced SERS. Reprinted from [72] with permission of AAAS. (b) A plasmonic inverted opal structure. Reprinted from [73] with permission. Copyright 2012 OSA.

However, recently, a quantum mechanical model has been developed for a plasmonic nanostructure that offers a huge local field concentration. This model claims that the Raman radiation of a molecule located in a plasmonic hot spot can be additionally enhanced by an opto-mechanical resonance of the plasmonic surface coupled to the vibrating molecule by electrostatic forces. Roughly speaking, the surface of the plasmonic nanostructure vibrates in resonance with the molecule [74]. This model apparently explains the literature data on the short-range Raman gain exceeding 10^9 in advanced SERS techniques (see, e.g., in [31]). In this case, vibrations of different molecules can be synchronized by some nonlinear mechanisms and the aforementioned stimulated Raman scattering can arise. It is a coherent radiation and may even result in the regime of the so-called *Raman laser* [75].

Often, high Raman gain is required for as many molecules as possible. This is the case when tracking of separate molecules is not needed because one studies the chemical content of a macroscopic substance. In this case, in sacrificing possibly huge values of the local Raman gain, one maximizes the volume in which sufficiently high Raman gain is obtained (usually, one needs $G_R > 10^3$). For a liquid analyte, the most advanced design solution of this problem is the replacement of the nanocorrugated surface by a bulk plasmonic structure, penetrable for liquids. Here, the term SERS is not fully adequate since the enhancement holds for molecules located not on the surface but in the voids of a porous structure. One of the best structures is a plasmonic photonic crystal structured as an inverted opal [76, 77]. The structure and the technology of its fabrication are shown in Figure 9.8(b). At some points of this inverse opa, $G_R > 10^8$, whereas G_R averaged over the volume of any nanovoid is of the order of 10^4. This structure is an excellent tool for chemical analysis of liquids because the plasmonic framework is substantially thick and the volume filled with the analyte experiencing the Raman gain is large. Additionally, the structure is mechanically robust and allows liquid flow through it. More details can be found in works [76, 77].

Probably, all-dielectric SERS substrates in which the high Raman gain is achieved beyond the plasmon resonances can be also referred as advanced SERS schemes. In fact, both gold and silver are destructive for the Raman detection of several molecules, such as glucose and saccharose, and toxic for many biological cells. Therefore, for sensing such analytes, plasmonic SERS substrates were conventionally covered by a passivating nanolayer that significantly decreased the Raman gain. Recently, it was understood that high local field enhancement and, therefore, the high Purcell factor are achievable due to resonances in dielectric nanoparticles with sufficiently high refractive index. In the optical range, it implies semiconductor materials [78].

9.7 Conclusions

In this chapter, two important topics in the area of optical nanosensing have been considered – namely, the technique of molecular detection and tracking, called SERS, and the technique of biological sensing and monitoring, called PEF. Both techniques are based on the fundamental physical effect called Purcell's effect, which has been discussed in detail. Both SERS and PEF can be described and understood using simple equations of classical electrodynamics. In this theory, only a few scalar parameters entering the main equations need to be obtained from quantum mechanics. For SERS, these are the Raman polarizability α_R and the Raman shift Ω. For PEF, besides the fluorescence frequency ω_0 and the lifetime of the fluorescence state τ_1 of a molecule in free space (or the free-space decay rate $\gamma_1 = 1/\tau_1$) it is also the magnitude of the dipole moment of the optical transition d_{10} in free space. Once we know these parameters, the remaining story is nothing but a classical dipole–dipole interaction, though it occurs between a quantum emitter and a plasmonic nanoparticle. These classical models allow the use of the well-established antenna theory to understand and optimize nanophotonic structures for optical sensing and other applications.

Problems and Control Questions

1. The Purcell effect in both PEF and SERS refers to weak coupling. The weak-coupling approximation implies that the initial dipole moment of the quantum emitter $\mathbf{d}_1 = \mathbf{d}_1^{(0)}$ oscillating at the fluorescence frequency ω_0 does not change due to interaction. Strictly speaking, the interaction always changes it, though in the case of weak interaction the dipole moment induced in the quantum emitter by the nearby resonant dipole is negligibly small.

   Consider a dimer formed by two dipole particles: quantum emitter 1 with Lorentzian polarizability α_1 and scatterer 2 with Lorentzian polarizability α_2. For simplicity, let them be parallel, as in Figure 9.9, or collinear. Then the vector notations are not relevant since the dipoles are oriented along

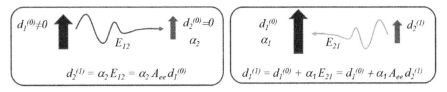

Figure 9.9 Dipole–dipole interaction changing the dipole moment of the quantum emitter can be presented as a set of the interaction events. The first event comprises the direct action (Purcell's effect) and the backward action. The steady emission spectrum corresponds to the final stage.

the same axis. The dipole–dipole interaction changing the dipole moment of the quantum emitter can be presented as a set of interaction events. The first event (step 1) comprises the direct action and the backward action. The direct action (Step 1.1) is the same as Purcell's effect: the creation of the dipole moment in the scatterer, $d_2 = d_2^{(1)} = \alpha_2 E_{12} = d_1^{(0)} A_{ee}$, where A_{ee}, as before, is the interaction factor of the dimer. The backward action is the creation of an additional dipole moment in the quantum emitter: $d_1 = d_1^{(1)} = d_1^{(0)} + \alpha_1 E_{21} = d_1^{(0)} + \alpha_1 d_2^{(1)} A_{ee} = d_1^{(0)} + \alpha_1 \alpha_2 A_{ee}^2 d_1^{(0)}$. Continue to consider this set of events, and find the dipole moment of the dimer in the steady regime. Present the result for the total dipole moment in the form $d_{\text{tot}} \equiv d_1 + d_2 = d_1^{(0)} F$, where F is the function (of α_1, α_2, and A_{ee}) you have to derive.

2. Take into account that the quantum emitter emits not a monochromatic light but a decaying pulse whose spectrum is Lorentzian. Therefore, the dipole moment of the quantum emitter is not constant over the frequency axis but a Lorentzian function:

$$d_1^{(0)}(\omega) = \frac{d}{1 - \frac{\omega^2}{\omega_0^2} + j\frac{\omega \gamma_1}{\omega_0^2}}. \qquad (9.35)$$

Here, d is called the matrix element of the dipole transition (assumed to be known as well as γ_1). Similarly,

$$\alpha_1(\omega) = \frac{\alpha_1(0)}{1 - \frac{\omega^2}{\omega_0^2} + j\frac{\omega \gamma_1}{\omega_0^2}}. \qquad (9.36)$$

Notice that the static polarizability of the molecule is known: $\alpha_1(0) = 2d^2/\hbar\omega_0$ [9]. Let the scatterer polarizability have the same resonance frequency as that of the molecule:

$$\alpha_2(\omega) = \frac{\alpha_2(0)}{1 - \frac{\omega^2}{\omega_0^2} + j\frac{\omega \gamma_2}{\omega_0^2}}, \qquad (9.37)$$

where $\alpha_2(0)$, γ_2 are also known parameters, and $|\alpha_2(0)| \gg |\alpha_1(0)|$.

Neglecting the frequency dispersion of A_{ee} and having in mind the solution of Problem 1, derive the conditions for A_{ee} under which the dipole–dipole interaction at the resonance reduces to the Purcell effect – i.e., $d_1(\omega) \approx d_1^{(0)}(\omega)$. Calculate the Purcell factor F_P for this case analytically.

What is the condition of the inequality $|F(\omega_0)| \gg 1$? Does it obviously mean a strong Purcell effect or something else?

3. Assume that dipoles 1 and 2 are collinear and distanced from one another by the gap R. Adopt the quasi-static approximation:

$$A_{ee} = \frac{1}{2\pi\epsilon_0 R^3}. \tag{9.38}$$

Having in mind the solution of Problem 1 and formulas (9.35)–(9.37), prove that the spectral maximum of $d(\omega)$ in presence of the nanoantenna is different from the unperturbed fluorescence frequency ω_0. You may perform calculations in MATLAB to see it. For a reasonably chosen set of Lorentzian parameters of both dipoles 1 and 2, find the interval of distances R for which this difference (called the Lamb shift) becomes noticeable. Is the Lamb shift a blueshift, or a redshift?

4. Show that further decrease of R results in a split of the unique resonance of $d(\omega)$ into two maxima. One of them is sharp and high; another one is low and spread. These maxima are separated by a deep minimum. You may implement this calculation in MATLAB. If your calculation is correct, the high resonance peak holds at a higher frequency compared to the low peak. This regime is called the Fano resonance.

5. Show (e.g., numerically in MATLAB) that further decrease of R results in symmetrization of the Fano resonance – both peaks acquire a nearly equal height and width. This regime is called Rabi splitting.

6. Prove analytically that under condition $A_{ee}(\omega_0) \gg [\alpha_1(\omega_0)]^{-1}$ the total dipole moment at ω_0 becomes much smaller than $d_1^{(0)}(\omega_0)$ (fluorescence quenching). How must the phase of the resonant dipole moments $\mathbf{d}_1(\omega_0)$ and $\mathbf{d}_2(\omega_0)$ be related in this regime? What can you say about the absolute values?

7. Show numerically – e.g., in MATLAB – that under condition $A_{ee}(\omega_0) \gg [\alpha_1(\omega_0)]^{-1}$ the fluorescence is suppressed not only at the resonance frequency but in the whole resonance band.

8. Is it possible to find F from Problem 1 analytically for a nonreciprocal dimer, when $E_{12} = A_{ee}^{12}d_1$ and $E_{21} = A_{ee}^{21}d_2$ and $A_{ee}^{21} \neq A_{ee}^{12}$? If possible, please find it.

9. Why are Problems 1–7 relevant for PEF, whereas in conventional SERS, we may restrict by the regime of weak coupling when $F = 1 + \alpha_2 A_{ee}$? Which peculiarity of conventional SERS makes the strong coupling not feasible?

10. Why local field enhancement is absent in PEF?

9.8 Appendix: Additional Information on Plasmon-Enhanced Fluorescence and Surface-Enhanced Raman Scattering

9.8.1 On Fundamental Linear Effects in Plasmon-Enhanced Fluorescence

Let us consider the enhancement of fluorescence with the use of core–shell plasmonic nanoparticles (see Figure 9.6[b]) in more detail. The regime of plasmon-enhanced fluorescence, which is not accompanied by a noticeable distortion of the fluorescence spectral line, implies relative weakness of coupling between the quantum emitter and the plasmonic core of the core–shell PNP. This weakness means that the "feedback" of the plasmon resonance changes neither the resonant value of d_1 nor the spectral shape S_1 of the primary emission. The emission spectrum remains Lorentzian, and only its level increases [58, 61]. The resulting emission is the sum of the fields produced by two dipoles: the quantum emitter with spectrum $d_1(\omega) = d_{10}S_1(\omega)$ and the plasmonic nanoparticle with the spectrum $d_2(\omega) = d_{20}S_2(\omega)$, where the dominant contribution to radiation is that of d_2. Therefore, the width of the spectral line is determined by the spectrum S_2 of the dipole moment d_2 induced in the PNP. Therefore, the Purcell effect in PEF results in enhancement of fluorescence intensity and in a certain broadening of its band.

For a PNP with a fixed core radius ρ and a silica shell of thickness $\delta = r - \rho$, the maximal possible Purcell factor corresponds to the minimal δ for which the backward action of the PNP to the emitter is still negligible. For silver cores with radius ρ in the interval 60–70 nm the weak-coupling regime corresponds to $\delta > 8{-}10$ nm. For $\rho = 60$ nm (silver) and $\delta = 10$ nm (silica), the maximal (resonant) Purcell factor for radially polarized molecules is nearly equal to 50 [79], whereas the estimate (9.24) gives $F_P = 58$. If we can vary the thickness of the core, the maximal Purcell factor of such PNP is maximal namely for this δ.

When δ is smaller than the optimal value, a stronger coupling results in a modification of the dipole moment spectrum $d_1(\omega)$ compared to (9.13). In order to see this effect, one can use the model resulting from the solution of Problems 1–7 of this chapter. If we do it and analyze the spectrum of the total dipole moment, we will see that the emission spectral line slightly shifts for $\delta < 10$ nm, and when $\delta < 5$ nm, it reshapes. The concept of "Purcell factor" is not well defined in this case. The Purcell factor refers to the enhancement at the same frequency ω_0, whereas for small δ, the maximal enhancement holds at a different frequency. Sometimes there may be two local maxima of the spectral line.

If coupling is not very strong ($5 < \delta < 10$ nm), the spectral line experiences a shift, whereas the spectrum $d_1(\omega) + d_2(\omega)$ centered at the shifted resonance frequency ω_L remains practically Lorentzian. This effect is called *Lamb shift* [80], and ω_L is called the Lamb frequency. In this case, the aforementioned model based on the assumption of weak coupling is not valid – formula (9.24) is not applicable. Lamb shift is accompanied by a decrease of the spectral gain of fluorescence compared to its maximal value obtained for the optimal thickness δ. The regime with Lamb shift was called *intermediate coupling* in work [56]. For a radially polarized rhodamine molecule, the Purcell factor referred to the spectral maximum reduces from 50 (when $\delta = 10$ nm) to 12 (when $\delta = 5$ nm).

If the shell thickness reduces to $\delta = 3{-}4$ nm, coupling of the quantum emitter and the silver core becomes strong and results in an asymmetric splitting of the spectral line. In this regime, a narrowband spectral minimum arises in the spectrum of $|d_1(\omega) + d_2(\omega)|$. This spectrum has a sharp minimum below ω_0 and a Lorentz-like maximum above ω_0. This maximum is called *Fano resonance* [81]. The spectral maximum of the fluorescence gain in this case still equals 10–12, whereas at the minimum (Fano's spectral hole) F_P is nearly zero.

Further qualitative implications of increased coupling occurs when $\delta = 1$–3 nm. Here, the fluorescence spectrum reshapes once more. Two spectral maxima at $\omega_+ > \omega_0$ and $\omega_- < \omega_0$ arise at the frequency axis almost symmetrically with respect to ω_0, and at these maxima, the Purcell factor is of the order of unity or even smaller. These spectral maxima are separated by a substantial interval of frequencies. This effect is called *Rabi splitting* [9, 82]. In the abundant literature, it is treated as a purely quantum effect. However, it is fully analogous to the splitting of resonant frequencies of two identical coupled resonant circuits (see more details in the next section). The radiation in this regime is much lower than dissipation that drastically grows compared to the case of the isolated emitter. For some quantum emitters such as rhodamine in the case $\delta = 1$ nm, one observes the fluorescence quenching. This effect is considered in details in the next section. Here we only notice that this situation is inverse to the Purcell effect. Instead of enhancing the emission, the nanoparticle suppresses it.

All aforementioned physical effects, including fluorescence quenching, had been known, and all of them (except the Purcell effect), were numerically and experimentally studied very well prior to publication of [56]. In general, the recognized theory of all these effects is based on a semiclassical approximation of quantum mechanics. In the semiclassical approximation of quantum mechanics, the classical electric field equations are combined with the Shrödinger equation of a quantum system. This approximation is fully suitable for modeling fluorescence (but hardly suitable to describe the Raman radiation).

The theory of fluorescence in presence or in absence of different plasmonic nanostructures can be represented by three known models. Qualitatively, they give the same result, but quantitatively, they are more or less accurate in some particular cases. One semiclassical model introduces the so-called *Langevin equation* [83, 84]. Another one deduces the so-called *Maxwell–Bloch's equations* [85]. The third one results in the so-called *Lindblad master equations* [86]. All of them describe evolution of the density matrix of the quantum system in real time (in the case of fluorescence, during the process of fluorescence emission) governed by the interaction of two quantum subsystems 1 and 2 in a pumping field. Here, a PNP is also considered to be a quantum subsystem whose localized surface plasmon is quantized [86, 87].[3]

However, as it was already mentioned, for a more or less accurate analysis of such linear effects as plasmon enhancement of fluorescence, all these complex models are not really necessary. Quantum emitter 1 and nanoantenna 2 interact with one another as two classical Lorentzian scatterers. Though the optical transition as such is a quantum process, the result of this transition manifests as a classical radiation with a Lorenztian spectral line. Moreover, the frequency spectrum of this transition is modified in the presence of a nanoantenna in accordance with the Lorentzian dispersion. This is so because the polarizability of the two-level quantum system, in accordance with its accurate semiclassical model [9], is a Lorenztian function of frequency. For the given central frequency ω_0 of the spectral line, the quantum model determines only the matrix element d_{10} of the optical transition dipole moment and the individual decay rate γ_1 determining the spectrum of the individual emitter. These values can be tabulated for molecules and represent the input data for further classical calculations.

The case of weak coupling – the Purcell effect – has been considered above in detail. Note, that this classical method is useful also for studies of Raman radiation. In SERS, the interaction of a quantum emitter and a plasmonic scatterer is always weak. For weak interaction, it does not matter if the emitter is a two-level quantum system or a macroscopic object. The polarizability of the molecule is not involved in calculations of the Purcell factor.

9.8.2 Circuit Model of Rabi Oscillations and Transition to the Spaser

We have already introduced a circuit model for a PNP that delivers a correct result for the polarizability of Lorentz-resonant particles. Now, let us use the fact that the fluorescent emitter is also a Lorentzian scatterer as well as a PNP. This property allows us to consider the interaction of the quantum emitter and a nanoantenna as that of two classical resonant scatterers, despite the quantum nature of fluorescence. Thus, we come to the model of coupled resonant circuits depicted in Figure 9.10. Here, the ability of a quantum emitter to produce fluorescence is described by a negative resistor R_{neg}^{QE}.

In the following paragraphs, we use the term *nanoantenna* (NA) instead of a PNP, implying that the role of a PNP can be played by a plasmonic dimer or an oligomer, which may grant a stronger coupling to our quantum emitter than a single silver or gold nanoparticle.

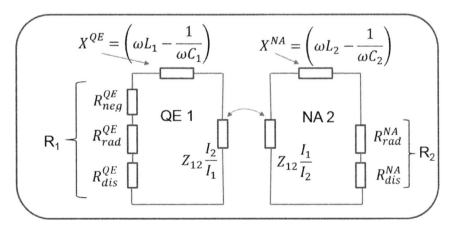

Figure 9.10   Circuit model of plasmon-enhanced fluorescence for the case when the nanoantenna is a resonant dipole [56].

The description of the emission by R_{neg}^{QE} is needed only in the case of strong coupling between the quantum emitter and the nanoantenna. In the case of weak coupling between them, the ability of the emitter to produce electromagnetic energy is properly described by a simple ideal current generator. Really, when the dipole moment $\mathbf{d}_1$ is not perturbed by the feedback from the NA, the radiating dipole moment is fixed, and we can express it in terms of an ideal current generator I_1 driving this dipole moment: $\mathbf{d}_1 = I_1 l_1 / j\omega$. The frequency spectrum of this generator is also fixed – it is the Lorentzian line of $\mathbf{d}_1$.

In the case of strong coupling, the fluorescence can be, in principle, modeled by a imperfect current generator – that with a finite shunt resistance or a imperfect voltage generator – a voltage source connected in series with some internal resistance. However, finding these equivalent shunt or series resistances is a difficult task. Therefore, in work [88], it was suggested to describe the generation of electromagnetic energy by a quantum emitter using *negative resistance*. For many radio-frequency generators, this concept is relevant in the regime of steady generation when the negative resistance is a differential resistance resulting from the linearized Volt-Ampere characteristics of a device. In the present case, the meaning of the negative resistor is different. It models the source of electromagnetic energy in the frequency range of emission sufficient for compensation of radiative and dissipative losses in the regime of steady fluorescence (called the continuous-wave regime). The advantage of this concept of fluorescence compared to the concepts of effective current or voltage generators is the possibility of finding this (a priori unknown) parameter very simply [88] and [64].[4]

If we admit this approach, it is possible to model and understand interactions of quantum emitters and resonant nanoantennas using the equivalent *LRC*-circuit presented in Figure 9.10. Emission of electromagnetic energy by quantum emitter 1 to free space is modeled by its radiation resistance and coupling to the adjacent NA by mutual impedance Z_{12}. Here, $R_{\text{neg}}^{QE} < 0$, whereas the radiation R_{rad}^{QE} and dissipation R_{dis}^{QE} resistances are positive.

We should understand that the model of negative resistance R_{neg}^{QE} cannot be used for calculation of radiated power. The value of the effective parameter R_{neg}^{QE} depends not only of the properties of the primary emitter, but also on the parameters of the nanoantenna and on coupling between the two particles (in the circuit model, on the parameters of the second circuit and on coupling between the two circuits). Really, for an isolated emitter, its total resistance $R_1 = R_{\text{neg}}^{QE} + R_{\text{rad}}^{QE} + R_{\text{dis}}^{QE}$ should vanish at the resonance, so that a non-zero current is supported. Thus, for this case, we have $R_{\text{neg}}^{QE} = -(R_{\text{rad}}^{QE} + R_{\text{dis}}^{QE}) \approx -R_{\text{rad}}^{QE}$. However, this value of negative resistance is not sufficient in the presence of the nanoantenna – in this case, it must compensate its radiative and dissipative losses, too. This important limitation of the model comes from the fact that the complete circuit model of a power source needs two parameters (source voltage or current and internal impedance) while in this model, we use only one parameter: the effective negative resistance.

However, this model is still very useful because it allows us to find eigenfrequencies of the system, where the total reactances of both circuits vanish. Assuming that the primary source has enough available power, negative resistance can compensate the total loss, and non-zero currents may flow in the effective circuits. Thus, the model is not suitable for studying the shape of fluorescence spectral lines and the fluorescence amplitude, but it is useful for finding the eigenmodes.

As to nanoantenna 2, its self-impedance comprises the resistance R_2 of two positive components R_{rad}^{NA} and R_{dis}^{NA} and reactance $X^{NA} = \omega L_2 - 1/\omega C_2$. Field coupling between objects 1 and 2 is modeled by the mutual impedance $Z_{12} = Z_{21}$ that results also in impedance $Z_i = Z_{12}I_2/I_1$ induced in the emitter. Here, I_1 and I_2 are the effective currents in the emitter and nanoantenna, respectively. When we calculate Z_i, we describe the system as a single circuit 1, because the impact of particle 2 is fully accounted for by Z_i. The induced impedance Z_i was derived in [56] in terms of the constant A_{ee} describing the dipole–dipole interaction of objects 1 and 2 (index ee means that both particles 1 and 2 are electric dipoles):

$$Z_i = \frac{I_2}{I_1} Z_{12} = \frac{l_1^2 l_2^2 A_{ee}^2}{\omega^2 Z_2}, \tag{9.39}$$

where $l_{1,2}$ are the effective lengths of the emitter and nanoantenna, respectively. For the case when the NA is a simple plasmonic sphere, we may substitute into (9.39) expression (9.23) for A_{ee}. Emission process corresponds in the circuit model to free oscillations of the circuit at the frequencies where impedances satisfy $Z_1 + Z_i = 0$. This is a complex-valued equation, whose real part determines the value of the negative resistance (needed to compensate the radiative and dissipative loss), and the imaginary part determines the emission frequency that can be different from ω_0.

Instead of calculating Z_i, let us find an analytical expression for the mutual impedance Z_{12}. It is, by definition, the electromotive force $\mathcal{E}_{12}$ induced in the NA by the unit effective current $I_1 = 1$ of the emitter:

$$Z_{12} = E_{12}l_2 = A_{ee}d_1 l_2, \quad d_1 = I_1 l_1/j\omega, \tag{9.40}$$

and since $I_1 = 1$, we have

$$Z_{12} = \frac{l_1 l_2 A_{ee}}{j\omega}. \tag{9.41}$$

The currents in the two dipoles satisfy the *coupled Kirchhoff equations*

$$I_1 Z_1 + I_2 Z_{12} = 0, \qquad I_1 Z_{12} + I_2 Z_2 = 0, \tag{9.42}$$

and the eigenfrequencies are found by equating the determinant to zero:

$$Z_1 Z_2 - Z_{12}^2 = 0. \tag{9.43}$$

After the substitution of formula (9.41), this equation becomes

$$-\omega^2 Z_1 Z_2 = (l_1 l_2 A_{ee})^2. \tag{9.44}$$

Substituting $Z_{1,2} = R_{1,2} + j\omega L_{1,2} + 1/j\omega C_{1,2}$ into Eq. (9.44), we obtain an equation for the eigenfrequencies: the frequencies at which $I_{1,2} \neq 0$:

$$\left(1 - \frac{\omega^2}{\omega_0^2} + j\omega C_1 R_1\right)\left(1 - \frac{\omega^2}{\omega_0^2} + j\omega C_2 R_2\right) = \kappa^2. \tag{9.45}$$

Here, it is denoted

$$\kappa = \sqrt{C_1 C_2} l_1 l_2 A_{ee}. \tag{9.46}$$

This parameter is the *dimensionless coupling coefficient* of the dimer.

Within the resonance band, we may neglect frequency dispersion of the effective resistances and approximate $R_{1,2}(\omega) = R_{1,2}(\omega_0)$, in which case all the parameters in (9.45) are frequency independent. This equation coincides with the eigenmode equation of two inductively coupled circuits known in electronics or elastically coupled oscillators known in mechanics. In fact, coupling of these two circuits is not purely inductive because the mutual impedance expressed by (9.41) for complex A_{ee} has all possible components: inductive, capacitive, and resistive. However, in the case of coinciding individual resonances and small distance between QE 1 and NA 2, A_{ee}, l_1, and l_2 are real-valued quantities. Then, κ is real, and it is instructive to rewrite (9.45) as

$$\left(1 - \frac{\omega^2}{\omega_0^2}\right)^2 + j\frac{\omega}{\omega_0} a\left(1 - \frac{\omega^2}{\omega_0^2}\right) + \frac{\omega^2}{\omega_0^2}\kappa_0^2 = \kappa^2. \tag{9.47}$$

Here, it is denoted $a = \omega_0(C_1 R_1 + C_2 R_2)$ and $\kappa_0^2 = -\omega_0^2 C_1 C_2 R_1 R_2 = \omega_0^2 C_1 C_2 |R_1| R_2$. In practice, κ_0 is of the order of 0.01–0.1, whereas κ, depending on A_{ee}, may vary from nearly zero up to nearly 0.1 [56]. Much higher values of κ compared to κ_0 for a NA performed as a spherical PNP are not feasible. However, if the molecules are distanced from a plasmonic nanosphere by a small gap 1–2 nm, κ may become close to κ_0 [56].

Let us consider the case when the system is stable – that is, there is neither decaying nor growing oscillations. Then the eigenfrequencies of the system – solutions of (9.47) – must be real, and we have to let $a = 0$ – i.e., $C_1 R_1 + C_2 R_2 = 0$. Therefore, the effective negative resistor entering R_1 is determined and equals $|R_{neg}^{QE}| = R_{rad}^{QE} + (R_{rad}^{NA} + R_{dis}^{NA})C_2/C_1$. The negative resistance exactly compensates in this regime all losses (here, we have neglected R_{dis}^{QE}, which is small compared to R_{rad}^{QE}).

The case when $\kappa < \kappa_0$ in Eq. (9.47) corresponds to weak coupling. Then there is an only eigenfrequency (which is nearly equal to ω_0 with a small redshift). As we have already noticed, the concept of negative resistance is introduced for the regime of strong coupling. The coupling turns strong when κ approaches κ_0 and exceeds its value.

There are two scenarios when the ratio κ/κ_0 grows up to unity. In the first scenario, we fix the effective length of molecule l_1 entering (9.46) (i.e., fix the dipole moment d of the optical transition) and increase A_{ee}. This scenario corresponds to a molecule approaching to NA 2. In this scenario, the system fluorescence varies non-monotonously. When the coupling is still weak, the total dipole moment grows with the reduced distance as in plasmon-enhanced fluorescence. This is the regime of the Purcell effect. Further, when κ approaches κ_0, the total dipole moment starts to decrease. Simultaneously, the spectral line reshapes – instead of the Lorentz line, it becomes the so-called *Fano spectral line*. This circuit model cannot describe this reshaping because it only refers to the eigenfrequencies. To analyze the spectrum, we have to calculate the total dipole moment of the system versus frequency using the electromagnetic model of coupled dipoles. Finally, when κ attains κ_0 (and, perhaps, slightly exceeds it), the dipole moment of the system at ω_0 vanishes. This can be seen from the circuit model. In the case when $\kappa = \kappa_0$, the solution $\omega \approx \omega_0$ still exists. However, in this case, currents I_1 and I_2 have opposite directions. Using Eqs. (9.41) and (9.42), one can deduce that their values are related as $I_1 l_1 = -I_2 l_2$. It means that the total dipole moment of the system vanishes; this is why the system practically does not radiate at the eigenfrequency. The electromagnetic model shows that the total dipole moment is not exactly nullified at other frequencies. However, in the whole fluorescence spectrum, it is suppressed compared to $d_1^{(0)}$, the unperturbed Lorentzian dipole moment of a single emitter. This regime implies significant reduction of radiation in the whole spectrum of fluorescence. When we bring a fluorescent molecule to a NA so that the gap between them is as small as 1–2 nm, the fluorescence reduces significantly – it practically disappears. In this regime, the emitter not only transfers the whole emitted power to the NA, also this power is fully spent for heating the NA. This phenomenon is called *fluorescence quenching*.

In the second scenario, we fix A_{ee} (assuming that its absolute value is large enough so that κ is close to κ_0) and increase l_1. If we do it, we also increase κ_0 because l_1 enters R_{rad}^{QE}, which is a part of R_1 entering κ_0^2 as a factor. However, this increase is slight, whereas κ^2 is proportional to l_1^2. In this scenario, we may have in mind a fluorescent emitter performed as a semiconductor nanocrystal of sizes 0.5–5 nm. Such a crystal is called *quantum dot*. When we increase the radius of the quantum dot from 0.5 to 5 nm, its transition dipole moment d (and, consequently, l_1) increase quadratically. In this case, we may easily pass the limit $\kappa = \kappa_0$ and achieve strongly overcritical regime with $\kappa \gg \kappa_0$. In this regime, (9.47) simplifies to

$$\left(1 - \frac{\omega^2}{\omega_0^2}\right)^2 \approx \kappa^2, \qquad (9.48)$$

and its solutions are as follows:

$$\omega_\pm \approx \omega_0 \sqrt{1 \pm \kappa} \approx \omega_0 \left(1 \pm \frac{\kappa}{2}\right). \qquad (9.49)$$

Two frequencies, $\omega_\pm$, describe the aforementioned Rabi oscillations. The difference $\omega_+ - \omega_0 = \omega_0 - \omega_- = \omega_0 \kappa / 2$ is called the Rabi frequency shift Ω_R. In [56], it was shown that in the case when NA 2 is a simple plasmonic nanosphere, this Ω_R coincides with the correct result for the Rabi shift derived from the quantum model. In work [64], a similar result was obtained for a dimer-type NA. Thus, this simple circuit model correctly predicts the Rabi oscillations in a fluorescent system. These Rabi oscillations are nothing else but the normal modes of two coupled classical oscillators that have the same resonance frequency.

At the Rabi frequencies, the real part R_i of the impedance Z_i induced by the NA in the emitter is not some additional radiation resistance. The radiative Purcell factor $F_P^{\text{rad}} = |d_{\text{tot}}/d_1|^2$ at $\omega_\pm$ is much lower than the total Purcell factor $F_P^{\text{tot}} = R_i/R_1 + 1$. Their difference is called the non-radiative Purcell factor. The electromagnetic model shows that it may attain dozens, whereas the radiative one is of the order of unity [64]. Thus, in the regime of Rabi oscillations, the NA grants a drastic increase of dissipation, but the fluorescence is not quenched. Compared to the power radiated by an isolated emitter (at frequency ω_0) in the regime when $\kappa \gg \kappa_0$, the radiation (at ω_+ and ω_-) may be slightly enhanced or slightly suppressed, depending on the explicit parameters of the system.

Probably the most striking example of the use of this simplest circuit model is the explanation of the operation of the so-called *spaser*. Assume that a fluorescent molecule is oriented radially with respect to a plasmonic nanosphere, and the gap between the center of the molecule and the metal surface is as small as $g = 0.5–1$ nm (1 nm is the length of a rhodamine molecule). Then κ is slightly larger than κ_0. Practically, it means that the radially polarized molecule is located on the surface of a metal sphere. It implies the largest possible (and almost real) A_{ee}. It is the regime of slightly overcritical coupling when $\kappa^2 = \kappa_0^2(1 + \xi)$, where ξ is a small positive number.

Further, assume that the pumping of the fluorescent molecule is very strong and the balance between the negative and positive resistances in the system is broken. In other words, assume that $a \neq 0$ in Eq. (9.47). Then this equation has no purely real solutions. However, solutions of this equation though complex make sense if $|a| \gg \xi$. It is easy to check by a direct substitution that Eq. (9.47) has a root,

$$\left(\frac{\omega_s}{\omega_0}\right)^2 \approx 1 - j\frac{\xi}{2a}, \tag{9.50}$$

whose real part delivers the resonance frequency ω_0, and the imaginary part (called *increment*) is a small positive number. Solution (9.50) is an unstable eigenmode, whose increment describes broken equilibrium at the initial stage of accumulation of the electromagnetic energy in the system. This accumulation is nothing but the generation of the localized surface plasmon in the nanoantenna at frequency ω_0. In the linear regime, currents $I_{1,2}$ in the emitter and NA exponentially grow.[5]

Thus, we see that the generation of a localized surface plasmon occurs if the negative resistance of a molecule located at the surface of a plasmonic nanosphere is sufficiently high. The phenomenon of this generation, previously referred to quantum nanophotonics, was theoretically revealed by D. Bergman and M. Stockman in [89, 90]. Later, it was studied in a number of works theoretically and experimentally. Several models of spaser were developed in the literature, and the familiarization with any of them demands a profound knowledge of quantum mechanics. Meanwhile, this simple, classical, phenomenological model allows us to find the necessary and sufficient conditions of generation.

Notice that (9.50) is not an only solution of Eq. (9.47) for the case under study (when $a \neq 0$ and $\kappa^2 = \kappa_0^2(1 + \xi)$). It is possible to show that Eq. (9.47) has, in this case, two other solutions whose real parts nearly equal $\omega_+ = \omega_0(1 + \kappa_0/2)$ and $\omega_- = \omega_0(1 - \kappa_0/2)$, respectively, and whose imaginary parts are small negative numbers (decrements). It means that Rabi oscillations coexist in the system with the solution describing the plasmon generation. These Rabi oscillations are also not stable, but they decay in time instead of growth. The electromagnetic model allows one to see that the total dipole moment vanishes at all three eigenfrequencies: ω_0, ω_+, and ω_-. In other words, all these oscillations – one growing and two decaying – are non-radiative. On the initial stage of generation, the spaser does not radiate. All these observations, following from the circuit model, fit the results of strict quantum models and match available experimental data [64].

Notice that the regime when $|a| > \kappa^2 - \kappa_0^2$ is not realizable with a single molecule attached to a nanoantenna. The generation ability of one fluorescent molecule is rather weak because for a very strong pumping, the emission of a molecule saturates. However, if NA 2 is covered with N fluorescent molecules and $N \gg 1$, the situation changes. Then the negative resistance of a single molecule is effectively multiplied by N. This multiplication implics that N molecules emit incoherently. It is well known that the radiated powers of incoherent sources sum up. If N quantum emitters could emit light coherently, R_{neg}^{QE} should have been multiplied by N^2. In fact, fluorescence is spontaneous radiation, coherence is impossible in the linear regime, and R_{neg}^{QE} grows only N times. However, if the whole surface of the plasmonic nanosphere is covered by fluorescent molecules (as it holds in spasers), N is very large, and multiplication of R_{neg}^{QE} by N is sufficient to overcompensate for the radiative and dissipative losses of the NA. In other words, covering the whole surface of the nanopshere by attached fluorescent molecules and applying sufficient pumping, we engineer the generation of a localized surface plasmon in the NA. We can formulate the conditions of generation as follows:

$$a < 0, \qquad 0 < \kappa - \kappa_0 < |a|. \tag{9.51}$$

In work [64], the effective negative resistance per one fluorescent emitter was found, which models the same effects as the strict quantum theory [83]. The possibility of such fitting means that the probably simplest circuit model makes sense even for a seemingly purely quantum phenomenon of nanophotonics as generation of localized surface plasmons in spasers [89].[6]

Before we finish the discussion of fluorescence, we need to prevent the reader from believing that the previously discussed effects – Fano resonance, Lamb shift, Rabi oscillations, and emission quenching – are *always* classical effects. For the moment, it was proven that the classical model has predictive power only for usual PEF. In the usual PEF, the fluorescence is a purely linear effect – this is why it obeys the Stokes law of down conversion. Besides the usual fluorescence, there exists the so-called *up-conversion fluorescence*, also called anti-Stokes fluorescence. This is a nonlinear fluorescence, and it can be also enhanced by a nanoantenna [91]. This process can hardly be described as a classical one because here quantum emitter 1 is a multilevel quantum system. The spectrum of fluorescence in such quantum objects is not a classical Lorentzian line [92]. The polarizability is also not compatible with idea of a classical resonance. It is difficult to expect that our classical model could be helpful in this case.

9.8.3 SERS: Physicists and Chemists

The model of the Raman gain resulting in formula (9.34) is in perfect agreement with most works on SERS [5, 31–35, 66, 93]. This formula was derived from the theory of the Purcell effect and includes the Purcell factor, which equals the local field intensity enhancement L^2. As it was already mentioned, neither in [5, 32–35, 66] nor in any other available work on SERS can the name of Purcell be found.

The main difference between PEF and SERS is that in PEF, the local field acting on the molecule is not enhanced by the resonant nanostructure because the pumping radiation has the frequency spectrum located above the plasmon resonance band. If the nanoantenna in PEF would have two plasmon resonances, one at the frequency of the pumping radiation and one at the frequency of fluorescence, the gain in PEF would be as huge as the gain in SERS. So, in what concerns the mechanism of gain, there is no fundamental difference between PEF and SERS. However, in the literature dedicated to PEF, almost all authors refer to the Purcell effect and deal with the Purcell factor. Why is the name of the nobelist Edward Purcell, well known to experts in PEF, not familiar to experts in SERS?

In a recent work [94], the author, who is an expert in nanophotonics, calculated the Raman gain for a molecule located in an open plasmonic cavity and obtained this gain as a product of the factors both proportional to the resonator quality Q and inverse proportional to the cavity mode volume V (as in the famous Purcell formula (9.10)). However, the *cavity enhancement factor* introduced in [94] is a product of the Purcell factor and the local intensity enhancement granted by the cavity. Therefore, even the author of [94] has not noticed that this Raman gain is nothing but the square power of the Purcell factor.

It can be noticed that the equivalence of the Purcell factor F_P to L^2 is not a specific property of a spheroidal scatterer. It holds for whatever reciprocal substrate – with clusters of plasmonic nanoparticles, with nanowires, nanocones, or nanopyramides [31, 34, 35, 72, 95]. How important is it to mention that the second factor L^2 in (9.34) is the Purcell factor? Perhaps it would be not important if this factor were uniquely called by the experts in SERS and uniquely explained. However, even in three famous initial overviews of SERS theory [5, 32, 33], this factor is explained in such a complex manner that the equivalence of all three explanations is not evident. Following these works, many authors correctly wrote that besides the local field enhancement at the source frequency, there is a second enhancement factor of electromagnetic nature. However, this second electromagnetic factor is explained in different ways in different works.

And in different works, the second factor is called different names: dipole coupling enhancement in [5, 32, 35], nanocavity enhancement in [34], analytical (why analytical?) enhancement in [31], surface-plasmon enhancement in [96], and single-substrate enhancement in [93]. Sometimes, this second factor (which is, in fact, a single factor F_P) is presented as a product of two electromagnetic factors called either image-field enhancement or first-layer enhancement and surface electron-photon coupling factor, respectively (see, e.g., in [95, 97, 98]).

There are works in which the equivalence of the second factor in the Raman gain to L^2 was assumed to be a consequence of reciprocity. However, that was proven in none of these works, and F_P was also called by different, vague names, multiplying possibilities of terminological confusion. Therefore, we think that the statement that the second factor in the Raman gain is the Purcell factor is methodologically useful.

It is not surprising that many electrochemists reading all these works would not understand why the second factor in the electromagnetic Raman gain is equal to L^2 when the molecule is located on the surface of a plasmonic protrusion or a nanoparticle. Why does this factor, similarly to the local intensity enhancement, sharply drop when the molecule is taken away from the surface to the distance of a few nm, and why is the further decrease with the distance not so sharp? Unclear explanations of the electromagnetic model by experts/physicists motivated chemists to search for alternative explanations to G_R. This is why the theory of charge-transfer complexes (see, e.g., in [36–41]) was developed. From 2007 to 2011, it became clear that although the charge-transfer complexes really exist on metal substrates and may modify the shape of the Raman spectral lines, they cannot drastically change the power of the Raman signal (see, e.g., in [99–101]). The situation is different only for semiconductor substrates with protrusions resonating in the infrared range, and in that case, the electrochemical mechanism of the Raman gain can even be dominating over the electromagnetic enhancement [102]. However, for plasmonic SERS substrates operating in the visible range, the Raman gain is practically equal to the electromagnetic gain. This conclusion has been made in recent works [101–103] written by interdisciplinary teams of experts. These works present the modern consolidated point of view of SERS on metal plasmonic substrates by both chemical and physical communities. Only for some complex-shape and relatively large organic molecules, the charge-transfer complexes contribute into the Raman gain. However, even for these molecules, the electrochemical gain does not exceed 4, whereas the electromagnetic gain attains 10^6 and more. For the majority of molecules, the electrochemical gain on plasmonic substrates equals unity [101–103]. So, in contrast to Wikipedia (https://en.wikipedia.org/wiki/Surface-enhanced_Raman_spectroscopy), recently, the consensus between the chemists and physicists about the relationship between the electromagnetic and electrochemical factors in the Raman gain has been reached. Now, the time demands to clear up the terminological mess in the electromagnetic theory of SERS, and for it, the present book can be helpful.

Bibliography

[1] A. Smekal, "Zur Quantentheorie der Dispersion," *Die Naturwissenschaften* **11**, 873–875 (1923).

[2] C. Raman and K. Krishnan, "A new radiation," *Nature* **121**, 501–502 (1928).

[3] L. Brillouin, "Diffusion de la lumière et des rayons X par un corps transparent homogène," *Annale de Physique* **17**, 88–122 (1922).

[4] G. Landsberg and L. Mandelstam, "Über die Lichtzerstreuung in Kristallen," *Zeitschrift für Physik* **50**, 769–780 (1928).

[5] V. I. Emelyanov and N. I. Koroteev, "Giant Raman scattering of light by molecules adsorbed on the surface of a metal," *Soviet Physics Uspekhi* **24**, 864–891 (1981).

[6] D. F. Walls, "Quantum theory of the Raman effect," *Zeitschrift für Physik* **237**, 224–233 (1970).

[7] D. Lin-Vien, N. B. Colthrup, W. G. Fateley, and J. G. Grasselli, *The Handbook of Infrared and Raman Characteristic Frequencies of Organic Molecules*, J. Wiley and Sons (1991).

[8] J. R. Ferraro and K. Nakamoto, *Introductory Raman Spectroscopy*, Academic Press (1994), pp. 107.

[9] L. Novotny and B. Hecht, *Principles of Nano-Optics*, Cambridge University Press (2006).

[10] B. N. J. Persson, "On the theory of surface-enhanced Raman scattering," *Chemical Physics Letters* 82, 561–565 (1981).

[11] P. P. Shorygin and L. L. Krushinski, "A propos de la théorie de la diffusion combinatoire de la lumiére," *Comptes rendus de l'Académie des Sciences de l'URSS* (French translation of Russian journal *Doklady Akademii Nauk SSSR*) **133**, 337–340 (1960).

[12] D. G. Rea, "On the theory of the resonance Raman effect," *Journal of Molecular Spectroscopy* **4**, 499–506 (1960).

[13] A. C. Albrecht, "On the theory of Raman intensities," *Journal of Chemical Physics* **34**, 1476–1483 (1961).

[14] R. Boyd, *Nonlinear Optics* (3rd edn.), Academic Press-Elsevier (2008), pp. 13–15.

[15] D. C. Harris and M. D. Bertolucci, *Symmetry and Spectroscopy: An Introduction to Vibrational and Electronic Spectroscopy*, Oxford University Press (1978), pp. 198–200.

[16] P. Hendra, C. Jones, and G. Warnes, *Frequency-Time Raman Spectroscopy*, Ellis Horwood Ltd (1991).

[17] A. Fadini and F.-M. Schnepel, *Vibrational Spectroscopy: Methods and Applications*, Ellis Horwood Ltd (1989).

[18] N. B. Colthrup, L. H. Daly, and S. E. Wiberley, *Introduction to Infrared and Raman Spectroscopy* (3rd edn.), Academic Press (1990).

[19] J.-X. Cheng and X. S. Xie, eds. *Coherent Raman Scattering Microscopy*, CRC Press (2013).

[20] R. L. McCreery, *Raman Spectroscopy for Chemical Analysis*, John Wiley and Sons (2000).

[21] D. M. Adams, *Metal-Ligands and Related Vibrations*, Edward Arnold Ltd (1967).

[22] E. Smith and G. Dent, *Modern Raman Spectroscopy: A Practical Approach*, J. Wiley and Sons (2005).

[23] T. Ichimura, N. Hayazawa, M. Hashimoto, Y. Inoue, and S. Kawata, "Tip-enhanced coherent anti-Stokes Raman scattering for vibrational nanoimaging," *Physical Review Letters* **92**, 220801 (2004).

[24] J. P. Ogilvie, M. Cui, D. Pestov, A. V. Sokolov, and M. O. Scully, "Time-delayed coherent Raman spectroscopy," *Molecular Physics* **106**, 587–594 (2008).

[25] M. Cui, J. P. Skodack, and J. P. Ogilvie, "Chemical imaging with Fourier transform coherent anti-Stokes Raman scattering microscopy," *Applied Optics* **47**, 5790–5798 (2008).

[26] S. Yampolsky, D. A. Fishman, S. Dey, et al., "Seeing a single molecule vibrate through time-resolved coherent anti-Stokes Raman scattering," *Nature Photonics* **8**, 650–656 (2014).

[27] M. Fleischmann and S. Pons, "Electrochemically induced nuclear fusion of deuterium," *Journal of Electroanalytical Chemistry* **261**, 301–308 (1989).

[28] D. C. Rislove, "A case study of inoperable inventions: Why is the USPTO patenting pseudoscience?" *Wisconsin Law Review* **4**, 1302–1307 (2006).

[29] M. Fleischmann, P. J. Hendra, and A. J. Mc-Quillan, "Raman spectra of pyridine adsorbed at a silver electrode," *Chemical Physics Letters* **26**, 163–166 (1974).

[30] S. Nie and S. Emory, "Probing single molecules and single nanoparticles by surface-enhanced Raman scattering," *Science* **275**, 1102–1106 (1997).

[31] K. Kneipp, M. Moskovits, and H. Kneipp, *Surface-Enhanced Raman Scattering: Physics and Applications*, Springer (2007).

[32] A. M. Brodskii and M. I. Urbakh, "The effect of the microscopic structure of metal surfaces on their optical properties," *Soviet Physics Uspekhi* **25**, 810–832 (1982).

[33] M. Moskovits, "Surface-enhanced spectroscopy," *Reviews of Modern Physics* **57**, 783–826 (1985).

[34] F. J. Garcıa-Vidal and J. B. Pendry, "Collective theory for surface-enhanced Raman scattering," *Physical Review Letters* **77**, 1163–1166 (1996).

[35] E. Zeman and G. Schatz, "An accurate electromagnetic theory of surface enhancement factors for silver, gold, copper, lithium, sodium, aluminum, gallium, indium, zinc, and cadmium," *Journal of Physical Chemistry* **91**, 634–643 (1987).

[36] D. L. Jeanmaire and R. P. van Duyne, "Surface Raman electrochemistry. Part I: Heterocyclic, aromatic and aliphatic amines adsorbed on the anodized silver electrode," *Journal of Electroanalytical Chemistry* **84**, 1–20 (1977).

[37] J. R. Lombardi, R. L. Birke, T. Lu, and J. Xu, "Charge-transfer theory of surface enhanced Raman spectroscopy: Herzberg-Teller contributions," *Journal of Chemical Physics* **84**, 4174–4175 (1986).

[38] A. B. Myers, "Resonance Raman intensities and charge-transfer reorganization energies," *Chemical Reviews* **96**, 911–926 (1996).

[39] T. E. Furtak and D. Roy, "Nature of the active site in surface-enhanced Raman scattering," *Physical Review Letters* **50**, 1301–1304 (1983).

[40] J. R. Lombardi and R. L. Birke, "A unified approach to surface-enhanced Raman scattering," *The Journal of Physical Chemistry C* **112**, 5605–5617 (2008).

[41] E. Burstein, Y. J. Chen, S. Lundquist, and E. Tosatti, "Giant Raman scattering by adsorbed molecules on metal surfaces," *Solid State Communications* **29**, 567–570 (1979).

[42] Y. Chu, M. G. Banaee, and K. B. Crozier, "Double-resonance plasmon substrates for surface-enhanced Raman scattering with enhancement at excitation and Stokes frequencies," *ACS Nano* **4**, 2804–2810 (2010).

[43] M. Cox, D. R. Nelson, and A. L. Lehninger, *Lehninger's Principles of Biochemistry*, W.H. Freeman Publishers (2008).

[44] H. Sahoo, "Fluorescent labeling techniques in biomolecules: A flashback," *RSC Advances* **2**, 7017–7029 (2012).

[45] J. S. Paige, K. Y. Wu, and S. R. Jaffrey, "RNA mimics of green fluorescent protein," *Science* **33**, 642–646 (2011).

[46] M. Kang, P. Xenopoulos, S. Munoz-Descalzo, X. Lou, and A.-K. Hadjantonakis, "Live imaging, identifying, and tracking single cells in complex populations in vivo and ex vivo," *Methods in Molecular Biology* **1052**, 109–123 (2013).

[47] A. Stallmach, C. Schmidt, A. Watson, and R. Kiesslich, "An unmet medical need: Advances in endoscopic imaging of colorectal neoplasia," *Journal of Biophotonics* **4**, 482–489 (2011).

[48] T. Fujii, M. Kamiya, and Y. Urano, "In vivo imaging of intraperitoneally disseminated tumors in model mice by using activatable fluorescent small-molecular probes for activity of cathepsins," *Bioconjugate Chemistry* **25**, 1838–1846 (2014).

[49] S. Sensam, C. L. Zavaleta, E. Segal, et al., "A clinical wide-field fluorescence endoscopic device for molecular imaging demonstrating cathepsin protease activity in colon cancer," *Molecular Imaging and Biology* **18**, 820–829 (2016).

[50] Z. Bai, R. Chen, P. Si, et al., "Fluorescent pH sensor based on Ag-SiO_2 core-shell nanoparticle," *ACS Applied Materials* **5**, 5856–5860 (2013).

[51] F. Tam, G. P. Goodrich, B. R. Johnason, and N. J. Halas, "Plasmonic enhancement of molecular fluorescence," *Nano Letters* **7**, 496–501 (2007).

[52] J. R. Lakowicz, K. Ray, M. Chowdhury, et al., "Plasmon-controlled fluorescence: A new paradigm in fluorescence spectroscopy," *Analyst* **133**, 1308–1346 (2008).

[53] S. A. Camacho, P. H. B. Aoki, P. Albella, et al., "Increasing the enhancement factor in plasmon-enhanced fluorescence with shell-isolated nanoparticles," *Journal of Physics Chemistry C* **120**, 20530–20535 (2015).

[54] M. Bauch, K. Toma, Q. Zhang, and J. Dostalek, "Plasmon-enhanced fluorescence biosensors: A review," *Plasmonics* **9**, 781–799 (2014).

[55] R. Y. He, C. Y. Lin, K. C. Chu, et al., "Imaging live cell membranes via surface plasmon-enhanced fluorescence and phase microscopy," *Optics Express* **18**, 3649–3659 (2010).

[56] C. Simovski, "Circuit model of plasmon-enhanced fluorescence," *Photonics* **2**, 568–593 (2015).

[57] E. M. Purcell, "Spontaneous emission probabilities at radio frequencies," *Physical Review* **69**, 681–687 (1946).

[58] A. Kavokin, J. Baumberg, G. Malpuech, and F. Laussy, *Microcavities*, Clarendon Press (2006).

[59] W. Demtröder, *Atoms, Molecules and Photons: An Introduction to Atomic, Molecular and Quantum Phyiscs*, Springer, Berlin (2006).

[60] Q. Gu, B. Slutsky, F. Vallini, et al., "Purcell effect in sub-wavelength semiconductor lasers," *Optics Express* **21**, 15603–15617 (2013).

[61] A.E. Krasnok, A.P. Slobozhanyuk, C.R. Simovski, et al., "An antenna model for the Purcell effect," *Scientific Reports* **5**, 12956–12978 (2015).

[62] V. Klimov, *Nanoplasmonics*, Pan Stanford Publishing (2014).

[63] E. Dulkeith, M. Ringler, T. A. Klar, and J. Feldmann, "Gold nanoparticles quench fluorescence by phase-induced radiative rate suppression," *Nano Letters* **5**, 585–589 (2005).

[64] C. Simovski, "Circuit theory of metal-enhanced fluorescence," *Photonics and Nanostructures: Fundamentals and Applications* **36**, 100712 (2019).

[65] C. Simovski, "Point dipole model for metal-enhanced fluorescence," *Optics Letters* **44**, 2697–2700 (2019).

[66] J. P. Goudonnet, J. L. Bigeon, R. J. Warmack, and T. L. Ferrell, "Substrate effects on the surface-enhanced Raman spectrum of benzoic acid adsorbed on solver oblate microparticles," *Physical Review B* **43**, 4605–4612 (1991).

[67] S. I. Maslovski and C. R. Simovski, "Purcell factor and local intensity enhancement in surface-enhanced Raman scattering," *Nanophotonics* **8**, 429–434 (2019).

[68] M. Rybin, S. Mingaleev, M. Limonov, and Y. Kivshar, "Purcell effect and Lamb shift as interference phenomena," *Scientific Reports* **6**, 20599 (2016).

[69] M. I. Stockman, "Electromagnetic Theory of SERS," in *Surface-Enhanced Raman Scattering*, K. Kneipp, M. Moscovits and H. Kneipp, eds., Springer (2006).

[70] S. Xiao, N. A. Mortensen, and A. P. Jauho, "Nanostructure design for surface-enhanced Raman spectroscopy: Prospects and limits," *Journal of the European Optical Society* **3**, 08022 (2008).

[71] J. B. Jackson and N. J. Halas, "Surface-enhanced Raman scattering on tunable plasmonic nanoparticle substrates," *Proceedings of the National Academy of Sciences of the United States of America* **101**, 17930–17935 (2004).

[72] K. A. Willets and R. P. Van Duyne, "Localized surface plasmon resonance spectroscopy and sensing," *Annual Review of Physical Chemistry* **58**, 267–297 (2007).

[73] L. D. Tuyen, A. C. Liu, C.-C. Huang, et al., "Doubly resonant surface-enhanced Raman scattering on gold-nanorod decorated inverse opal photonic crystals," *Optics Express* **20**, 29266–29275 (2012).

[74] M. K. Schmidt, R. Esteban, A. Gonzalez-Tudela, G. Giedke, and J. Aizpurua, "Quantum mechanical description of Raman scattering from molecules in plasmonic cavities," *ACS Nano* **10**, 6291–6298 (2016).

[75] H. Rong, R. Jones, A. Liu, O. Cohen, D. Hak, A. Fang, and M. Paniccia, "A continuous-wave Raman silicon laser," *Nature* **433**, 725–728 (2005).

[76] D. M. Kuncicky, B. G. Prevo, and O. D. Velev, "Controlled assembly of SERS substrates templated by colloidal crystal films," *Journal of Materials Chemistry* **16**, 1207–1211 (2006).

[77] D. M. Kuncicky, S. D. Christesen, and O. D. Velev, "Role of the micro- and nanostructure in the performance of surface-enhanced Raman scattering substrates assembled from gold nanoparticles," *Applied Spectroscopy* **59**, 401–410 (2005).

[78] A. Krasnok, M. Caldarola, N. Bonod, and A. Alú, "Spectroscopy and biosensing with optically resonant dielectric nanostructures," *Advanced Optical Materials* **6**, 1701094 (2018).

[79] Y. Pang, Z. Rong, R. Xiao, and S. Wang, "Turn on and label-free core-shell $Ag-SiO_2$ nanoparticles-based metal-enhanced fluorescent (MEF) aptasensor for Hg_{2+}," *Scientific Reports* **5**, 9451 (2015).

[80] P. Bharadwaj and L. Novotny, "Spectral dependence of single molecule fluorescence enhancement," *Optics Express* **15**, 14266–14271 (2007).

[81] S. Savasta, R. Saija, A. Ridolfo, et al., "Nanopolaritons: Vacuum Rabi splitting with a single quantum dot in the center of a dimer nanoantenna," *ACS Nano* **4**, 6369–6376 (2010).

[82] G. Khitrova, H. M. Gibbs, M. Kira, S. W. Koch, and A. Scherer, "Vacuum Rabi splitting in semiconductors," *Nature Physics* **2**, 81 (2006).

[83] E. S. Andrianov, A. A. Pukhov, A. V. Dorofeenko, A. P. Vinogradov, and A. A. Lisyansky, "Rabi oscillations in spasers during nonradiative plasmon excitation," *Physical Review B* **85**, 035409 (2012).

[84] R. Zwanzig, *Nonequilibrium Statistical Mechanics*, Oxford University Press (2001).

[85] F. Arecchi and R. Bonifacio, "Theory of optical maser amplifiers," *IEEE Journal of Quantum Electronics* **1** 169–178 (1965).

[86] V. Gorini, A. Kossakowski, and E. C. G. Sudarshan, "Completely positive dynamical semigroups of N-level systems," *Journal of Mathematical Physics* **17**, 821 (1976).

[87] G. Lindblad, "On the generators of quantum dynamical semigroups," *Communications in Mathematical Physics* **48**, 119 (1976).

[88] J.-J. Greffet, M. Laroche, and F. Marquier, "Impedance of a nanoantenna and a single quantum emitter," *Physical Review Letters* **105**, 117701 (2010).

[89] M. I. Stockman, "Spasers explained," *Nature Photonics* **2**, 327 (2008).

[90] M. I. Stockman, "The spaser as a nanoscale quantum generator and ultrafast amplifier," *Journal of Optics* **12**, 024004 (2008).

[91] D. M. Wu, A. Carcia-Etxarri, A. Salleo, and J. A. Dionne, "Plasmon-enhanced upconversion," *Journal of Physical Chemistry Letters* **5**, 4020–4031 (2014).

[92] F. Auzel, "Upconversion and anti-Stokes processes with f- and d-ions in solids," *Chemical Reviews* **104**, 139–174 (2004).

[93] J. R. Lombardi and R. L. Birke, "A unified view of surface-enhanced Raman scattering," *Accounts of Chemical Research* **42**, 734–742 (2009).

[94] S. A. Maier, "Plasmonic field enhancement and SERS in the effective mode volume picture," *Optics Express* **14**, 1957 (2007).

[95] G. C. Schatz and R. P., van Duyne, "Image field theory of enhanced Raman scattering by molecules adsorbed on metal surfaces: Detailed comparison with experimental results," *Surface Science* **101**, 425–438 (1980).

[96] E. C. Le Ru, E. Blackie, M. Meyer, and P. G. Etchegoin, "Surface-enhanced Raman scattering enhancement factors: A comprehensive study," *Journal of Physical Chemistry C* **111**, 13794–13803 (2007).

[97] A. Otto, I. Mrozek, H. Grabhorn, and W. Akemann, "Surface-enhanced Raman scattering," *Journal of Physics D: Condensed Matter* **4**, 1143–1212 (1992).

[98] I. Pockrand and A. Otto, "Surface enhanced and disorder induced Raman scattering from silver films," *Solid State Communications* **37**, 109–112 (1981).

[99] V. Giannini, J. A. Sanchez-Gil, J. V. Garcia-Ramos, and E. R. Mendez, "Electromagnetic model and calculations of the surface-enhanced Raman-shifted emission from Langmuir-Blodgett films on metal nanostructures," *Journal of Physical Chemistry* **127**, 044702 (2007).

[100] P. Alessio, C. J. L. Constantino, R. F. Aroca, O. N. Oliveira, "Surface-enhanced Raman scattering: Metal nanostructures coated with Langmuir-Blodgett films," *Journal of the Chilean Chemical Society* **55**, 469–478 (2010).

[101] S. Schlucker, ed. *Surface Enhanced Raman Spectroscopy: Analytical, Biophysical and Life Science Applications*, Wiley-VCH (2011).

[102] A. Musumeci, D. Gosztola, T. Schiller, et al., "SERS of semiconducting nanoparticles: TiO_2 hybrid composites," *Journal of the American Chemical Society* **131**, 6040–6041 (2009).

[103] M. Thomas, S. Mühlig, T. Deckert-Gaudig, et al., "Distinguishing chemical and electromagnetic enhancement in surface-enhanced Raman spectra: The case of para-nitrothiophenol," *Journal of Raman Spectroscopy* **44**, 1497–1505 (2013).

Notes

1 An English electrochemist of Czech origin, best known as one of the two fathers of cold nuclear fusion (the second is Stanley Pons, an American physicist). His work [27] evoked one of the scientific sensations of the twentieth century. Initially, the results of [27] were confirmed by several research groups, but further studies have shown that they were misinterpreted. Nowadays, the whole concept of cold nuclear fusion is treated by the majority of specialists in nuclear physics between a disputable science and pseudoscience. No practical yield has been obtained in this field in spite of numerous experiments, and as a result, cold fusion patents were disclaimed [28].

2 SERS is sometimes interpreted as surface-enhanced Raman spectroscopy.

3 The Langevin equation was transposed to quantum mechanics from the classical theory of Brownian movement. The Maxwell–Bloch's equations were called in the name of J. C. Maxwell because the classical electric field vector enters them and in the name of F. Bloch because one of these equations resembles the well-known *Bloch equation* describing the phenomenon of nuclear magnetic resonance. As to the Lindblad equations, they were derived simultaneously by an international team [86] and, independently, by G. Lindblad [87].

4 In seminal work [88], this value was found wrongly and resulted in some nonphysical features of the plasmon-enhanced emission. Therefore, this approach had been dropped until the mistake was corrected in work [64].

5 In the next stage, which cannot be described by this linear model, their amplitudes will be restricted by nonlinear effects.

6 Strictly speaking, the spaser implies not only generation of a plasmon. In the steady nonlinear regime it becomes a laser – a generator of light at frequency ω_0 [89]. In the steady regime of generation, the fluorescence of all N molecules turns synchronized – their dipole moments sum up in phase (this effect refers to quantum optics and is called super-radiance). The array of molecules radiates almost coherent light at frequency ω_0. For our spaser – that performed as a spherically symmetric system – this radiation is weak and has a very narrow (non-Lorentzian) spectral line.

10   Nanostructures for Enhancement of Solar Cells

10.1   Introduction

The Sun is the main source of energy on Earth. Mankind has created passive and *active* technologies in order to use this energy. Active solar technologies are those that convert the sunlight's energy into other forms. One of the most important active solar technologies is *solar photovoltaics*. Photovoltaic devices convert the electromagnetic energy of sunlight into electricity. The mechanism of this conversion is the *photovoltaic (PV) effect*. This effect is a special case of the *photoelectric effect*, also called photoelectricity and photoinduction. The photoelectric effect is the generation (induction) of conductivity charges in a semiconducting or conductive medium by absorption of incident light. The absorbed light increases the concentration of free charge carriers compared to the dark state of the same medium. In a semiconductor, photoelectricity is the absorption of a photon, resulting in the excitation of an electron-hole pair. An electron transits to the conduction band, and in the valence band, a hole appears. In other words, photoelectricity in a semiconductor is the same as optical excitation.

The photovoltaic effect is the creation of voltage and electric current due to spatial separation of the photoelectric (photoinduced or photogenerated) charges. The *charge separation* prevents the otherwise inevitable *recombination* of these charges. Charge separation is granted by an intrinsic electrostatic field that can exist in the illuminated medium. Separation of the photoinduced charges by the intrinsic field creates voltage (electromotive force) at the electrodes connected to the device. If an illuminated sample of the PV medium with an intrinsic field is connected to an external circuit, the positive and negative photoinduced charges (moving in opposite directions) create *photocurrent*. This mechanism of light-to-electricity conversion is potentially the most efficient one [1].

10.1.1   A Bit of History

In the year 1839, Edmond Becquerel (father of Henri Becquerel, who discovered radioactivity) was a student and experimented in the laboratory of his father, Antoine-César Becquerel, a pioneer of a new science called *electrochemistry*. Edmond illuminated a circuit comprising a silver halide electrode submerged into an electrolyte by a collimated light beam. The illumination changed the current in the circuit. Then the battery was disconnected from the circuit. However, some short-circuit current arose when the light beam illuminated the contact [2]. It was historically the first *PV cell*, though not efficient enough for practical generation of electricity. In 1879, W. Adams and R. Day – studying the PV effect in different media, substituting silver halide – revealed for *gray selenium* much a higher photocurrent than that observed in all previous studies [3]. In 1883, Charles Fritts found that IR light does not produce photocurrent in these elements, that an electrolyte is not necessary, and that the best illumination is normal to the surface of the gray selenium layer. He achieved for a PV cell impinged by collimated solar light an unprecedented *overall efficiency* of about 1%. Overall efficiency is the ratio of electric power delivered in the external circuit to light power incident on the cell. This ratio depends on the loading resistance – maximal power is delivered to the load whose resistance is equal to that of the cell. The overall efficiency is the ratio of this maximal output power to the power of the incident light.

The PV cell suggested by Fritts as a source of electricity was implemented as a millimeter-thick gray selenium layer on a platinum substrate whose top surface was covered by a strongly submicron (partially light penetrable) gilding [4].[1] The work by Fritts allowed scientists to understand that the PV effect is frequency selective and to assume that the photocurrent arises in the bulk of gray selenium and not only in the contact area. In 1888, Aleksander Stoletov (Stoletow) proved the direct proportionality of the photocurrent in this PV cell to the absorption of visible light in the layer of gray selenium [5]. It was understood that the absorbed light converts into electricity namely in the semiconductor. All these studies were only prerequisites for the breakthrough in photovoltaics done by Albert Einstein. In work [6], Einstein explained both photoelectricity and its implication, the PV effect. His explanation paved the road to really efficient PV cells – giving birth to solar photovoltaics. For this theory, Einstein was awarded by his second Nobel Prize (1921).

10.1.2 On Enhancement of Solar Cells by Passive Nanostructures

Solar cells (SCs) are **PV** cells whose excitation is done by the Sun. Sunlight has a very broad frequency spectrum. On the Earth's surface, the main part of the direct sunlight spectrum comprising nearly 90% of the whole power covers a part of the near-IR range (770–900 nm) and the whole visible range ($\lambda = 400$–770 nm). Ten percent of the power refers to the UV tail (320–400 nm) and three infrared sub-bands ($\lambda = 950$–1050 nm, 1100–1320 nm, and 1500–1800 nm) [7]. Though the operation band of practical SCs is more narrow that this consolidated ultra-broad band, an efficient SC is always a very broadband converter of light into electricity. This property determines the main difference between SCs and relatively narrowband converters such as photodiodes used in sensing and other similar applications. Here, most of our attention will be given to nanostructures that enhance the operational characteristics of SCs. Nanostructures that we will discuss can be classified as *antireflecting coatings* (ARCs) or *light-trapping structures* (LTSs). The purpose of both is the same: to maximize the useful absorption of solar light – that is, the absorption of sunlight converting into electricity inside the PV layer. The maximization of this absorption is the same as the minimization of optical losses of an SC.

ARCs are used to minimize reflection loss. LTSs are used to minimize transmission through the device and parasitic scattering from the internal elements of the PV cell, which results in the leakage of the light energy from the cell. In this chapter, we explain how nanostructured ARCs and LTSs enhance the SC performance. However, in order to understand it, one has to know what needs to be enhanced. Therefore, a basic understanding of physics of an SC is necessary. Here, we restrict the discussion by the most important class of SCs: solar photodiodes.

10.2 Basics of the Diode Solar Cell Physics

The following brief overview of operation of PV diodes and their main characteristics basically follows popular textbooks [7, 8]. Here, we assume that the readers are familiar, at least in general, with the physics of a usual *semiconductor diode*.

10.2.1 Operation of a Diode Solar Cell

The PV effect can be illustrated by the zone diagram of a p–n junction, shown in Figure 10.1 (left panel). This sketch also shows the energies of photons that are absorbed in the diode. Photons whose energy $\hbar\omega$ is lower than that of the semiconductor bandgap $W_g = \hbar\omega_g = W_c - W_v$ are absorbed without creating photoelectric charges i.e., are lost (Loss 1). Photons whose energy $\hbar\omega$ exceeds that of the semiconductor bandgap $\hbar\omega_g$ generate electron–hole pairs in the material bulk. These charge carriers are born with some initial velocity (due to the pulse of the absorbed photon) and spatially separate, as it was pointed out by Einstein, before their recombination occurs. The lifetime of photoinduced charges in a semiconductor is much longer than the lifetime of the excited state in a fluorescent molecule of a dye. It can be sufficient to let these charges to pass though a rather thick layer of semiconductor. However, in the absence of intrinsic static field **E** inside the semiconductor volume, it is useless to illuminate the semiconductor sample. All photoinduced electrons and holes after their relaxation to the edge states of the valence and conductivity zones, respectively, will recombine.

If a static electric field **E** is present in bulk of a semiconductor, the generated electron and hole drift along **E** in opposite directions. Then the photoelectricity is accompanied by the PV effect. For semiconductor PV materials, the best way of getting the needed static field "for free" is building a diode. In the diode, static electric field is obviously present due to the junction of two parts of the sample that are doped differently. Recall that at the contact of the p- and

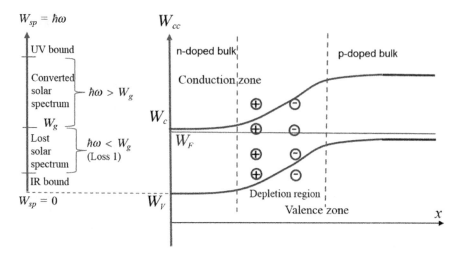

Figure 10.1 Left: Zone diagram of a dark semiconductor diode (no illumination). In this diagram, the energy of absorbed solar photons W_{sp} is shown for comparison with the energy levels W_v and W_c of the semiconductor. Sub-band solar light ($\hbar\omega < W_g = W_c - W_v$) is absorbed without producing the electron–hole pairs. Photons of higher frequencies are efficiently converted into electron–hole pairs. Right: Without light illumination, electrostatic field inside the depletion region of a semiconductor diode creates only the built-in voltage: No electromotive force is formed in the junction. The Fermi level W_F is uniform across the whole PV layer (this uniformity means the energy equilibrium of an unbiased junction).

n-doped parts of a semiconductor, positive and negative charges form a bilayer called the *depletion region* with some static electric field inside, like in a charged capacitor. The voltage V_c corresponding to this static field, called the *contact voltage* across the p–n junction, is close to the value of the semiconductor energy gap recalculated into a voltage $V_g = \hbar\omega_g/e$ ($e = -1.6 \cdot 10^{-19}$ C is the electron charge). This value in practice may attain 1–2 V. The difference V_g-V_c is relatively small and depends on the doping level.

The energy diagram of a p–n junction is shown in Figure 10.1 together with the energy diagram of solar photons. The energy of absorbed photons is counted from the level W_c, the top edge of the conductivity zone of a semiconductor. In the absence of illumination, the *Fermi level* W_F is the same in both bulk p- and n-doped regions, meaning that if we connect the extremities of these regions with a wire, no current will flow. The diode is obviously not a power source and cannot discharge like a capacitor. In fact, the contact voltage is compensated by the outer voltages V_b created by weak electrostatic fields (of the opposite sign with respect to **E** in the depletion region). These fields exist in the bulk of the p-doped and n-doped parts called the quasi-neutral regions of the diode. As a result, the contact voltage is not an electromotive force, and in the theory of p–n junction, V_c is called *built-in voltage*. So, in the dark regime, there is no voltage between the extremities of the p-doped and n-doped regions of a diode.

Now imagine that the diode (made of a PV material) is illuminated by light. The processes in an illuminated diode are illustrated in Figures 10.2 (zone

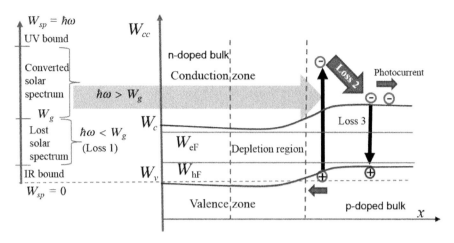

Figure 10.2 Zone diagram of an illuminated semiconductor diode and sketch of intrinsic losses of an SC. Illumination works as a direct biasing and destroys the uniformity of the Fermi level. Beyond the equilibrium, one may introduce quasi-Fermi levels of electrons (W_{eF}) and holes (W_{hF}), uniform across the junction. Their difference determines the open-circuit voltage. Loss 1 is the percentage of the sub-band sunlight. Losses 2 and 3 are illustrated for a photoinduced electron–hole pair in the p-doped region. A part of its energy $\hbar\omega - W_g$ is lost due to the relaxation process (Loss 2). Loss 3 is the probability of this pair to meet free charges of the opposite sign and to recombine. Losses 1–3 do not include the loss of useful photons ($\hbar\omega > W_g$) due to their possible reflection, scattering, and absorption beyond the PV layer.

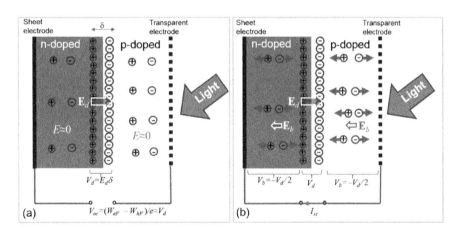

Figure 10.3 The static field in the bulk of a loaded photodiode results from the p–n junction. (a) Open-circuit voltage V_{oc} is practically equal to the voltage in the depletion region $V_d = E_d\delta$. These is no electric field in the bulk beyond the depletion region; no current flows; all photoinduced charges recombine. (b) In the short-circuit regime, voltage V_d cancels out with two voltages, V_b, applied to the bulk regions and corresponding to the field E_b. This field urges the photocurrent to flow oppositely to the dark diode current.

diagram) and 10.3 (comparison of the *open-circuit* regime and *short-circuit* regime). The main absorption mechanism of useful photons ($\hbar\omega > \hbar\omega_g$) in a PV material is the generation of electron–hole pairs. These photoinduced charges break the equilibrium in the depletion region, decreasing the electrostatic field in it – i.e., acting like a direct bias. They also break the equilibrium in the quasi-neutral regions, and the Fermi level loses its physical meaning everywhere in the diode. Instead, one introduces the quasi-Fermi levels: that of electrons (W_{eF}) and that of holes (W_{hF}). Their difference implies an *open-circuit voltage* V_{oc} which arises across the diode.

Really, since the current cannot flow in the open-circuit diode, in the regime of steady illumination, the left extremity (that of the n-doped region) turns out to be charged positively, and the right extremity (that of the p-doped region) is charged negatively, as shown in Figure 10.3(a). This way, the built-in voltage in the illuminated regime extends between these extremities and transforms into the open-circuit voltage, which is smaller than V_g but has the same order of magnitude. One can say that the illumination transforms the built-in voltage V_c into electromotive force V_{oc}. Photoinduced charges in the quasi-neutral regions screen the charges of the depletion region, reducing the static field beyond it, and V_{oc} turns out to be close to $E_d\delta$, where E_d is the absolute value of the mean electrostatic field in the depletion region and δ is its thickness.[2] This situation is illustrated by Figure 10.3(a).

Now let us connect the electrodes of our SC by a wire, as it is shown in Figure 10.3(b). In this short-circuit regime, two drops of voltage V_b arise across the bulk n-doped and p-doped regions of the PV layer. These voltages compensate V_d, which is possible because the static field $\mathbf{E}_b \neq 0$ arises in the quasi-neutral regions (like it did in the dark regime). Again, this field is opposite to the built-in field of the depletion region. In this case, the current in the external circuit is maximal and equal to the photocurrent $I = I_{sc}$ (i.e., the *short-circuit current* and photocurrent are equivalent). From Figure 10.3(b), it is clear that the photocurrent in both bulk regions is formed by the minority charge carriers: holes in the n-doped part and electrons in the p-doped part. This means that the photocurrent is directed oppositely to the usual diode current (called in the theory of solar cells *dark current*).

10.2.2 About Optimization of Solar Cells

In order to optimize solar cell efficiency, one should minimize both the intrinsic losses and optical loss resulting from absorption of sunlight beyond the *PV layer* and scattering to free space. This chapter mainly discusses the reduction of the optical loss, which is attained in modern SCs with the use of nanostructures. However, these nanostructures are not intrinsic parts of the PV layer. The main component of the SC is the PV layer, and it must be designed separately. How is it designed?

First of all, it must be made of a *PV material* – a material that efficiently converts photons into electron–hole pairs. The percentage of useful photons producing photoinduced charges in the total amount of photons absorbed in the PV material is called *photoelectric quantum efficiency* or *photoelectric spectral response* of the material. PV materials suitable for solar photovoltaics are by definition materials for which this value is close to 100% at the maximum of the solar spectrum (green range) and weakly decreases with detuning to be significant in the whole useful part of the solar spectrum. However, this parameter for realistic materials is dispersive, and its integral value (averaged over the useful spectrum) can be a value within the 50%–80% range, depending on the explicit PV material and its doping level (if the doping level is reasonable, because in heavily doped materials photoelectric spectral response drops). Choosing among all semiconductors specifically a PV material and optimizing its doping level, we guarantee a rather high value of the integral photoelectric response.

Optimization of the PV layer implies the maximal possible reduction of intrinsic losses of an SC. The intrinsic power loss of an SC comprises losses of three types. Loss 1 – *sub-bandgap loss* – arises because the sunlight obviously comprises photons of energy $\hbar\omega < \hbar\omega_g = W_g$. The bandgap frequency ω_g of a PV semiconductor usually lies in the so-called inter-band range (wavelengths $\lambda = 770$–1100 nm). The portion of sunlight power in the long-wavelength part of the spectrum ($\lambda > 1100$ nm) is rather small (17%), and this power is most often sacrificed. For crystalline silicon, which is the most popular and one of the most efficient PV materials, $\lambda_g = 1050$ nm, and the value of Loss 1 related to the PV cutoff at frequency $\omega_g = 2\pi c/\lambda_g$ is about 18%. Another reason for Loss 1 is a parasitic effect called the *Franz–Keldysh effect* [9, 10]. This is nonlinear transfer of visible-light photons into the IR range that takes place in semiconductors in presence of high static electric field **E**. We will see that this field is present in SCs and more or less significantly distorts the band structure. In obsolete SCs, this static field was too high, and the Franz–Keldysh effect increased Loss 1 quite noticeably. In modern SCs, the Franz–Keldysh effect is very small. The value $\eta_{sp} = (1 - \text{Loss 1})$ is called the *spectral efficiency* of the SC. For modern SCs based on c-Si, $\eta_{sp} \approx 81\%$ [11–14].

The dissipative loss of the power for a given photon equals $\hbar\omega - \hbar\omega_g$ because the charge carriers obviously relax toward the edges of the bandgap before they attain the electrodes. The integral parameter describing the relaxation losses averaged over the whole useful spectrum is usually called Loss 2 or *super-bandgap loss*. Since Loss 2 is dissipative, it can be modeled as an additional resistance connected in series to that of Ohmic loss, which arises due to the non-ideal photoelectric efficiency. The value of Loss 2 depends on the solar spectrum shape and the statistics of photogenerated charges in the PV material. Since the solar spectrum is ultra-broadband and its maximum on the surface of the Earth is in the green range, for practical PV materials, a significant percentage of solar photons have energy higher than $\hbar\omega_g$. It makes Loss 2 a large value; for practical SCs, it is within the interval 30%–60%. The explicit value of Loss 2 for a given material depends on several parameters: the doping level, the thickness of the PV layer, the geometry of the SC, etc. Notice, however, the rather high value of Loss 2 (close to 30%), is inevitable for a simple or *single-junction SC* (the term "single-junction" also refers to *heterojunction* SCs such as modern silicon SCs). In a third-generation SC – in a monolithic *multijunction SC* and in an SC with parallel decomposition of the solar spectrum – solar spectrum is effectively split onto sufficiently narrow bands converted by different parts of the SC separately. In these advanced and very expensive SCs, Loss 2 is reduced to 10%–15%.

For single-junction SCs, the problem of high Loss 2 is responsible for the inevitable choice of such PV materials, which have noticeable Loss 1. The infrared part of the incident light power is sacrificed (in SCs based on crystalline Si, GaAs, Ge, and other practical PV materials), and there is an optimum trade-off between Loss 1 and Loss 2. If we choose the bandgap frequency ω_g at the IR edge of the solar spectrum $\lambda = 1850$ nm (this is the bandgap wavelength of InSe), we get rid of Loss 1 almost completely but broaden the operational band of our SC so that 90% of solar photons have energy higher than W_g. Then, Loss 2 will increase nearly twofold (from nearly 30%, corresponding to SCs based on c-Si, to nearly 60%, corresponding to indium selenide SCs). This increase will definitely overcompensate for the 18% gain in the spectral efficiency.

Thus, practical PV materials for solar photovoltaics are not only materials with high photoelectric spectral response in the solar spectral range. They are also materials whose ω_g is selected as a trade-off between Loss 1 and Loss 2. Notice that Loss 2 for a given PV material can be somewhat reduced by the proper design of an SC. For obsolete silicon SCs with a simple p–n junction, Loss 2 $\approx$ 40%–45% [7]. For advanced silicon SCs, Loss 2 = 30%–35% [12].

Loss 3 – *recombination loss* – is the percentage of photoinduced charges that recombine before they attain the electrodes. For our purposes, it is enough to know that the recombination holds everywhere: at the surface illuminated by the sunlight, in the bulk of the semiconductor, in the depletion region of the p–n junction, and at the surfaces of the metal contacts. This effect is especially important for crystalline SCs with a thick PV layer for which the struggle against this effect drives advanced technical solutions.

The recombination in the bulk is reduced by the reduction of the PV layer thickness. Here, it is worth mentioning that the static electric field **E** in the bulk of PV layers with a thick depletion region is low because the modest contact voltage turns out to be applied to a thick layer. In this situation (that holds in the crystalline silicon SCs), the main mechanism of the separation of the photogenerated charges is their diffusion. Field **E** is needed only to ensure the directional drift of the photogenerated charges toward the electrodes. Roughly speaking, the electric field makes the diffusion of charges directional. Since the mean speed of the carriers is mainly determined by diffusion, the effectively maximal path of the minority carrier in the PV layer (that of the electrons in the p-doped layer and that of the holes in the n-doped layer) is their diffusion length. It is a parameter of the PV material with a given level of doping. Thicknesses of the p-doped and n-doped sublayers of the PV layer should be smaller than the diffusion length. Otherwise, most of the generated charges recombine in the bulk before they attain the electrodes.

However, if these layers are much thinner than the mean free path of the minority carrier, such a PV diode will be also inefficient. Mean free path is of the same order as the thickness of the depletion region in a simple p–n junction. Very high strength of electric field in the depletion region of a thin p–n junction implies a rather low photoelectric quantum efficiency of this region. The light absorption in quasi-neutral regions is more efficient. One needs high light absorption in the depletion region for higher electromotive force (the open-circuit voltage), but this absorption holds with a low quantum efficiency and, therefore, results in a high Ohmic loss. Recall also that in the region of high electric fields, the Franz–Keldysh effect further reduces the cell efficiency.

This dilemma has been resolved without a trade-off between Ohmic loss and electromotive force of the SC by replacement of simple p–n junctions by a heterojunction. Modern silicon SCs comprise a relatively thick, weakly n-doped c-Si layer and a heavily n-doped nanolayer on top of it. The bottom layer of the PV layer is p-doped. The main light absorption holds in the thick, weakly doped layer because the thin n^+ layer on top is sufficiently transparent for the sunlight. The photoelectric quantum efficiency of the central layer is high because the electric field strength is not too strong. The light energy is effectively collected in this sublayer, since it plays the role of an effective depletion region, occupying most of the volume of the solar cell. Of course, the layer thickness is chosen so that it does not exceed the diffusion length of the charge carriers; otherwise, the recombination loss – Loss 3 – will be too high. Thus, the thickness of the region where the photocurrent is mainly generated is increased without damaging the quantum efficiency. For the optimal thickness of the PV layer chosen in this way (for c-Si, its total thickness is of the order of 200 μm), the recombination of the photogenerated charges in the bulk is reasonably small and is within 5%–10% [7, 8, 15]. The surface recombination in such SCs became the main issue since 1970s, when heterojunction silicon SCs were created.

The surface of a crystal lattice is a kind of a crystal defect that results in the appearance of defect states. These states, called *surface states*, arise inside the bandgap, and their presence drastically enhances the probability of recombination. Therefore, the lifetime of the charge carriers generated near the open surface is reduced, and they recombine fast. In modern SCs, this problem is partially resolved by using the so-called *chemical passivation* of the top surface – covering it with a material that eliminates surface states. It can be a dielectric whose lattice is compatible with that of the semiconductor. Alternatively, it can be a semiconductor with a large bandgap (e.g., with ω_g in the UV range) that weakly absorbs sunlight.

However, passivation resolves the issue of surface recombination only for the open surface of the PV layer. Recombination remains high on the interfaces between the optimally doped PV material and the metal of the contacts. The level of this surface recombination can be reduced by preparing heavily doped areas near metal contacts. In this way, the contrast of two contacting materials is reduced, the transition of photoinduced charges across the interface is not drastically slowed, and they penetrate into metal before they would recombine. Thus, the heavily n-doped layer on top of c-Si SC serves not only to form of a heterojunction. It also reduces surface recombination. As to the p-doped layer on the bottom, in modern silicon SCs it forms a p–p^+ junction with a heavily p-doped nanolayer. This layer grants the reduction of surface recombination in the bottom part of the SC.

In the described layered design, the heavily n-doped layer is on the top, and parasitic absorption in it is tolerable but not negligibly small. Moreover, the transparency of the top contact is granted by making it into a wire mesh. We will discuss in the next section why this mesh produces a significant shadow, making it so that a noticeable part of the PV layer is under the wires and also stays in the dark. Finally, the presence of metal strips on top of c-Si hinders its chemical passivation, increasing the cost of this procedure.

Therefore, in more advanced silicon SCs, both contacts are located on the bottom side of the PV layer. In this design, the two contacts usually form an interdigital structure of metal microstrips making contact alternating heavily doped n- and p-regions on the bottom of the PV layer. Since these heavily doped regions are rather thin, only the recombination in the bulk of the PV layer is important. In this design, the intrinsic electric field (in multiple depletion regions) is not vertical anymore, and the paths of photoinduced carriers elongate. Therefore, the bulk recombination and Ohmic loss (Loss 2) increase. However, this harm (about 10% for the overall efficiency) is overcompensated by gain in the optical efficiency and lower cost of passivation of the fully open top surface of these SCs [12], called *interdigital-back-contact* SCs.

To conclude this discussion, let us compare a semiconductor PV diode with the historically first SCs. These SCs were also PV diodes, but they used the contact voltage at electrolyte-semiconductor (Becquerel) or metal-semiconductor (Fritts) interfaces. Such structures refer to the so-called *Schottky diodes*. SCs based on Schottky diodes are inefficient because the contact voltage is applied not to a quite substantial depletion region but to the sharp physical interface of two media. At this interface, the intrinsic electric field is very high (theoretically singular for an abrupt junction), and the junction represents a potential barrier. The absolute majority of photogenerated charges are stuck on this barrier and are not separated before they recombine. Therefore, photocurrent in an illuminated Schottky PV diode is lower by an order of magnitude than that in a p–n diode or a heterojunction diode.

10.2.3   Volt-Ampere Curve of a Diode-Type Solar Cell

The usual operation regime of a PV diode is in the middle between two extreme regimes: the open-circuit one and the shortcut one. Let us briefly recall their definitions. If the external load resistance R_L is infinite, the illumination of the diode results in the maximal possible drop of voltage between the electrodes, which is equal to V_{oc}, but the current is zero: all photogenerated pairs recombine in the same regions where they are born. In the short-circuit regime ($R_L = 0$), the current is maximal and equal to the photocurrent, whereas the drop of voltage is zero.

The maximal power in the external circuit corresponds to the matching condition when the load resistance is equal to R_{iSC} of the illuminated SC. Since the photocurrent is the additive part of the diode current I and since it is directed oppositely to the dark current, the equivalent scheme of the loaded PV diode is a parallel connection of a current generator and a usual (not light-illuminated)

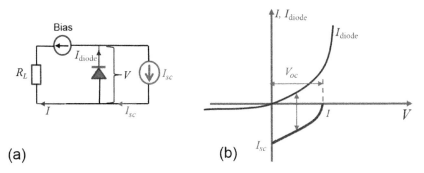

Figure 10.4 (a) Equivalent scheme of an illuminated and biased PV diode. (b) An $I–V$ plot of the PV diode (blue curve) is obtained from the $I–V$ plot of the same diode without illumination (magenta) by adding $-I_{sc}$ to I_{diode}. The solar-cell regime implies no external bias.

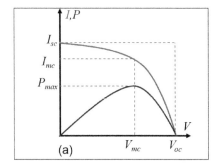

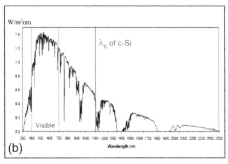

Figure 10.5 (a) $I–V$ and $P–V$ plots of a typical SC. Values I_{mc} and V_{mc} correspond to the matched regime when the delivered power attains its maximum P_{max}. The fill factor is equal to $FF = I_{mc}V_{mc}/I_{sc}V_{oc}$. (b) Solar spectrum at the sea level. Loss 1 corresponds to the range $\lambda > \lambda_g$.

diode, as shown in Figure 10.4(a). An $I–V$ dependency of an unbiased PV diode is represented by the bottom curve in Figure 10.4(b). When $V > V_{oc}$ or $V < 0$, some external biasing is implied. However, it is not relevant for SCs. The Volt-Ampere curve of a SC corresponds to the case when the biasing is internal; that is, the bias is due the voltage drop across the load resistor R_L.

10.2.4 Main Parameters of Diode-Type Solar Cells

To discuss the main parameters of a SC, it is instructive to consider such the $I–V$ plot where the positive current direction is chosen so that I is positive. A typical $I–V$ plot of a c-Si SC is shown in Figure 10.5(a) together with the $P–V$ plot showing the output power versus V, which practically implies variation of R_L. The maximal electric output P_{max} corresponds to the matched circuit (mc) regime. Corresponding values of the voltage and current are denoted as $V = V_{mc}$ and $I = I_{mc}$, respectively. The target is to maximally approach the $I–V$ plot curve to a rectangle. Then, $V_{mc} \rightarrow V_{oc}$ and $I_{mc} \rightarrow I_{sc}$. In this case, the value of P_{max} attains its absolute maximum $P_{max}^{max} = I_{sc}V_{oc}$, but it is achievable only theoretically. The ratio P_{max}/P_{max}^{max} for a realistic PV material is called the *fill factor* (*FF*). In some advanced SCs, the fill factor is close to unity – e.g., for some cadmium telluride SCs $FF_{max} = 0.93$ [12]. However, the most popular SCs for civil terrestrial applications are SCs based on c-Si. Panels of modern silicon SCs combine sufficient efficiency with reasonably low fabrication and exploitation costs and long life. Nowadays, their production does not imply any toxic waste. For silicon, SCs the fill factor is in the interval $FF = 0.7$–0.85, depending on design specifics. This value enters the *overall efficiency of the SC* as a multiplicative factor.

Usually, in the analysis of the SC conversion efficiency, one assumes that the bandgap of the semiconductor is chosen optimally as a compromise between Loss 1 and Loss 2, as explained earlier. In Figure 10.5(b) the solar spectrum on the Earth's surface is shown and the wavelength λ_g corresponding to the bandgap of c-Si is marked. Loss 1 is the area below the curve corresponding to $\lambda > \lambda_g$ divided by the area under the whole plot. Once the PV material is selected, this loss is determined. In calculations of the PV conversion efficiency, one considers the incident power as that of the useful solar spectrum – only the range $\lambda_{min} < \lambda < \lambda_g$ is taken into account. Here, $\lambda_{min} = 320$ nm (937 THz) is the UV bound of the solar spectrum. Thus, the conversion efficiency of the SC is defined as the efficiency of conversion of the *useful power* of the sunlight into electricity in a system composed by a maximally illuminated SC and its matched load $R_L = R_{iSC}$.

Not all the incident power can convert into electricity due to many factors, such as non-ideal photoelectric quantum efficiency of the PV material, non-ideal fill factor, intrinsic losses of the PV layer, and, finally, optical loss. Let us first define the optical loss (*OL*). It is the difference between the integral absorption coefficient A_{PV} of the PV layer and unity. A_{PV} is the integral of the time-harmonic absorption coefficient $A_{PV}(\lambda)$ over the band $\lambda_{min} < \lambda < \lambda_g$, which takes into account the solar spectrum shape as a weight function. Since the number of photogenerated charge carriers is proportional to the number of absorbed photons, the conversion efficiency is directly proportional to A_{PV}. The latter is often called *optical efficiency of the SC*. Optical loss $OL = 1 - A_{PV}$ is also called *matching loss* because many researchers identify the optical loss with the reflection loss (the main loss factor for silicon SCs) and deduce the last one from the impedance mismatch of the illuminated surface and free space.

Intrinsic losses of the PV layer are called Losses 1, 2, and 3. Loss 1 has been discussed in the preceding paragraphs. Losses 2 (relaxation loss) and 3 (recombination loss) are taken into account when calculating the so-called *internal quantum efficiency of the SC*. Previously, we have defined quantum efficiency of a PV material as a ratio of the number of electron–hole pairs induced in the PV material to the number of useful (above-bandgap) solar photons absorbed in the PV material during the same time. This quantum efficiency is a *photoelectric* parameter and sometimes is called *photoelectric spectral response*. Meanwhile, the internal quantum efficiency *of the SC* η_i is a PV parameter. By definition, it is equal to the ratio of the number of electron–hole pairs *reaching the electrodes of the SC* to the number of the absorbed useful photons.[3]

Both relaxation and recombination losses reduce the values $P_{\max}^{\max} = I_{sc} V_{oc}$ and η_i proportionally. In work [11], by Shockley and Queisser, it was shown that $\eta_i = (1 - \text{Loss}\,2)(1 - \text{Loss}\,3)$. In the absence of optical loss for an SC with the ideal fill factor, internal quantum efficiency of the SC is equal to $P_{\max}^{\max}/P_{\text{inc}}$. For advanced silicon SCs, Loss 2 = 0.33 and Loss 3 = 0.05, which gives $\eta_i \approx 0.63$. Since c-Si SCs became available on the market, when the typical value of the internal quantum efficiency was equal, $\eta_i \approx 0.59$ [7], in 36 years it has increased only slightly. We may conclude that the internal quantum efficiency value is, nowadays, close to its practically achievable limit.

Now, let us estimate the maximal achievable value of the *overall efficiency* of any SC based on a given PV material. In the literature, there are different definitions of this parameter. So-called *ultimate efficiency of an SC* introduced by Shockley and Queisser normalizes the electric power produced by the unit area of an SC not to the real power flux of the incident sunlight but to the *useful* (high-frequency) part of the solar power spectrum *absorbed in the PV layer*. In other words, the ultimate efficiency does not take into account Loss 1 and optical loss. Also, it does not take into account the Ohmic loss that arises in the contact areas, due to realistic impurities of materials and other imperfections of a real SC. For advanced silicon SCs having $FF_{\max} = 0.85$, we have $\eta_u = FF_{\max} \cdot \eta_i \approx 0.5$.

The ultimate efficiency is not a realistic target value for a designer of an SC based on a given PV material. *Maximal achievable efficiency of an SC* in its most actual definition takes into account all inevitable losses omitted in η_u. Among them, the optical loss is the main one. A solar panel even in a big power station called *solar power plant* cannot rotate after the Sun. Such rotation in the terrestrial conditions would take too significant of a part of the output power. Therefore, the angle of sunlight incidence may vanish only at midday and varies during the daytime up to the nearly grazing incidence. Optical efficiency OE is the mean value of the integral absorption coefficient of power in the PV layer. Mean value implies the averaging over the incidence angles during the daytime. Therefore, in the definition of the maximal achievable overall efficiency,

$$\eta_{\max} = \eta_{sp} \cdot \eta_u \cdot VF \cdot CCF \cdot OE_{\max}, \qquad (10.1)$$

$OE_{\max}$ is obviously smaller than unity. Also, in (10.1), *VF* (*voltage factor*) and *CCF* (*current collection factor*) are factors that take into account the aforementioned voltage and current losses, respectively. For most advanced SCs operating on standard terrestrial conditions OE is nowadays equal to 0.85 (see Section 10.3). It is reasonable to assume that it can hardly be made higher than 0.90. For silicon SCs with the optimized geometries of the PV layer and electrodes, *VF* and *CCF* can be as high as $VF \approx CCF \approx 0.95$. Since, for silicon SCs, $\eta_{sp} = 0.81$ and $\eta_u = 0.5$, by adopting $OE_{max} = 0.9$ and $VF = CCF = 0.95$, we obtain that the maximal achievable overall efficiency of silicon solar photovoltaics in civil terrestrial applications is equal to $\eta_{max} = 0.3$ [12]. In the literature, this value is sometimes also called the Shockley–Queisser limit. Some authors, in order to stress that this value is different from the ultimate efficiency (originally calculated by Shockley and Queisser), call η_{max} *practical* Shockley–Queisser limit.

Practical limit efficiency $\eta_{max} = 0.3$ is not the highest one determined for usable PV materials. For SCs based on purified germanium and for those based on so-called *metal-organic perovskites*, the practical Shockley–Queisser limit was estimated as nearly equal to 40% [16–18]. For SCs based on amorphous semiconductors – the most popular PV material used in flexible thin-film SCs – this limit is estimated as 13%–15% [15, 19]. In the following paragraphs, we will explain why this low limit still keeps a good market perspective for amorphous PV materials.

In 2009, an advanced silicon SC manifested the record overall efficiency $\eta = 0.25$ [20]. However, this SC was fabricated in a laboratory. For similar samples industrially adapted for mass production, this record value could not be achieved. Moreover, this record efficiency was measured for the normal incidence of sunlight when A_{PV} was close to unity. The mean (daytime-averaged) optical efficiency OE of that SC was not published but should have been of the order of 0.8. Potential improvement of that SC has been mainly related to the possibility to further reduce the daytime-averaged optical loss. The other possibilities to enhance it without a drastic increase of the fabrication cost were exhausted [21]. This why, since 2009, most efforts of researchers exploring this area have been invested specifically in nanostructured ARCs decreasing the reflection loss of silicon SCs for the oblique incidence of sunlight. However, nanostructures decreasing the optical loss are topical not only for silicon solar photovoltaics. They are especially topical nowadays for *thin-film solar cells*. In Section 10.5, we will see why it is so.

10.2.5 On Multijunction Solar Cells

As it was already mentioned, SCs with an effective split of the solar spectrum can beat the restriction of 30% minimum for Loss 2 and 19% minimum for Loss 1. Therefore, in this way, one may overcome the Shockley–Queisser limit for silicon SCs. A multijunction (cascaded) SC represents an effective series connection of relatively narrowband PV cells (called PV cascades), separated from one another by nanometer-thick layers of an insulator through which the photocurrent flows practically without losses due to the tunnel effect. Each partial cell of this monolithic structure absorbs a dedicated sub-band of the solar spectrum. In Figure 10.6(a), we show an example of multijunction SCs reviewed in [15]. Physically, it comprises three PV diodes connected in series by *tunnel junctions*. The top SC converting the high-frequency part of the solar spectrum is based on InGaAs.

In Figure 10.6(b), the solar spectrum is depicted as split into three parts, corresponding to three cascades of the multijunction SC. The gray area

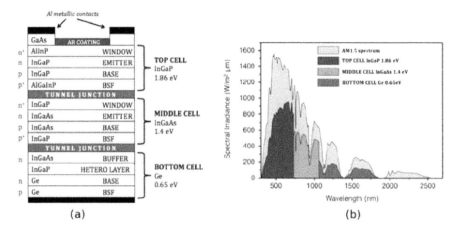

Figure 10.6 (a) Multijunction SC with a monolithic design and multilayer antireflective coating in between the microstrips forming the top contact. (b) Three bands of the solar spectrum (those centered at $\lambda = 520$ nm, $\lambda = 770$ nm, and $\lambda = 1{,}440$ nm, respectively) are converted into electricity by three effective single-junction SCs entering the multijunction structure. Reprinted from [12] with permission of the editors.

corresponds to the solar spectrum marked as $AM1.5$ (this notation corresponds to a typical orbit of the Earth's satellites). The part of the solar spectrum highlighted by dense gray is the part of the power spectrum absorbed by the top cascade. GaAs beams are added here for a better Ohmic contact of the top cell with the metal microstrips. GaAs forms a good contact with both InGaP and metal (Al), preventing the formation of parasitic Schottky diodes on the interfaces Al/InGaP (which would reduce V_{oc} and increase surface recombination). So-called *window layer* and so-called BSF layer (*back-surface-field layer*) suppress surface recombination reducing Loss 3 to its minimal value, about 5%. The second effective PV diode converting the middle part of the spectrum is based on InGaAs. The part of the spectrum absorbed by the middle cascade is highlighted by light grey. The third cascade converting the low-frequency part of the spectrum (the absorbed portion is highlighted by dark gray) is a *heterojunction* PV diode based on three PV layers – InGaAs, InGaP, and Ge – that form a unique *polycrystalline* structure. It is clear that the overall efficiency of this SC is mainly restricted by parasitic absorption of the sunlight in the undedicated cascades. The useful (PV) absorption is lowest in the bottom PV diode whose optical efficiency is close to 0.5 due to parasitic absorption of the red light in two PV cells located above.

In this SC, Loss 2 reduces from the 35% to 40% typical for a silicon SC to about 10%. Moreover, the consolidated operation band now covers the IR part of the solar spectrum, and Loss 1 is reduced so that the spectral efficiency exceeds 90%. The optical efficiency of this SC is close to 0.7 (due to high parasitic absorption in the cascades), and the overall efficiency is about 0.25 [15]. Figure 10.6(a) shows the composition of the SC described in [12, 15]. Its conversion efficiency, 0.25, was the record value for SCs in the mid-1970s.

Monolithic multijunction SCs are very expensive and, therefore, are not competitive as electricity sources for civil terrestrial applications even in huge solar power plants. However, in space (and military) applications, the price of electricity is not the main issue. Therefore, the development of such SCs since 1970s has continued and has resulted in 46% of their modern overall efficiency [22].

Notice that their present-day optical efficiency exceeds 90% because the satellite SCs (arrayed in big panels) rotate to be orthogonal to the sunlight flux all the time when the satellite is illuminated (beyond the shadow of the Earth). This rotation offers low reflection losses but is specific for the space orbit where the rotation of a large mass does not consume an important part of the produced electric energy. The terrestrial SCs we consider in the remainder of the chapter are stationary, as we have already discussed.

Here we finish the introductory section, and in the next section, we will mainly concentrate on the issue of reflection losses. This issue is either fully omitted or very weakly concerned in available books on SCs.

10.3 Optical Efficiency of Solar Cells with an Optically Thick Photovoltaic Layer

If the PV layer is optically thick, meaning that it is thicker than the penetration depth of the sunlight, the light is absorbed before it reaches the bottom electrode. In this case, there are two components of the optical loss. One component is parasitic (non-PV) absorption in the heavily doped contact layers. This type of absorption almost is absent in interdigital-back-contact (IBC) SCs. The second (and main) loss component is the reflection or backscattering loss. Let us consider the problem of this loss by revisiting Figure 10.6(a), where an obsolete multijunction SC is depicted. First, we see that the top surface of this SC is not completely illuminated. A part of the surface is shadowed by the current-collecting electrode representing a mesh of metal microstrips. The minimal width of the microstrip is restricted by the mesh impedance, which should be low enough for effective carriage of current. This impedance growth reduces for very tiny contacts [8], and, practically, the microstrips should have the width of the order of $w = 10\,\mu m$ or more [7]. The mesh impedance depends also on the mesh period a. If this period is too large, the effective paths of the charges (the lines of the photocurrent bulk density) condensing around the microstrips elongate too much, and their lengths become large compared to the thickness of the PV layer. It results in growth of both bulk recombination and Ohmic losses in the SC.

Therefore, there is an optimal ratio w/a, representing a compromise between the minimal optical loss and minimal bulk recombination and Ohmic losses. For example, when $a = 10w = 100\,\mu m$, CCE and VF are both close to 0.95 (high enough), whereas the bulk recombination and Ohmic losses in the SC are nearly the same as if the photocurrent lines were vertical. Therefore, a typical design of an SC with a front metal contact implies an array of parallel microstrips of the width $w = 10$ μm and the period about 0.1 mm. This design implies an irreducible component (10%) of optical loss due to the shadow. Special solar concentrators were developed between the 1960s and the 1980s to focus the sunlight into the open areas between these microstrips [15]. However, the cost of these concentrators made the solar electric energy too expensive. Therefore, the term $w/a \approx 0.1$ practically inevitably enters the reflection loss of SCs with metal top electrode. For silicon SCs with a front contact mesh, irreducible optical loss is as high as 15% due to the parasitic absorption in the heavily n-doped silicon in the top part of the PV layer.

In modern silicon SCs, there is no front mesh. In the case of the IBC design, the optical loss related to parasitic absorption in heavily doped layers is negligible. The PV layer of a crystal-silicon SCs is optically thick, and the sunlight is absorbed before it reaches the heavily doped nano-regions [23]. Since the factor w/a is removed from the optical loss of the SC with back contacts, only the reflection loss of the open surface remains.

As we have mentioned, this reflection loss factor grows versus the incidence angle θ. However, before inspecting this angular dependence, we need to understand how the antireflection coating (ARC) operates for the normal incidence. Basically, the antireflection coating is a matching device whose purpose is to ensure that the input impedance at the input surface is equal to the free-space impedance. As is well known, broadband matching can be realized with multilayered stacks of low-loss dielectrics. However, we should take into account that multilayer ARCs of 10 or more stacked nanolayers (used in satellite SCs because they should be ultra-broadband and very efficient) are never used in civil terrestrial applications. The first reason why multilayer ARCs are replaced in commercial SCs by single-layer antireflectors is simple: the high costs of multilayer ARCs. The second reason is not so simple. It is the presence of the so-called *front glass* or another macroscopically thick laminate protecting the surface of the ARC from atmospheric abrasion. In the following section, we will see how this factor erases the advantages of multilayer ARCs.

10.3.1 Single-Layer Antireflection Coating and the Impact of the Front Glass

In an obsolete silicon SC developed in work [24], the ARC was an 80-nanometer-thick layer of MgF (refractive index $n = 1.37$). If MgF is not covered by a front glass, the open area of this SC reflects only 8% of the normally incident sunlight power [24]. When this SC is covered by a front glass layer ($n_{rmglass} = 1.48$), this reflectance increases to 12%. In the following paragraphs, we deduce both these values.

Let a layer (1) of thickness d_1 with refractive index $n_1 = \sqrt{\varepsilon_1}$ (here, ε_1 is its relative permittivity in the optical range) be placed on a semi-infinite substrate (2) with refractive index $n_2 = \sqrt{\varepsilon_2}$. For a normally incident plane wave of frequency ω (the wavenumber in free space is $k_0 = \omega/c$), the surface impedance on top of this structure can be found from the transmission-line equation (e.g., [25]):

$$Z_s = Z_1 \frac{Z_2 + jZ_1 \tan \gamma_1 d_1}{Z_1 + jZ_2 \tan \gamma_1 d_1}, \tag{10.2}$$

where γ_1 and $Z_{1,2}$ are the wavenumber in medium 1 and the wave impedances of media 1 and 2, respectively:

$$\gamma_1 = k_0 n_1, \quad Z_{1,2} = \sqrt{\frac{\mu_0}{\varepsilon_0 \varepsilon_{1,2}}} = \frac{Z_0}{n_{1,2}}. \tag{10.3}$$

Here, Z_s means the input impedance of a transmission line (whose role is played by the layer d_1), and Z_2 is the load impedance of this transmission line. In general, Z_2 is the surface impedance of the bottom interface, but the substrate is assumed to be semi-infinite, and the surface impedance of a half-space for the normal incidence is equal to the wave impedance of the medium [25].

If the incidence is oblique ($\gamma_1 = k_0 \sqrt{\varepsilon_1 - \sin^2 \theta}$, where θ is the incidence angle), formula (10.2) is still applicable with replacements $Z_{1,2} \rightarrow Z_{1,2}^{TE,TM}$ for TE-polarized (or s-polarized) and TM-polarized (or p-polarized) incident waves, respectively. Here,

$$Z_{1,2}^{TE} = Z_{1,2} \frac{n_{1,2}}{\gamma_{1,2}}, \quad Z_{1,2}^{TM} = Z_{1,2} \frac{\gamma_{1,2}}{n_{1,2}}, \tag{10.4}$$

and, of course, $\gamma_2 = k_0 \sqrt{\varepsilon_2 - \sin^2 \theta}$ [25].

If medium 2 is a layer with a finite thickness d_2, formula (10.2) is generalized by the replacement of the impedances $Z_2^{TE,TM}$ with the surface impedance of layer 2. The last impedance refers to the interface of media 1 and 2 and is calculated by the same transmission-line formula, where the surface impedance of the substrate (layer 3) replaces that of layer 2. This approach allows us to deduce the surface impedance of an arbitrary number of stacked layers [25, 26].

For the normal incidence, the amplitude reflection coefficient ρ and power reflectance R are easily found from the surface impedance:

$$\rho = \frac{Z_s - Z_0}{Z_s + Z_0}, \quad R = |\rho|^2. \tag{10.5}$$

Formula (10.5) generalizes for the oblique incidence by substitutions $Z_0 \rightarrow Z_0^{TE,TM}$, where $Z_0^{TE} = Z_0/\gamma_0$, $Z_0^{TM} = Z_0 \gamma_0$, and $\gamma_0 = k_0 \sqrt{1 - \sin^2 \theta} = k_0 \cos \theta$ [25].

The normal-incidence reflectance in (10.5) vanishes under the matching condition $Z_s = Z_0$ that occurs in accordance with (10.2) when $n_1 = \sqrt{n_2}$ and $\gamma_1 d_1 = \pi/2$. The layer with such thickness and refractive index is called a *quarter-wave plate* (also referred to as a *simple antireflector*). For frequencies slightly lower than $\omega_0 = \pi c/2 n_1 d_1$, the normal-incidence reflectance remains very small.

A simplified model of the silicon refractive index (adequate in the red and inter-band frequency ranges) neglects frequency dispersion of its refractive index, assuming that it equals $n_2 = 3.61$. For this medium, the best simple antireflector is a quarter-wave plate of silicon nitride with $n_1 = 1.9$. In the red band, R of the c-Si half-space covered with an antireflector of SiN is practically zero, and R attains 0.1 only in the blue band of the solar spectrum. This property of an antireflector makes the operation of this simplest ARC quite broadband. In fact, it is true, in spite of the frequency dispersion of silicon (or any realistic PV material underlying the antireflector). Though this dispersion is detrimental for the antireflection operation of a quarter-wave plate (it increases the mean value of R), small dielectric losses accompanying this dispersion are favorable for this operation [27]. In fact, only reasonably small losses with $\text{Re}(n_2) \gg \text{Im}(n_2)$ are favorable. If the dielectric loss factor in the PV material is too high, reflectance increases [27]. Obviously, losses in the antireflection layer are always harmful for optical efficiency [28, 29].

Thus, for silicon, $n_2 \approx 3.6 - j0.3$, without a front glass, the practically optimal simple antireflector is a layer of silicon nitride of thickness $d_1 = \pi/2\gamma_1 = 76$ nm [28]. If we average R with the weight $S(\omega)$ over the whole operation band of the silicon SC $[\lambda_{\min}, \lambda_g]$, we obtain an even smaller value, $R_{\text{int}} \approx 0.04$; i.e., the optical loss in the open area of the SC for the normal incidence is only 4%. Taking into account the shadow caused by the front contact mesh, we obtain the optical loss factor for the normal incidence as $OL = 10\% + R = 14\%$. If the antireflector is made of MgF, as in [24], we obtain from (10.5) $R_{\text{int}} \approx 0.08$ because MgF has the refractive index $n = 1.37$, which is not optimal. Then, for the normal incidence, we have $OL = 18\%$.

Notice that the value $R_{\text{int}} \approx 0.04$ is achieved by using an antireflector made with nanometer precision. Making one layer with such precision is not very expensive for mass production. Costs drastically increase as the number of stacked nanolayers increase [21, 29, 30]. However, even for a single nanolayer, it is impossible to maintain this precise thickness during many years of operation because the solar cell is under the influence of such atmospheric factors as dust, rain, and even hail. Atmospheric abrasion decreases the thickness of the antireflector, resulting in about 30% degradation of the SC overall efficiency per year [31]. That is why silicon SCs utilizing antireflectors are either encapsulated by a protective laminate or covered by a macroscopically thick layer of glass. This protective layer is prepared without micron or submicron precision; i.e., its thickness D has statistic deviations from the mean value that are larger than λ.

Let us elucidate the role of the front glass in the optical loss. First, let us see that the front glass reduces R, even in the absence of an antireflector. Really, the normal reflectance of bare Si at $\lambda = 600$ nm in accordance with (10.5) is equal to

$$R_{Si} = |\rho_{Si}|^2 = \left| \frac{n_2(\lambda) - 1}{n_2(\lambda) + 1} \right|^2 = 0.33, \qquad (10.6)$$

whereas the amplitude reflection coefficient of a bilayer formed by a glass layer of an arbitrary thickness D and typical refractive index $n_{\text{glass}} = 1.5$ on the Si half-space can be calculated as

$$\rho = \rho_{\text{glass}} + (1 - \rho_{\text{glass}}^2)\rho_{Si/\text{glass}} e^{-2j\gamma_{\text{glass}}D}. \qquad (10.7)$$

Formula (10.7) is the so-called *two-ray approximation* [32]. The first term in (10.7) corresponds to the ray reflected from the protecting glass to free space. The second term in (10.7) corresponds to the ray transmitting into glass with the amplitude transmission coefficient $1 + \rho_{\text{glass}}$, passing the optical path $\gamma_{\text{glass}}D$ toward the Si substrate, reflecting from Si with the reflection coefficient

$$\rho_{Si/\text{glass}} = \frac{n_{Si} - n_{\text{glass}}}{n_{Si} + n_{\text{glass}}}, \tag{10.8}$$

passing after this reflection the backward optical path $\gamma_{\text{glass}}D$ and, finally, transmitting to the air with the amplitude transmission coefficient $1 - \rho_{\text{glass}}$. Multiple internal reflections in the glass layer of thickness D can be neglected because the macroscopic layer cannot be perfectly flat – its interfaces are wavy [28]. Thus, formula (10.7) gives, for the power reflectance $R = |\rho|^2$,

$$R = \rho_{\text{glass}}^2 + (1 - \rho_{\text{glass}}^2)^2 \rho_{Si/\text{glass}}^2 + 2\rho_{\text{glass}}^2(1 - \rho_{\text{glass}}^2)\cos(2\gamma_{\text{glass}}D). \tag{10.9}$$

Since the front glass is a macroscopic layer and its thickness randomly varies over the area, parameter D is stochastic. The statistically averaged value for R implies the averaging over possible values of D. Since the mean value of the cosine function is zero, the statistic averaging of (10.9) does not contain a term that describes interference effects between rays 1 and 2. Omitting the last term in (10.9) and substituting $\rho_{\text{glass}} = (n_{\text{glass}} - 1)/(n_{\text{glass}} + 1)$, we deduce for the reflection loss:

$$R = \left(\frac{n_{\text{glass}} - 1}{n_{\text{glass}} + 1}\right)^2 + \left[\frac{2n_{\text{glass}}\rho_{Si/\text{glass}}}{(n_{\text{glass}} + 1)^2}\right]^2. \tag{10.10}$$

The substitution of (10.8) at $\lambda = 600$ nm into (10.10) gives $R \approx 0.21$, which is smaller than $R_{Si} = 0.33$. This improvement refers to all frequencies, and it was confirmed by comparison with experiments in [28].

Introducing a nanolayer with the optimal refractive index $n = \sqrt{n_{\text{glass}}n_{Si}} = 2.3$ (e.g., zinc oxide) and the optimal thickness $d = 82$ nm (a quarter-wave plate at $\lambda = 600$ nm), sandwiched between silicon and glass media, we nullify the second term in (10.10) at $\lambda = 600$ nm and make it rather small in the whole operation band. Really, in this case, we should replace $\rho_{Si/\text{glass}}$ by the reflection coefficient of an antireflector (10.5), where Z_0 is replaced by $Z_{\text{glass}} = Z_0/n_{\text{glass}}$, with Z_s is given by (10.2) with substitutions $Z_1 = Z_{\text{ZnO}}$, $n_1 = n_{\text{ZnO}}$. ZnO layer of thickness $d = 82$ nm grants us $R_{\text{int}} \approx R_{\text{glass}} \approx 0.05$ – i.e., almost the same value of the open-area reflection loss as that obtained for a silicon nitride antireflector without glass. For an antireflector of MgF utilized in [24], the second term in (10.10) does not vanish at any λ. In this case, after the averaging of R over the solar spectrum, we obtain $R \approx 0.12$; i.e., an antireflector of MgF is not recommended in the presence of front glass.

The reflectance of the front glass, given by the first term in (10.10), definitely suffers from the atmospheric dust and other factors. Therefore, the reflectance of the front glass increases with time. However, this reflectance can be further decreased if we make the refractive index of the front glass nonuniform in the vertical direction. A slab of porous glass whose porosity varies with the depth can be modeled as a composite slab whose reflective index varies across the slab. Practically, one can achieve $n_{bot.} = 1.5$ near the bottom interface and $n_{top} = 1.4$ near the top interface [28, 32]. In this case, the front glass reflectance reduces from 0.04 to 0.028, and the antireflector remains the same because $n_{bot.}$ does not change. In this case, the integral normal reflectance of the whole SC is nearly 0.03 – almost equal to that of the front glass. Self-cleaning functionality engineered for such porous glasses is described in [33].

In 1973, an ARC called the *moth-eye texture* was suggested in work [34] for better suppression of reflection from optical glasses. Instead of a porous glass plate, one can use a specially prepared composite layer whose effective refractive index changes smoothly versus the vertical coordinate. On the top surface, it is equal to $n = 1$; i.e., the front reflection is presumably eliminated. In the following section, we will inspect this issue in more detail studying the *moth-eye ARCs*.

10.3.2 Oblique Incidence and Daytime-Averaged Optical Loss

In the case of the incidence under an angle θ, the reflectance of an air-glass interface can be found as

$$\rho_{glass}^{TE} = \frac{\cos\theta - \sqrt{n_{glass}^2 - \sin^2\theta}}{\cos\theta + \sqrt{n_{glass}^2 - \sin^2\theta}}, \qquad (10.11)$$

which gives $(\rho_{glass}^{TE})^2 \approx 0.17$ for $\theta = 60°$ and $(\rho_{glass}^{TE})^2 \approx 0.32$ for $\theta = 70°$. For the TM-polarization, the reflection coefficient is not monotonous versus θ, due to the Brewster effect:

$$\rho_{glass}^{TM} = \frac{n_{glass}^2 \cos\theta - \sqrt{n_{glass}^2 - \sin^2\theta}}{n_{glass}^2 \cos\theta + \sqrt{n_{glass}^2 - \sin^2\theta}}. \qquad (10.12)$$

From these formulas (applicable at any optical frequency), we find the averaged power reflectance $R_{glass} = \left(|\rho_{glass}^{TE}|^2 + |\rho_{glass}^{TM}|^2\right)/2$ corresponding to the unpolarized light. R_{glass} as well as the polarized reflectances $[\rho_{glass}^{TE}]^2$ and $[\rho_{glass}^{TM}]^2$ are shown in Figure 10.7(a). The curves corresponding to the case when R_{glass} is replaced by the integral (over the solar spectrum) reflectance R_{int} of the front glass with an ARC inserted between the front glass and the PV layer do not visually differ from the curves obtained for a semi-infinite glass at any λ [35]. The physics of antireflection is based on the mutual cancelation of partial reflections from the top and bottom interfaces of the ARC layer, as shown in Figure 10.7(b). Realization of the effect of destructive interference demands the equivalence of the magnitudes of the two reflected waves and their opposite phases. Their exact cancelation is possible only at one wavelength λ_0 where the reflectance of the whole structure is equal to that of the front glass. However, for the normal incidence, the reflectance turns out to be close to that of the front glass in a rather broad band of wavelengths. For the oblique incidence, the performance of this simple ARC is not so broadband.

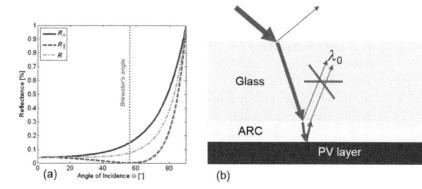

Figure 10.7 (a) Integral reflectance of sunlight from a usual glass layer: TE-polarization (solid curve), TM-polarization (dashed curve), and unpolarized light (dash-dotted curve). Reprinted from [36]. Unrestricted use of this open-access article is permitted. (b) The physics of antireflection is based on the mutual cancelation of partial reflections from the top and bottom interfaces of the ARC layer.

To perform the averaging over the incidence angle θ, which varies during the daytime, we should specify the maximal incidence angle θ_{max} corresponding to the daytime edges. There is no agreement in the solar photovoltaic community which value of θ_{max} should be adopted. In the following paragraphs, we assume that $\theta_{max} = 80°$ – in this case, our calculations better fit the literature data than if we choose $\theta_{max} = 90°$. With our choice of θ_{max}, the daytime-averaged reflectance of the front glass turns out to equal 0.095. This value describes the effective averaged reflectance of the open area of an SC. Also, in the daytime-averaged optical loss factor, the shadow term w/a must be present, which does not depend on θ. Thus, the expression for the mean (or effective) optical loss factor is as follows:

$$OL = \frac{1}{\theta_{max}} \int_{0}^{\theta_{max}} R(\theta)\,d\theta + \frac{w}{a}. \qquad (10.13)$$

For a c-Si SC with a front contact mesh $w/a = 0.1$, formula (10.13) gives $OL \approx 0.2$; i.e., the optical efficiency of a silicon SC with a front glass and a contact mesh equals 80%, in spite of the presence of an ARC. The use of a bilayer and even a multilayer ARC instead of a simple antireflector cannot give a real improvement because the presence of the front glass erases the advantages of a multilayer ARC. For SCs with interdigital back contacts, the shadow factor w/a disappears. Then the mean optical loss reduces by nearly 50%, and we have $OL \approx 0.1$ [28, 35] (also, the recombination loss decreases, though the Ohmic loss increases, and current collection efficiency decreases, as we discussed earlier).

To further reduce the optical loss, researchers continue developments of transparent composite glasses (see, e.g., in [37]). However, the progress in this field is modest, and other technical solutions that imply absence of the front glass have been found.

10.3.3 Moth-Eye Antireflectors

If we remove the front glass from the top of a solar panel, the lifetime of the latter reduces, due to atmospheric abrasion. Meanwhile, the optical loss also reduces because the first term in (10.13) is removed. Today, the companies working in solar photovoltaics, where the competition is very strong, struggle for every decimal point of a percent of the overall efficiency [18, 21, 22]. Therefore, a possibility of getting rid of an extra 10% weighs more than the theoretical decrease of the solar panel lifetime that will occur in the future. This practical need stimulated studies of moth-eye ARCs, the modern successors of the glass cover on top of an SC.

This dielectric texture is quite similar to the moth-eye photonic crystal of cylindrical protrusions (see Chapter 5). However, in contrast to the natural and photonic-crystal moth eyes, which have periods of the texture within the interval 250–300 nm, the period b of the moth-eye ARCs cannot exceed 150–200 nm because the key prerequisite of homogenization of any composite structure is smallness of its constituents compared to the wavelength. Practically, the period of protrusions is equal to 80–100 nm, whereas their height is 200–500 nm. Next, the protrusions are not simple nanorods but have the shape of cones with smoothed tops [38–43].

In 1973, when the first moth-eye ARC was suggested in [34] as an ARC for optical glasses, the target was to achieve a vertical gradient of the effective refractive index from 1.5 to unity. However, the necessary manufacturing technology did not exist. Therefore, only a microwave model experiment confirming the governing idea was made. Another implementation of this idea was suggested in work [44], and it was also realized at millimeter waves, at frequencies much lower than those of sunlight.

Today, a moth-eye ARC can be fabricated, for example, by using the so-called *nanoimprint lithography*. This method is based on the use of a hard stamp of quartz, fabricated once. Before losing the required quality, this stamp can replicate many times, molding a lot of much softer surfaces than that of quartz, producing needed textures. This technology is affordable for the solar PV industry on condition of mass production. It allowed one to drop the idea of challenging optical composites suggested in [34] in favor of simple arrays of dielectric protrusions obtained by this lithography technique [39, 41, 42].

The operation of this array is qualitatively illustrated by Figure 10.8. In the absence of front glass, it is reasonable to return to the consideration of a multilayer ARC, illustrated by Figure 10.8(a). Its refractive index is stepwise along the vertical coordinate, as depicted in Figure 10.8(b). The large number of layers gives a set of frequencies and angles for which the partial waves reflected from all the interfaces interfere destructively. The thinner that these layers are and the greater that their number is, the more broadband and more angularly stable the antireflective operation is. However, this way also implies an increase of fabrication costs. Finer precision is required as the nanolayers become thinner. As more layers are required, efforts necessary to ensure the layers' flatness increase.

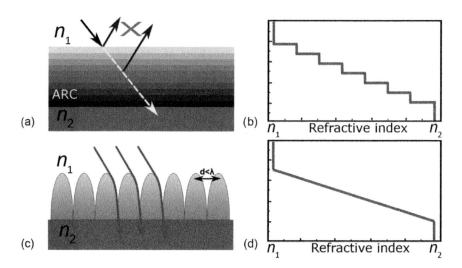

Figure 10.8 (a) A multilayer ARC grants several wavelengths at which the wave reflected by the top interface cancels out with that reflected by one of intermediate interfaces. Similarly, for several incidence angles, the reflection can be suppressed. (b) The refractive index profile of the multilayer ARC is stepwise. (c) Moth-eye ARC performed as an optically dense array of properly profiled protrusions suppresses the reflection better than the multilayer ARC. (d) The effective refractive index of the moth-eye ARC smoothly varies in the vertical direction from n_1 to n_2. These four figures are drawn by the author, after [43].

This is the reason why the moth-eye geometry shown in Figure 10.8(c) was chosen for open-air ARCs. Roughly speaking, a moth-eye ARC behaves as a layer of an effective medium whose refractive index at a given horizontal plane is determined by the cross-section area occupied by the medium having the refractive index n_2. In the bottom part of the ARC, it occupies the whole area, and the effective refractive index is equal to n_2. In the top part of the structure, it occupies a small area, and the effective refractive index approaches n_1. Thus, the effective refractive index of the moth-eye ARC can be thought to be the limit case of a multialyer ARC. The moth-eye ARC has no optical contrast at the effective top and bottom interfaces. If the situation illustrated by Figure 10.8(d) were applicable to all frequencies and all incidence angles, this structure would be perfectly matched with free space – i.e., reflection-free for all frequencies and incidence angles. Of course, this plot is an idealization – a moth-eye ARC is not an all-frequency and all-angle device, though it is much better than a simple antireflector.

In principle, a moth-eye ARC can be made as a flexible nanotexture of a transparent resin attached to the transparent top electrode of the SC by a submicron layer of a polymer glue with nearly the same refractive index n_2 as that of the electrode. However, this nanotexture is very vulnerable and needs to be covered by a macroscopically thick polymer laminate in order to operate in the open air. Such laminates, nowadays, have the refractive index $n = 1.34$ and grant the lifetime of a solar panel up to 1–3 years [45]. For a moth-eye ARC of acrylate resin located on a silicon SC and protected in this way, one obtained in

[39] the mean reflectance nearly equal to 7%. Together with the shadow factor $w/a = 10\%$, we obtain $OL = 17\%$ for the optical loss in the sandwiched geometry of the SC. However, the improvement of the optical efficiency from 80% to 83% is not sufficient when considering the reduced lifetime.

If we want to either reduce the reflection loss or extend the life of a solar panel to 5–10 years, we should get rid of the laminate and construct the ARC using a hard material. For silicon SCs, a month-eye ARC can be fabricated of silicon nitride [28, 38] or built as modified surface of c-Si itself [40]. The moth-eye ARC of silicon nanocones exhibits a very low reflectance for the normal incidence and small incidence angles (less than 2% in accordance with [40]). However, for large angles θ, the reflectance increases, and the mean value of the reflection loss is close to 10% (i.e., $OL \approx 20\%$, and there is no improvement compared to the simple ARC). Why is this so?

Unfortunately, the explanation of the operation of a moth-eye ARC in terms of a smoothly varying refractive index is simplistic and ignores the spatial dispersion effects. The spatial dispersion the dependence of the effective material parameters on the wavevector – may arise in regular arrays if the period is of the order of $\lambda/4$ or larger. This is the case of practically achievable moth-eye ARCs. Due to spatial dispersion, the effective refractive index of a moth-eye ARC calculated for the normal incidence differs from that calculated for the oblique incidence. For small incidence angles θ, this difference is minor. If the phase shift between two adjacent protrusions, $k_0 b \sin \theta$, is smaller than unity (here, $k_0 = 2\pi/\lambda$ is the free-space wavenumber), the interference of electromagnetic waves scattered by two adjacent protrusions is still negligible. In this case, the array can be homogenized the same way as in the case of normal incidence. However, for large incidence angles, when $k_0 b \sin \theta > 1$, the electromagnetic interference of the waves scattered by any two adjacent protrusions arises. This interference is not as strong as that resulting in the Fraunhofer diffraction maxima for the scattered field because $b < \lambda/2$. However, it is not negligible and disables the simplistic model of the gradient refractive index. Practically, the concept of a smooth transition from n_1 to n_2 becomes meaningless for $\theta > \pi/4$.

The analysis of the spatial dispersion in moth-eye ARCs shows that the reflectance of the unpolarized light for typical periods $b = 80$–150 nm grows versus θ in the same way as for a flat interface [39]. The only difference is the lower value of this reflectance as compared to the flat surface. If the material of the ARC has a low refractive index, the mean reflectance is low. This is the reason why, for the ARC based on the acrylate resin having the refractive index 1.4, the mean reflectance equals 0.07 and is mainly determined by the laminate. Silicon has the refractive index varying in the solar spectral range from 3.5 to 5 ($n_{Si} \approx 4.5$–5 in the blue and UV parts of the solar spectrum). Therefore, a silicon SC with a moth-eye ARC is not more efficient during the daytime than a silicon SC with a simple antireflector and a front glass (which is more robust to the atmospheric factors).

However, not all wafer-type SCs are based on silicon. Epitaxial multicomponent semiconductors – such as GaInP, AlGaP, and GaAs – are PV materials with similarly a high photoelectric spectral response and a lower refractive index than that of c-Si. For these PV materials, the use of hard laminate-free moth-eye ARCs turns out to be justified [41, 42]. Optimal moth-eye ARCs for such SCs require materials with the refractive index 1.7.

SCs based on epitaxial multicomponent semiconductors and gallium arsenide have a drastic difference in their optimal design as compared to silicon SCs. These semiconductors form a good Ohmic contact with transparent conductive oxides (TCOs). The conductivity of TCOs such as aluminum-doped zinc oxide (AZO), fluoride-doped tin oxide (FTO), and indium-doped tin oxide (ITO) is sufficiently high so that a nanolayer of thickness $\Delta = 80$–100 nm made of such TCOs can collect the photocurrent flowing from the PV layer and redirect this current to the contact wires with the same current collection efficiency as a wire mesh having $w/a \approx 0.1$. The contact wires are located on the perimeter of the SC and create no shadow for the PV layer. In design, the optical loss OL does not contain the term w/a, which was one of two main arguments against the sandwiched geometry of a silicon SC.

Unfortunately, TCOs are not ideally transparent within the solar spectrum, and a multiplicative factor describing the parasitic power absorption in a TCO layer replaces in (10.13) the term w/a. This factor, for a given frequency $\omega \equiv k_0 c$, is equal to $\exp(-2\alpha\Delta)$, where $\alpha = k_0 \mathrm{Im}\sqrt{\varepsilon_{TCO}})$ is the amplitude wave attenuation in a material with complex permittivity ε_{TCO} in the optical range, and Δ is its thickness. The thickness value is chosen as a compromise between the Ohmic loss, due to a finite DC conductivity of a TCO and the aforementioned parasitic power absorption averaged over the solar spectrum. For optimal Δ, the impact of the Ohmic loss in the TCO is negligible, whereas the parasitic optical absorption adds an extra 5%–6% to the mean reflection loss (that for advanced SCs is close to 10%. So, the total optical loss in these SCs is equal to 15%–16%).

SCs based on GaAs can be implemented with two technical solutions: with back contacts (like in modern silicon SCs) and with a top electrode. Both variants have nearly the same cost and efficiency. In the second variant, GaAs is covered by a nanolayer of transparent material serving as its chemical passivation (the window layer) and a front mesh of metal microstrips. Unlike chemical passivators used in silicon solar photovoltaics, this material of the window layer is electrically conducting. Materials of the window layer can be aluminium indium phosphate or gallium indium phosphate. They do not form surface states on the interface with GaAs because their crystal lattices allow a robust electric contact with GaAs. The optimal conductance of the window layer is achieved by its doping. Electric conductivity of the window layer is rather low in order to avoid high parasitic absorption of sunlight. Therefore, on top of it, there is a contact wire mesh. However, the period a of this mesh is much larger than that of a similar silicon SC, and the shadow factor w/a is much lower. In the optimal design, the parasitic absorption and the contact mesh shadow together result in the same optical efficiency as that for multicomponent SCs with a moth-eye ARC [41, 42].

Thus, the mean optical loss of practical SCs utilizing moth-eye ARCs is nearly 15%, where nearly 10% refers to the mean reflection loss and 5% refers to the parasitic absorption (or to the parasitic absorption and shadow). Though the improvement of the optical efficiency is modest (from 80% to 85%), it is easy to visually distinguish an SC covered with a laminate (e.g., front glass) from an SC with a moth-eye ARC on top. In the first case, the surface is bleached and looks blue from large observation angles. In the second case, the solar panel does not bleach and looks dark from any observation angle.

10.3.4 Black Silicon

We have already mentioned that, for silicon SCs, a moth-eye ARC is not as efficient as for some other practical SCs because, for large incidence angles θ, regular silicon protrusions are not efficient. However, the governing idea of the effective refractive index varying in the vertical direction obviously does not obviously require a regular array of protrusions. An alternative – random – implementation of this idea for Si-based SCs is called *black silicon*.

In [46], the authors suggested to use electrochemical etching or metal-assisted chemical etching that results in a stochastic roughening of the surface. In Chapter 9, we mentioned a similar technique for metal substrates, when it usually results in nearly hemispherical protrusions. Properly chosen technological parameters of electrochemical etching result for a semiconductor surface in a random but optically dense array of nearly conical protrusions. The mean values of the height and width of the cones can be 0.5–1 µm and 100–200 nm, respectively. Typical SEM pictures of such random arrays are presented in Figures 10.9 and 10.10. The effective layer formed by such protrusions on top of a layer of c-Si can be described as an effectively continuous medium whose refractive index decreases from the bottom to the top due to the conical shape of the protrusions. When this structure is illuminated by obliquely incident light, spatial dispersion effects, very strong for periodical arrays, are negligible because in subwavelength random arrays there are neither constructive nor destructive interference effects.

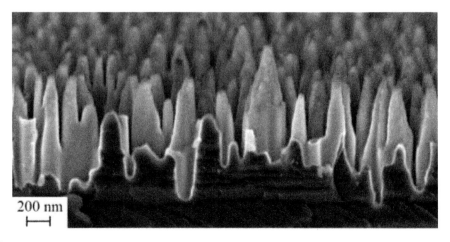

Figure 10.9 SEM micrograph of a black silicon surface developed in Aalto University. Reprinted from [47] with permission of the IEEE.

The structure shown in Figure 10.9 is a less expensive and more efficient ARC than the moth-eye arrays studied in papers [38, 40]. A demonstrator SC with this surface was fabricated in 2011 in Aalto University (the group of Prof. H. Savin). Here, the surface of c-Si including all nanocorrugations was passivated by a 2–3 nm thick layer of Al_2O_3 using atomic layer deposition. The material (corundum) resulting from this deposition does not contain even sub-nanometer pores, and a 2 nm thick layer ensures high-quality passivation. Therefore, a drastic increase of the surface area is not accompanied by an increase of surface recombination. The daytime-averaged reflectance of this black Si does not strongly differ from the normal reflectance and is nearly equal to 4% [47]. In 2014, this achievement resulted in a record for the daytime-averaged efficiency of silicon solar cells, $\eta = 22.1\%$ [48]. The record for the mean reflectance stood until 2017 [49]. However, the technical solution of [49] demands more expensive technologies.

Sometimes, very elaborate patterns of black Si surfaces can be obtained using affordable nanochemistry methods, complementing nanophysical techniques. Combining nanophysics and nanochemistry, the authors of [50] obtained a two-dimensional photonic crystal of free-standing 200 nm thick Si nanorods of micron height prepared with the period $b = 500$ nm on top of a c-Si SC. This array was optimized via numerical simulations to minimize the reflectance at all angles from 0 to θ_{max}, corresponding to the daytime edge. It turned out to be feasible in the range $\lambda = 400$–700 nm. Later, the angular stability of the reflection loss factor for this regular type of black silicon was later confirmed experimentally [51]. However, the antireflective operation of this regular black silicon as well as the moth-eye silicon SC is not sufficiently broadband. The reflectance is high in the inter-band range, above 700 nm, where c-Si is still an efficient PV medium, and the solar spectrum is significant until $\lambda = 10, 50$ nm. An experiment has shown that this arrayed ARC does not reduce reflection at $\lambda > 700$ nm. On the contrary, in this band this ARC increases the reflection compared to the bare flat interface of c-Si [51].

Black silicon ARCs arranged as random nanostructures are cheaper and very broadband. Profiles of two samples of black silicon SC with self-cleaning properties developed in [52] are presented in Figure 10.10. The mean reflectance for one sample is nearly 10%, and for the other one, it is about 7%. However, it is just a plate of silicon with antireflective properties, and it is unclear which overall efficiency a corresponding SC could have.

Judging from news media, the record of the daytime-averaged conversion efficiency in the industrially adapted silicon-based photovoltaics in 2018 was $\eta_{PV} = 22.8\%$. It belonged to the Russian company *Hevel* [53]. This result was previously claimed by an experimental demonstrator tested in a laboratory [54]. However, it was achieved not with a conventional black silicon SC but with a heterojunction SC based on different forms of silicon – polycrystalline (epitaxial) Si sandwiched between thin layers of c-Si (bottom) and amorphous silicon (top). In accordance with [54], this technology turned out to be affordable for commercialization. However, details of the fabrication process have not been published and, therefore, cannot be discussed here.

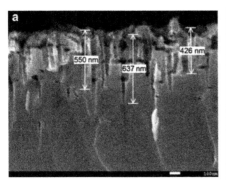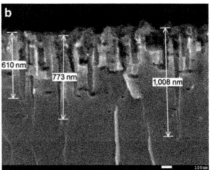

Figure 10.10 TEM micrographs of two samples (a and b) of black silicon obtained in Chengdu University by using the metal-assisted chemical wet etching. Reprinted from [52] with permission of Springer Nature.

In these works, the antireflective properties are granted by a nanostructure similar to black Si. This random nanotexture is located on top of this SC and represents an array of ZnO protrusions. The textured layer of zinc oxide has two functionalities – that of the current collection and the antireflective one [55]. The trade-off between parasitic absorption in the top electrode and its Ohmic losses is resolved using an additional 70 nm layer of ITO; i.e., the photocurrent is collected by a bilayer electrode. The electrode is backed by a nanolayer (thickness 20 nm) of a-Si, which has no PV functionality. Its purpose is to reduce surface recombination and to ensure the Ohmic contact of ITO and epitaxial silicon. The PV layer consists of epitaxial silicon and crystalline silicon sublayers, whose total thickness is sufficient for complete absorption of transmitted sunlight in one passage. For the normal incidence OL of this SC is close to 10% [55], where 6% refers to the parasitic absorption in the ITO layer and 4% corresponds to the contact wires $w/a = 0.04$ made of silver. This sparse wire mesh is designed to decrease the Ohmic surface resistance like it was done in gallium arsenide SCs, discussed earlier. For oblique incidence, OL grows only slightly, and the mean optical efficiency of the SC equals nearly 85% [54] (a 5% improvement compared to the conventional design).

Designers of this SC refer their heterojunction SCs covered by a roughened ZnO layer to the class of black silicon SCs (see [56]). There are four commonalities between these SCs: (1) the PV layer is made of solid silicon; (2) there is no laminate or a front glass; (3) the surface is randomly nanotextured with a subwavelength period and is not bleaching for all angles; (4) the nanostructured surface of ZnO is similarly hard and robust to the atmospheric factors, promising several years of operation without noticeable degradation.

Notice, however, that the exploitation of such solar panels is not very cheap due to the necessity of daily wet cleaning. It is especially important in polluted air where the dust contains so-called *particular matter* (particles that are only a few dozens of μm or even smaller), which attach to the zinc oxide and result in its abrasion [57]. The black silicon structures with *self-cleaning functionality*, suggested in [52], suffer high surface recombination. For such black silicon SCs, Loss 3 exceeds 15%, which does not allow them even to approach to the record mean efficiency 22%. Review papers [51, 58] discuss a trade-off between the surface recombination level and the self-cleaning effect for black silicon. The dilemma was only resolved by using expensive nanotechnologies [58]. Therefore, it appears that self-cleaning black silicon SCs are hardly promising for commercialization [51].

At this time, information on solar power plants built by Hevel Solar in different countries is available only from mass media (see, e.g., [59]) and confirms the daytime-averaged overall efficiency of 22% on sunny days. The same mean efficiency was achieved earlier (in 2016) by the American company Silevo.[4] This company used SCs based on so-called *oxide tunneling junctions* (see, e.g., in [60–62]). Here we do not discuss the physics of these pioneering SCs, which would deserve a separate monograph. It is enough to mention that these SCs utilize composite PV layers and do not refer to the class of black silicon SCs, though they also comprise nanostructures.

Finally, note that in the scattered sunlight (cloudy weather), the electric output of black silicon SCs drops by an order of magnitude. This refers to all SCs (see, e.g., [63]). The benefits of black silicon solar photovoltaics and of that based on oxide tunneling junctions will be clear when the lifetime of solar panels in recently launched solar power plants will be known, and corresponding exploitation costs will be counted.

10.4 Antireflecting Coatings Formed by Arrays of Small Spheres

An interesting affordable technique for manufacturing nanostructured ARC is presented in paper [64]. A densely packed array (monolayer) of micron- or submicron-sized spheres of silica (or another transparent dielectric) can be easily prepared on a flat dielectric or semiconductor surface because spheres attach better to the surface than to one another and may even self-assemble on the surface submerged in a colloidal suspension of such spheres (such colloids are inexpensive and available on the market with a great variety of particle diameters – from 100 nm to 10 μm). In order to prevent an occasional formation of multilayers of spheres, one often uses the spin-coating technique. Then a monolayer structure results from the centripetal force, which overcomes the mutual attachment of the spheres to one another.

After the monolayer has been obtained and water has been dried out, the whole surface can be encapsulated by a submicron film of a glue. In work [64], this glue was the spin-on glass; however, it can be also a polymer. This technique enables cheap fabrication of large-area ARCs with a regular nanostructure.

However, in what concerns reflection losses, these structures did not stand comparison with black silicon. Being placed on the surface of c-Si, these spheres are inefficient due to a high optical contrast between silica and silicon: the effective refraction index of the monolayer of silica spheres is nearly 1.4, whereas the refractive index of c-Si in the green range of the solar spectrum is about

4.0. We will discuss the applications of these ARC in the next section, which is dedicated to *thin-film solar cells* (TFSCs).

10.5 Thin-Film Solar Cells

In the previous section we saw that modern practical silicon SCs have a high fill factor that achieves $FF = 0.8–0.85$, the daytime-averaged optical efficiency achieves $A = 0.85–0.9$, and the mean overall efficiency is as high as $\eta_{PV} = 22\%$. We also saw that the current research in this field is not fundamental but applied, and this applied research is a competition of numerous scientific groups for every decimal fraction of a percent. This struggle is mediated by the obvious necessity of keeping fabrication costs low, maintaining safety or fabrication, and preventing toxic waste.

However, there is an alternative branch of solar photovoltaics – that based on flexible sheets of PV materials. In this alternative area, the potential for further improvement is huge, since in this technique, SCs have an optically thin PV layer, insufficient to absorb the sunlight in one passage and even two passages. Techniques that prevent the passage of the incident light across the PV layer before it has been absorbed are called *light trapping* techniques. Also, the high potential of improvement stems from the fact that basic PV materials of these SCs are either very inefficient (such as amorphous silicon) or not sufficiently stable in time (such as synthetic perovskites).

10.5.1 Drawbacks, Advantages and Practical Perspective of Thin-Film Solar Photovoltaics

Let us understand why TFSCs are practically interesting. Though advanced silicon SCs, such as those produced by Silevo and Hevel, justify their cost in one year (whereas their lifetime is longer), they are not very attractive for domestic use. They are heavy and hard, requiring mechanically robust supports. Moreover, they are quite fragile and can occasionally (or intentionally) be broken. Meanwhile, TFSCs can be fabricated as lightweight, flexible films. Their flexibility makes them compatible with the so-called *roll-to-roll fabrication technique*, which minimizes fabrication costs as the market grows. Additionally, this technique allows one to fabricate variable-area solar panels [19]. Finally, a flexible film is conformable to curved surfaces and does not spoil the appearance of a tile roof.

A very important type of TFSCs commercialized in 1970s (as a charging source for pocket calculators) was the SC based on amorphous silicon [19]. The Shockley–Queisser overall efficiency and the ultimate efficiency of a-Si TFSCs (13% and 42%, respectively, in accordance with [65]) are noticeably worse than those of their crystalline counterpart (30% and 50%, respectively). This is due to high intrinsic losses (especially Loss 1) and low open-circuit voltage.

Loss 1 is high because the bandgap wavelength of a-Si is almost half the length of that of c-Si; it lies within the interval $\lambda_g = 600\text{–}700$ nm depending on the explicit design (λ_g depends on the static electric field in the depletion region). If $\lambda_g = 650$ nm, as it occurs in most a-Si SCs, about one-half of the incident solar power is lost, and the spectral efficiency is only $\eta_{sp} = 50\%$. Next, the mobility of the minor charge carriers in a-Si is much lower than their mobility in the usual (crystalline) silicon. This drawback implies low drift velocity and small diffusion length of the carriers. Recall that the time in which most of minor carriers pass through the doped bulk part to the corresponding electrode must be shorter than their lifetime. Otherwise, most photocurrent carriers will recombine, which will result in very large values of Loss 3. Therefore, the low drift velocity of the carriers requires a smaller thickness of the PV layer. Usually, the thickness for the PV layer of a-Si is no more than 400 nm, and the PV diode cannot be a p–n one. First, in the doped a-Si layer minor charges are very slow, and high recombination restricts the total thickness of these layers by maximum of 100 nm. Second, in the p–n junction of a-Si, the harmful Franz–Keldysh effect (see Section 10.2.2) is too strong. Therefore, instead of a p–n junction one fabricates p–i–n diodes, where the typical thickness of the doped parts is as small as 30–60 nm, and the intrinsic layer at the center has the thickness 200–350 nm. The total thickness of the PV layer, 300–400 nm, is not sufficient for absorption of solar light before it reaches the bottom electrode. Therefore, in the past, one fabricated an intrinsic a-Si layer with the thickness of the order of 1 μm (see, e.g., in [66]). In these SCs, Loss 3 exceeded 60%, and the overall efficiency was about 2.5%.

If the thicknesses of all three sublayers are chosen to maximally reduce Losses 2 and 3 (and the doping level in the p- and n-sublayers is also optimal – namely, the minor carriers concentration is equal to $(2\text{–}3) \cdot 10^{18}\ cm^{-3}$ in these sublayers), Loss 2 and Loss 3 are reduced so that their sum does not exceed 30% [19]. However, the fill factor of a-Si in these best SCs attains 0.52. That restricts the ultimate efficiency, and the practical Shockley–Queisser limit calculated using formula (10.1) by the values $\eta_u = 0.42$ and $\eta = 0.21$, respectively [67]. At a first glance, these numbers are promising. However, in these calculations, the optical efficiency is assumed to be 100%. Moreover, even neglecting the optical loss, this estimate was too optimistic because Ohmic losses, which are especially high in the contact area, can hardly be negligible. Unlike its crystalline counterpart, a-Si does not form a good electric contact even with polished metal. Notice that the polished metal cannot be flexible and is prohibited if flexibility is required. For flexible TFSCs, the back electrode is made of a flexible metal foil electrically connected to a-Si by an intermediate contact material, such as *conductive paste* – carbon or graphite paste with copper inclusions. Alternatively, it can be a layer of a semitransparent conductive metal oxide such as doped ZnO. These materials possess noticeable optical losses. The thickness of the contact material is chosen as a compromise between parasitic absorption of sunlight in the back electrode and its Ohmic resistance. Therefore, the overall efficiency suffers both Ohmic loss and optical loss in the area of the back electrode.

The top electrode in TFSCs is usually made of a TCO (AZO or ITO, in perovskite SCs – from FTO). If it is too thick, parasitic absorption makes the optical efficiency of the SC very low. If it is too thin, the Ohmic resistance of this electrode (recall that the current flows horizontally in it) strongly exceeds the internal resistance of the PV layer, and too much of the generated electric power is dissipated. The optimized TCO layer for a TFSC has the thickness of 70–80 nm, and the parasitic absorption in it subtracts 6%–8% from the optical efficiency. The Ohmic resistance of this optimized electrode is comparable to the internal resistance of the PV layer. In both bottom and top electrodes with the optimized parameters, 30%–35% of the output power is lost.

Therefore, the maximally achievable overall efficiency reduces practically from 0.21 to nearly 0.134–0.139. If we further take into account that the maximally achievable optical efficiency is about $OE = 0.94$ (see (10.1)), we obtain the result of $\eta_{PV} = 0.13$ [65]. This is the true Shockley–Quesser limit for a flexible TFSC based on a-Si. And even this low limit is not approached in commercially available samples of flexible TFSCs. In 2009, commercially available TFSCs with a micron-thick PV layer and SiN antireflector on top of the front electrode demonstrated the same daytime-averaged overall efficiency (about 3%) as the TFSCs with the optimized thickness of the p–i–n structure [66]. In six years, this value increased to 4%–5% [12]. So, in spite of the low efficiency, such advantages as low costs, light weight, and flexibility determined fast development of thin-film solar photovoltaics in the past decade. Fifteen years ago, flexible TFSCs based on a-Si had the cost about 5% of the price of silicon solar cells, whereas their overall efficiency was only 1/5 lower [15]. The overall comparison was evidently in favor of a-Si TFSCs.

Six years ago, TFSCs were at the peak of their development and represented a point of keenest interest for both industry and research institutes. However, in 2013–2014, the situation dramatically changed. During these years, the price of monocrystalline silicon dropped several times and became comparable to the price of the other components of wafer solar panels. This development resulted from the ramp-up in the production of semiconductor devices for consumer electronics. In 2011–2012, the industry started to prepare large-area silicon crystals from waste accompanying fabrication of cell phones, tablets, and other widespread gadgets. Crystallization of the melt waste did not imply an additional purification of Si. Fabrication of c-Si solar panels reduced the amount of industrial waste and became as nature-friendly as the production of a-Si TFSCs.

At the same time, the cost of a-Si also reduced similarly and became smaller than the cost of all other components of a thin-film panel. However, the cost of these components – antireflective coating, current-collecting electrodes, metal contacts, circuitry for matching the solar panel to the external load – stayed the same. Therefore, the difference in prices per unit area of a monolithic SC based on c-Si and a TFSC based on a-Si has drastically changed in favor of c-Si. This change resulted in a crash of companies specializing in thin-film solar photovoltaics, and it appears that not one of them survived in 2015. Consequently, the amount of funds allocated for research in this area over the world was recently drastically reduced [12]. Note that this situation was predicted in 2011 in work [68]. In this report, researchers working in the American solar power industry explained (properly but in vain) to their managers how to avoid the collapse of thin-film solar photovoltaics.

The industrial collapse of TFSCs does not mean that we should stop scientific research in this field. "Energy is needed where it is consumed," and this paradigm acts in favor of future TFSCs. The area of applicability of TFSCs is much wider than that of the silicon solar photovoltaics oriented to power plants. The main issue is the low efficiency of TFSCs that for flexible a-Si SCs fabricated by scientific groups, as a rule, does not exceed 5%, and this is mainly due to low optical efficiency [12]. The market of TFSCs is waiting for a scientific breakthrough – a drastic enhancement of the optical efficiency of TFSCs without a significant increase of their costs is needed. In the following section, we will show how this breakthrough may start.

10.5.2 ARC of Nanospheres for a-Si Solar Cells and the Issue of Transmission Losses

Although moth-eye ARCs are relatively inexpensive on condition of their mass production, they are considered by many authors as not affordable for TFSCs, based on a-Si whose idea is cheap electricity from flexible sheets. Therefore, alternative technical solutions for all-angle ARCs have been suggested for these TFSCs whose operation is presumably similar to that of moth-eyes. For example, one can take a honeycomb planar array of silica spheres with the diameter of 2 μm densely packed on the substrate and laminated by a glue of liquid glass, which was suggested in aforementioned work [64]. This monolayer is shown in Figure 10.11 together with its building block (encapsulated microsphere).

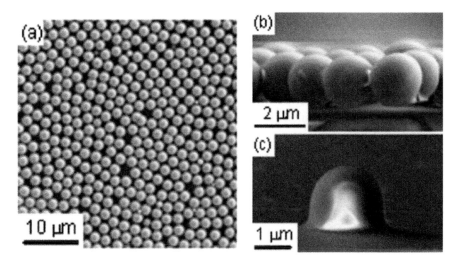

Figure 10.11 (a) Top view of an ARC of silica microspheres. (b) A SEM picture of the array in the vertical cross section. (c) A single microsphere encapsulated by a nanolayer of the spin-on glass illustrates the idea of a gluing laminate protecting the depicted nanostructure from abrasion and ensuring its mechanical robustness. Reproduced from [64] with permission of AIP Publishing.

The structure was obtained using self-assembly, and its fabrication is even less expensive than the nanoimprint lithography used for moth-eye ARCs. Instead of c-Si as in [64], these microspheres can be placed on the surface of the top electrode of a TFSC (a TCO nanolayer). Well, the period of this array is optically large and in the reflected field the Fraunhofer diffraction lobes arise. However, the same technique can be used with smaller spheres. Detailed and accurate studies have shown that the best suppression of reflection for such an ARC located on top of a TCO layer (top electrode) of an amorphous silicon TFSC and even of a gallium arsenide SC is granted by submicron spheres [69–71].

As it was shown in [70–72], the operation of the ARC on top of a TFSC is similar to that of a flat multilayer ARC. For TFSCs, any possibility to reduce the fabrication costs is important, and the fact that one can fabricate a cheap analog of multilayer ARCs (due to their high cost utilized only in satellite SCs) was a great finding. A TFSC based on a-Si and enhanced by an optimal ARC of nanospheres comprise nanospheres of silica ($n_{\mathrm{sil}} = 1.46$ in the visible range) with a 700 nm diameter fixed on top of the TCO by a 300 nm thick layer of a polymer adhesive [73]. Its operation is qualitatively illustrated in Figure 10.12 by an effective multilayer. This model is adequate for spheres with diameters smaller than 600–800 nm [70–73]. Effectively, a PV layer with $n_{\mathrm{a-Si}} \approx 3.2$ is covered by a layer of ITO with $n_{\mathrm{ITO}} \approx 1.9$, covered in its turn by a composite layer with $n_1 = 1.41$ (silica and glue) and finally by a composite layer with $n_2 = 1.38$ (silica and air). The mean integral reflectance of such TFSC reduces to 11% compared to 23%, corresponding to the bare TFSC. The mean integral optical loss in the 80 nm thick layer of ITO is 6%. This seemingly should have resulted in the daytime-averaged optical loss $OL = 17\%$. However, in reality, the optical loss is much higher. In these estimates, we have neglected the optical loss in the back electrode where a substantial part of the incident light transmits.

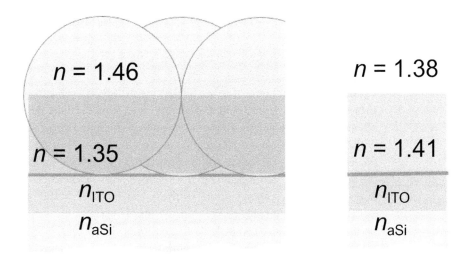

Figure 10.12 Multilayer model of the ARC of silica nanospheres of top of the a-Si TFSC. Left panel: the physical structure. Right panel: the model structure.

Though this multilayer ARC in view of its low cost is excellent, and 83% of the solar power flux transmits into the PV layer, the optical loss is very high. Namely, for this TFSC $OL \approx 59\%$, whereas for a bare TFSC without any ARC, we have $OL \approx 67\%$ [71]. The improvement granted to the optical efficiency by the ARC of silica spheres is too modest (8%) and hardly sufficient to justify even modest fabrication costs.

The p–i–n structure of the thickness 300–400 nm is optically thin – it absorbs nearly one-half of the solar power in one passage of the sunlight through the layer. The other half of the power transmitted into the PV layer passes to the area of the back contact and is dissipated there. An ARC is not very useful – it replaces the reflection loss by the transmission loss. Transmission loss is the key issue for all flexible TFSCs except the aforementioned perovskite TFSCs (to be discussed separately).

Since an increase of the thickness of the PV layer compared to the optimal one results in the same reduction of the efficiency as the reduction of the transmission loss, the only remedy is trapping the sunlight inside the PV layer. To achieve this goal, one should transform the incident plane waves into a package of such spatial harmonics that could stay inside the PV layer until the transmitted power is completely absorbed. It can be, for example, a package of evanescent waves – waves propagating in the horizontal direction and attenuating across the layer. It can be a package of Brillouin waves propagating in the PV layer like in a waveguide, or it can be any other waveform that would imply a better absorption of the solar energy inside the PV layer than that corresponding to a single plane wave propagating vertically (across the PV layer). This functionality is performed by specially engineered light-trapping structures.

10.5.3 Light Trapping in Epitaxial Silicon Solar Cells

The first LTSs were engineered not for *flexible SCs*, which on the microscopic level are flat multilayers, but for SCs based on epitaxial silicon. Epitaxial (poly-crystal with submicron grains) silicon TFSCs were developed in 1970s in order to reduce the amount of purified silicon needed for producing usual silicon SCs in those old times.

The optimal thickness of a c-Si PV layer is within 200–300 μm. This interval of values expresses the trade-off between the mechanical robustness, the necessity of minimizing surface recombination, and the maximal absorption of sunlight in the PV wafer. Meanwhile, the thickness of the PV layer of epitaxial Si cannot exceed 20 μm. The surface recombination for these SCs is not a key issue because the main PV conversion occurs in the p-doped layer sandwiched between heavily doped nanolayers (called $p+$ and $n+$ layers, respectively). In the epitaxial Si, both diffusion length and mean free path of minor carriers are smaller than those parameters of c-Si by two orders of magnitude. The critical bulk recombination prohibits for these SCs a PV layer with a thickness $d > 20$ μm. Moreover, if we ensure the maximal optical efficiency, it is better to reduce the thickness of the PV layer to $d = 200$–400 nm [19]. An amazing economy of purified silicon and a similar reduction of fabrication waste made the epitaxial SCs very attractive for the industry in 1970s.

However, even the thickness of the PV layer $d = 2$ μm was not sufficient for the absorption of solar light during two passages through the layer. When such flat PV layers are backed by a polished metal electrode, nearly 20% of the power reflected by the bottom electrode transmits back to the air. If the electrode is not polished and represents a metal foil and a contact layer, even more power is absorbed in it. In both situations, the ARC is not helpful at all because an ARC can only prevent the reflection from the top interface and can suppress neither the parasitic transmission of light downward nor the backward transmission of light reflected from the bottom electrode. It is schematically shown in Figure 10.13(a) for a bilayer ARC that practically eliminates reflectance for the normally incident light. In the photograph (top panel), this surface looks gray (not black) because it reflects a lot. Meanwhile, in the 1970s, SCs of epitaxial Si with $d = 200$–400 nm were developed; overall efficiency without LTSs was very low. Therefore, already in 1970s, light-trapping research started in solar photovoltaics.

The best-known technical solution (see, e.g., in [74–76]) is *pyramidal textures* the texturing of both top and bottom surfaces of the Si layer. The crystallography of Si allows such texturing with the pyramidal angle equal to 45°. In Figure 10.13(b) (top), an atomic force microscope image of such a surface is shown. The image shows the domain beyond the microstrips of the grid electrode. Here, the averaged height of the pyramids is 1.5 μm. The textured surface is covered by an antireflector (75 nm of MgF) that also serves for passivation of

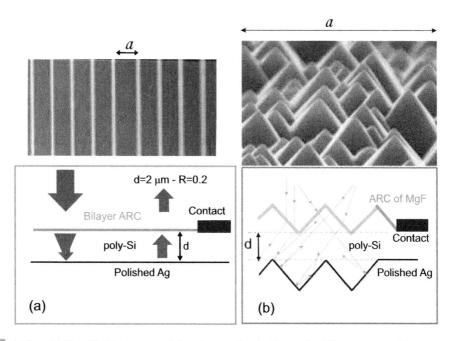

Figure 10.13 (a) Epitaxial silicon SC of the thickness of about 2 μm equipped with an optimal ARC (bottom panel) has the same reflective properties as the thick layer of silicon with the same contact mesh (top). (b) Pyramidal LTS for an epitaxial silicon SC: sketch of the cross section illustrating the idea of the ray-type light trapping (bottom panel) and the AFM micrograph of the surface (top). Micrographs have been prepared by M. Omelyanovich.

polycrystalline Si reducing surface recombination. The physical mechanism of this LTS is illustrated by the sketch in the bottom panel. The texture diverts the rays of the sunlight, and their oblique propagation in the PV layer makes the optical path of sunlight inside the PV layer sufficient for its full absorption. From the picture, it is clear that this enhancement of the path length holds for any incidence angle. Only the rays incident to the pyramidal apex or to the crevice are scattered.

It is also clear that this explanation based on geometrical optics is restricted by diffraction effects. For an individual pyramid with a height taller than $1.5-2\,\mu$m in the range of the visible light, these effects can be, in principle, negligible. However, the pyramidal texture of the top surface is obviously reproduced in the bottom surface of the PV layer – the substrate of polished silver that reflects the light [15, 19]. If the thickness of the PV layer is insufficient, the sunlight experiences strong diffraction on these two parallel textures and the expectations a of excellent light trapping and low scattering are not justified.

As it was shown in works [74–76], this ray light-trapping mechanism practically works only for $d > 1\,\mu$m. In fact, the condition of the applicability of the ray optics also restricts also the minimal ratio d/h. First, the height h must be much larger than the effective wavelength in Si ($\lambda_{Si} = \lambda/n_{Si}$); second, the ratio d/h must be larger than unity [75]. Under these conditions, light trapping using pyramidal or conical textures may enhance absorption in the PV layer (compared to the flat structure with the same thickness of all layers) up to $4n_{p-Si}^2$, where n_{p-Si} is the refractive index of the polycrystal Si. Of course, this ultimate gain can never be achieved, and this consideration only indicates great potential of light-trapping techniques.

For epitaxial TFSCs with $d < 1\,\mu$m, *pyramidal textures* do not work, and for such TFSCs, researchers have developed more sophisticated types of LTS, called photonic crystal LTSs (see, e.g., [12]). We do not discuss them here; it is enough to mention that they are too expensive for commercialization. Pyramidal textures have found application in wafer silicon solar cells as well, due to their antireflective action for large incidence angles. Since such textures are covered by an antireflector of MgF, the Fraunhofer backscattering from them is weak. Except black silicon SCs and SCs based on oxide tunneling junctions, all silicon solar photovoltaics imply pyramidal texturing, at least on top of the PV layer. The decrease in the mean integral reflectance granted by this texture is modest, but fabrication costs of the pyramidal texture are nowadays so low that this modification is justified.

10.5.4 Plasmonic Light Trapping in Thin-Film Solar Cells

For amorphous silicon TFSCs, pyramidal, conical, or prismatic textures are not efficient because the PV layer is too thin. Moreover, this texturing technique is not compatible with the target of flexibility. For flexible TFSCs, researchers have developed a lot of original LTSs. Most efforts of researchers were devoted to the development of so-called *plasmonic LTSs*. The history of this scientific direction started in 1990s by pioneering work [77], and since that time, it has resulted in hundreds if not thousands of journal articles reviewed in work [78]. This review, published in 2010, enthusiastically promoted the idea of plasmonic LTSs for amorphous solar photovoltaics. Meanwhile, work [79] (also published in 2010) criticized the whole approach because the light-trapping effect can be achieved only due to a plasmon resonance, which is obviously associated with resonant losses in the metal constituents of the LTS. Therefore, the increase of the useful PV absorption due to subwavelength concentration of the field in the PV layer is obviously accompanied by an increase of the parasitic absorption. The discussion has continued since that time. Nowadays, the attitude of the solar photovoltaic community to plasmonic LTSs for amorphous TFSCs is controversial [12]. The question is especially intriguing due to the fact that in the more than 20 years of its history, not one of hundreds plasmonic LTSs fabricated for amorphous silicon TFSCs has been commercialized.

As a rule, plasmonic light-trapping implies subwavelength concentration of solar light transmitted into the PV layer. Due to the subwavelength size of plasmonic nanoparticles, this concentration is not light focusing – a process when the incident plane wave is converted into a beam of converging waves. It is conversion of the incident plane wave into a package of evanescent waves, exponentially decaying in the vertical direction. There is a point of view promoted and shared by many authors that simple ray focusing is unsuitable for TFSCs for two reasons. First, a converging light beam experiences the so-called free-space diffraction or *Abbe diffraction*. This effect restricts the applicability of the geometrical optics for the domain of the focal spot and restricts the minimal size of the focal spot by the value 1.22λ (calculated by Lord Rayleigh and called the *Rayleigh limit*). Most part of TFSCs have the PV layer of the thickness smaller than the minimal focal spot size, and, therefore, focusing is not a suitable mechanism for light trapping. Second, even for those TFSCs which have sufficient thickness of the PV layer, focusing is not suitable because it will reduce the current collection efficiency. A lens located on top of the TFSC will focus light in a micron-sized area of the PV layer. All the area of the SC around the focal spot will remain dark and the photocurrent will be not formed there. Thus, the effective area of the TFSC squeezes to the microscopic values and the output power will be small. Such light concentration would be reasonable only if the unit area of TFSC were much more expensive than the unit area of the lens. However, the situation is the opposite. Solar light concentrators (Fresnel lenses and special mirrors) were strongly developed for solar photovoltaics 15–20 years ago (see e.g. in [15]) when wafer SCs were expensive and there was a market demand to decrease their area. Nowadays, silicon solar panels are cheaper than these lenses and mirrors, and TFSCs are much cheaper.

These authors conclude: Since the light focusing is not suitable for trapping light in a TFSC, one has to use nanostructured light concentrators, nearly uniformly covering the whole area of a TFSC [80–82]. The most suitable mechanism of such concentration appears to be the plasmon resonance (see similar speculations also in e.g. [83–85]). We may refer these plasmon resonances to two classes. The first class is that of localized surface plasmon resonances. These are resonances of an individual metal nanoparticle. In this book (Chapter 7) the dipole type of this resonance (that of a silver nanosphere) is analyzed. Meanwhile, for substantial in size or complex-shaped particle of a plasmonic metal (usually Ag or Au) there are also multipole resonances, among them the most important for light trapping is the quadrupole one. The second class of plasmon resonances relevant for LTSs is represented by collective surface plasmon resonances – resonances of the whole nanostructure covering the area of the SC. These resonances result in formation of packages of surface waves called surface-plasmon polaritons.

Individual plasmon resonances can be engineered in rather cheap, chemically-manufactured plasmonic arrays called *metal island films*. A metal island film is an optically dense random array of Au or Ag nanoparticles with the typical sizes of the particles 20–200 nm, whereas the dominant shape of the constitutive nanoparticles depending on the explicit technique can be different: from nanospheroids to nanopatches. As we have discussed in Chapter 9, too small nanoparticles manifest high losses and low magnitude of the plasmon resonance, thus particles smaller than 20 nm are not desirable. The maximal particle size is restricted by the requirement of low backscattering losses. The variation of the particle sizes and shapes offers overlapping plasmon resonances at different frequencies so that the resonance bands may cover an important part of the operation band of a TFSC (for those based on a-Si, up to one half of it [86, 87]).

However, broadband increase of absorption is not the only requirement for LTSs. The purpose of LTSs is broadband increase of the useful (PV) absorption and not the total absorption in the active layer into which the metal nanostructure is incorporated. However, many authors are happy to report the decrease of both integral reflectance and transmittances in the operation band, meanwhile the optical loss in their structures is high due to the resonant dissipation in the metal nanoparticles [79]. Moreover, beyond the range of the consolidated plasmon resonance such LTSs increase the reflection loss (because a non-resonant metal structure creates a shadow behind it). This effect is not very significant if the metal island film is incorporated sufficiently deeply into the PV layer where the sunlight intensity already decays due to the PV absorption. However, such structures are more expensive than similar plasmonic LTSs located on top of the active layer. There were many attempts to suppress parasitic reflection from plasmonic island films located on top of a TFSC, such as the use of overlapping dipole and quadrupole plasmon resonances [88]. However, all these attempts resulted only in a small (narrowband) enhancement. An efficient way to exploit localized surface plasmons in an LTS of a TFSC is to implement the bottom electrode of the SC as a metal island film located on a substrate of TCO [89, 90]. Then the light at frequencies beyond the plasmon resonance band is reflected into the PV layer. However, the hot spots resulting from the local field enhancement at the plasmon resonance still intersect with the metal nanoparticles.

Thus, it appears that cheap plasmonic nanostructures do not grant any significant improvement to any practical TFSC. Even in the case when gain in the PV absorption exceeds the increase of the parasitic absorption [89, 90], the gain for the overall efficiency is not sufficient to justify the fabrication costs [12]. In fact, these costs are higher than that initially expected because a metal island film incorporated into the PV layer or located on the back electrode surface needs to be isolated from the semiconductor. Otherwise the photoelectric conversion is spoiled by the atomic diffusion from the metal surface [91], and the gain in the optical efficiency is overcompensated by the decrease of internal quantum efficiency. In [12] it was concluded that all known attempts to enhance practical TFSCs using rather cheap, chemically manufactured plasmonic nanostructures have failed and the authors of [79] were right.

An alternative approach is exploitation of collective plasmon resonances which arise in periodic metal nanostructures. Utilizing collective plasmon resonances, one may minimize reflection losses, maximize the consolidated bandwidth of plasmon resonances and practically get rid of parasitic losses in metal elements, engineering evanescent wave packages so that the intensity maxima are located outside of metal. These advantageous operation regimes require a high regularity of metal elements dimensions and distances between them. Such elements prepared with nanometer precision are called *plasmonic nanoantennas* [92–94]. Figure 10.14(a) shows an example of such LTS from [95]. This type of nanoantenna arrays is applicable to different types of flexible TFSCs from those based on a multicomponent semiconductor called CIGS [96] to a TFSC based on an organic PV material [97, 98]. Organic TFSCs do not refer to the class of PV diodes, they are called exitonic solar cells [97, 98]. Nanoantennas depicted in Figure 10.14(a) grant a drastic reduction of both reflection and transmission losses being located on top of the TFSC, and the additional absorption holds not in their metal elements. Comparison with the optimal ARC shows a noticeable enhancement in the integral PV absorption for all types of TFSCs studied in works [96–98]. The collective plasmon resonances exploited in these nanostructures were studied in work [99], where it was shown that these resonances form a consolidated resonance band suitable for light trapping. This band may occupy more than a half of the operation band of an SC. Figure 10.14(b) shows the

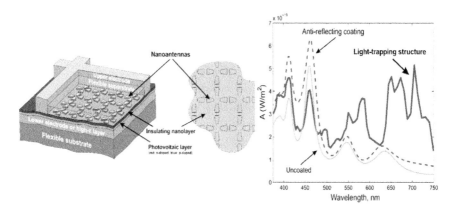

Figure 10.14 (a) Plasmonic LTS utilizing collective resonances of an array of specially designed nanoantennas.
(b) Spectral absorption of sunlight for normal incidence calculated for a bare TFSC (based on a multicomponent semiconductor called CIGS), for the same TFSC with an antireflector, and for the same TFSC enhanced by a LTS. Reprinted from [95] with permission of Aalto University.

spectrum of the PV (useful) absorption for three cases: bare SC, that enhanced by an ARC, and that enhanced by an LTS.

An array of nanoantennas can be fabricated also in the bottom part of the PV layer: at the interface with the bottom electrode. At the frequencies of the surface plasmon polariton resonances (which may occupy a significant portion of the operation band) an array of nanoantennas illuminated by a plane wave operates as a phase diffraction grating whose main diffraction maxima correspond to horizontally stretched wavevectors. The conversion of incident plane waves into such surface plasmon polaritons can be treated as conversion into waveguide modes propagating in the SC horizontally. The electric field of these modes almost does not penetrate into the back electrode (see e.g. in [100–103]).

However, the noticeable enhancement of the PV absorption (from 30% measured in [98] to 40% measured in [103]) accompanied by a corresponding enhancement of the overall efficiency from 18% (for organic solar photovoltaics in [98]) to 27% (for amorphous silicon solar photovoltaics in [103]) is not revolutionary. Recall, that the usual overall efficiency of a flexible TFSC based on a-Si is about 5% and the efficiency of a flexible organic SC is even lower. If we increase it up to 6.5% as in [103], it is not sufficient to justify the high fabrication cost of a regular plasmonic structure. When the back electrode is performed of polished silver and light can be absorbed after two passages through the PV layer, the optical efficiency is much higher and may attain 70%, whereas the light-trapping effect is wide-angle and broadband [101, 102]. Polished silver forms a good electric contact with the amorphous silicon and there is no need in conductive paste. Unlike Al foils, silver polished with submicron precision does not imply parasitic light absorption in the contact domain. Therefore, nowadays such TFSC enhanced by a moth-eye ARC or an ARC of nanospheres can have $\eta = 11.3 - 11.5\%$ approaching to their Shockley–Queisser limit of 13% [65, 104]. However, such TFSCs are hard and have no perspective for civil terrestrial applications. They may find applications in space technologies where TFSCs can be preferred to multijunction SCs due to their incomparably lower mass per unit area [15].

Returning to flexible TFSCs, we have to state that although regular LTSs are efficient optically, they are very inefficient economically. Even if the nanoimprint lithography is involved as it was suggested in [95, 96], it is combined with preparation of golden nanoparticles using some physical techniques. Fabricated of a metallic nanostructure with a nanometer precision in large areas is, nowadays, a rather expensive process. Therefore, TFSCs whose LTSs exploit collective plasmon resonances can be made flexible, but can find only some restricted applications beyond mass production.

To conclude this discussion, one can assert that plasmonic LTSs for TFSCs, in spite of the huge body of literature, are either inefficient or very expensive. Therefore, they do not promise a revival of the flexible TFSCs for commercial solar photovoltaics.

10.5.5 Dielectric Light-Trapping for Amorphous Silicon Solar Cells

The general advantage of dielectric LTSs compared to metal ones is the practical absence of optical losses. Recall that the same consideration is the main motivation behind development of all-dielectric metasurfaces (Chapter 4). Of course, even transparent dielectric materials are not ideally transparent: the imaginary part of the complex permittivity is non-zero, but for many solid dielectrics, this value is practically negligible in the optical range. There is a straightforward way for engineering all-dielectric LTSs. It is the transformation of a previously known ARC dedicated to TFSCs (an affordable large-area nanostructure) into an LTS. For example, it is possible to trap light with a monolayer of silica nanospheres, discussed earlier. To realize light trapping, we should properly tune their diameter. In work [73], the light-trapping operation was noticed for nanospheres of diameter 700 nm. However, the mechanisms of light trapping in the densely packed arrays of dielectric spheres can be different, and the dominant mechanism depends on the sphere diameter (for a given refractive index). The mechanism of light trapping in [73] corresponding to the increase of the sphere diameter from 500 nm (the best ARC) to 700 nm (ARC with light-trapping functionality) was related to the *whispering gallery resonances* that arise in substantial in size dielectric cavities. However, this mechanism is not optimal for a LTS dedicated to amorphous silicon TFSCs. For example, in [73], the corresponding effect resulted in a very slight decrease of the transmission loss accompanied by a 3% enhancement in useful absorption.

In works [105, 106], the authors suggested to transform an ARC of paper [64] into an LTS via a simple increase of the diameter of the spheres from 500 nm (optimal for antireflective operation) to 700 (see [105]) or 900 nm (see [106]). This modification increases the period of spherical particles. It exceeds the wavelength and the effect of the diffraction grid arises. Consequently, scattering back into air increases due to appearing diffraction lobes. However, the increase of the reflection loss (compared to the ARC of 500 nm large spheres) is small (about 10% for the integral reflectance) because the optical contrast of silica ($n_{silica} = 1.5$) and air is not very high. This increase is overcompensated by the significant light-trapping effect the reduction of the integral parasitic transmission by 18%. Thus, the gain in the PV absorption equals 8%.

When the diameter of silica spheres in the ARC of [64] is equal to 700 nm, the mechanism of light trapping is related to the generation of *whispering gallery modes* in the spherical cavity [105]. In the absence of a substrate, these modes are practically not excited by the incident wave. In the presence of the substrate, the eigenmodes of the cavity are coupled with the substrate via near fields. In the corresponding frequency range, this coupling gives rise to a whispering gallery mode representing a set of intensity maxima (hot spots) and minima alternating around the cavity. The lowest hot spot extends outside the cavity and partially intersects with the substrate. The presence of other (unnecessary) hot spots inside the cavity does not result in a noticeable increase of parasitic absorption because the cavity is practically lossless. Notice that the bottom hot spot extends into the PV layer through the nanolayer of TCO. The field concentration results in both increased parasitic absorption and PV absorption, whereas the last one prevails (since the absorption of the PV material is much higher). Within the operation band of a TFSC based on a-Si, a spherical cavity of silica with a diameter of 700 nm has three whispering gallery resonances. These resonances result in the relative bandwidth of the consolidated resonance band nearly equal to 10% [105]. The combination of this light-trapping effect and the antireflective properties of the LTS results in an increase of the overall efficiency of a flexible TFSC compared to the case when the monolayer of 500 nm large spheres is used [105]. Though the self-assembly technique is, in principle, affordable for preparing large-area nanostructures, the resonant regime demands nanometer precision for the nanospheres. Therefore, it is difficult to judge whether the 8% gain in the optical efficiency claimed in [105] justifies the corresponding fabrication costs.

In [106, 107], another mechanism of light trapping by silica spheres was suggested and exploited. This is the formation of a so-called photonic nanojet [108–112]. Roughly speaking, this effect means focusing the incident plane wave into an elongated focal spot by a sphere whose radius is noticeably larger than the wavelength but still comparable to it. In the present case, it implies the spheres with diameters $D > 0.8\,\mu$m. The larger the microlens is, the thinner the spot whose width can be estimated as $\lambda/2n$ (n is refractive index of the medium where the photonic nanojet propagates). A focal spot this thin is achieved when $D > 2\,\mu$m. An array of such microspheres produces a noticeable backscattering, which overcompensates the gain granted by the reduction of transmission loss. However, for light trapping in practical TFSCs based on a-Si, there is no need to enlarge the microspheres up to $D = 2\,\mu$m. Amorphous silicon is a highly refractive material, and even a microsphere with $D = 1\,\mu$m producing in the green range of the spectrum the focal spot of width $\lambda/n \approx 125$ nm (at $\lambda = 500$ nm, $n \approx 4$) grants the field concentration sufficient for the complete absorption of sunlight in a layer $h = 330$ nm of intrinsic a-Si. An effective light beam incident on a sphere has the same diameter, $D = 1\,\mu$m. This beam converts into a photonic nanojet whose diameter is eight times smaller. It means the 64-fold increase of the light intensity inside the nanojet. In reality, the squeezing of the wave beam width is not eightfold, because the microsphere is not submerged into a-Si, it is located above it. In practice, the cross section of the wave beam shrinks at the wavelength 500 nm only 10 times. However, even this tenfold increase of the light intensity is enough to increase the absorption sufficiently and the parasitic transmission in a practical TFSC turns suppressed.

It is worth stressing that the formation of the photonic nanojet is a kind of focusing – it is not a subwavelength light concentration granted by a localized surface plasmon or a Mie resonance. The non-resonant nature of the phenomenon is its advantage. A sphere whose diameter D is chosen as a compromise between the minimization of the reflection (backscattering) loss and that of the transmission loss grants the sufficient gain for the PV absorption in the whole operation band of an a-Si SC. For the normal incidence, this optimum corresponds to $D = 900$ nm [106]. The corresponding diffraction-limited light concentration is sufficient for practical absence of transmission loss. A more detailed analysis involving all angles of incidence typical for the daytime resulted in the optimal value $D = 1\,\mu m$ [107].

In fact, the advantage of this LTS compared to a simple antireflector arises only if we average the optical efficiency over all incidence angles corresponding to the daytime. Though, for the normal incidence, the parasitic transmission is prevented, and the reflection loss is reduced as well, a simple focusing has a drawback that was discussed earlier. Namely, the photocurrent is generated only within the area of photonic nanojets. The PV layer beyond the nanojets remains dark. Therefore, the gain of enhanced PV absorption is compensated by the decrease of the effective area of the SC. Fortunately, it is so only for the normal incidence. For obliquely incident waves, the effective area of the SC suffers not so significantly. The solar flux obliquely impinging one silica sphere of the array splits inside it onto two wave beams, shown in Figure 10.15(a). One wave beam (the lower ray) extends from the sphere into the air gap (where it becomes a photonic nanojet) and transmits into the substrate. Another wave beam transmits to the adjacent sphere because the spheres in the array touch one another. In the adjacent sphere, this second ray experiences three total internal reflections and finally transmits into the substrate in a different place. This mechanism of obtaining the second nanojet is called *cascade focusing*. In fact, there are more than two rays because our sphere touches several neighbors. Numerical simulations illustrated by Figure 10.15(b) confirmed the analytical model: we observed several light nanojets per unit sphere. As a result, the field concentration in the vertical direction prevails over the in-plane field concentration. For the angles larger than 40°–45°, almost the whole area of the PV layer is involved in the generation of photocurrent, and the parasitic transmittance is prevented like it is prevented for the normal incidence.

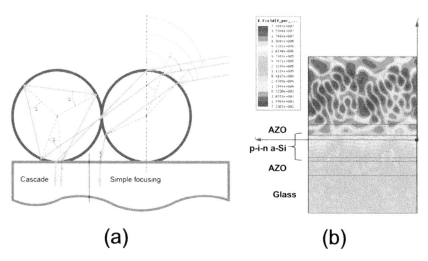

(a)   (b)

Figure 10.15 (a) Schematics of the simple and cascade focusing of an obliquely incident light beam illuminating a reference sphere. (b) Color map of the electric field simulated for the case $\theta = 60°$. Simulations are made by M. Omelyanovich.

For the daytime-averaged overall efficiency, this LTS grants the overall gain of 16% compared to the same TFSC covered by a simple antireflector of silicon nitride. The gain compared to the same TFSC enhanced by an ARC of nanospheres ($D = 500$ nm) is about 8%. A monolayer of densely packed micron-sized silica spheres operates as an efficient ARC for the reflected sunlight and as an efficient LTS for the transmitted sunlight [107]. Work [107] leaves no doubts about which type of LTSs is more promising for amorphous silicon TFSCs (dielectric or plasmonic).

However, it is still not an optimal technical solution. Though the self-assembly allows manufacturing regular monolayers in large areas almost for free, this regularity holds only at the scale of a few millimeters. The defects in these monolayers are significant for practical samples – those having the sizes 3×3 cm or more. For such samples, regions where the monolayer is not formed cover nearly 20% of the area. In these regions, microspheres either form clusters (monolayer structures of five to 10 densely packed spheres) or are isolated from one another by the micron gaps. The daytime-averaged integral reflectance of these defect areas is as high as 60%–70% due to high backscattering of microspheres. When only the defect areas are illuminated (by the collimated sunlight), the photocurrent per unit area turns twice as low than that measured for a bare TFSC. The gain in the photocurrent per unit area reported in [107] referred to the special illumination. In this work, a collimated wave beam of millimeter width was directed to a regular part of the SC. For practical applications of a-Si TFSCs, one should either eliminate these defective domains from the samples (which is hardly affordable if this technical solution is commercialized) or somehow reduce the impact of defects to the optical efficiency of LTS.

10.5.6 Light-Trapping Structures Integrated with the Top Electrode

A prototype of an affordable LTS in which the impact of defects is reduced can be also found among existing ARCs. In work [113], the authors revealed that nanopatterning of a flat antireflector by notches shaped as truncated cones improved its wide-angle operation. These notches of depths about 100 nm and width 200–300 nm were prepared using nanoimprint lithography with the period 400–500 nm in a layer of glass whose optimal thickness for antireflective operation on a-Si is about 110 nm. We have already seen that nanolithography is affordable for TFSCs being compatible with the roll-to-roll processing. Later, these ARCs were successfully applied for wide-angle suppression of reflection from optical glasses where similar notches were prepared in a polymer film of thickness 150 nm [114].

In work [115], light-trapping properties of similar notches when their horizontal sizes are increased up to 400 nm have been theoretically studied. Such notches still offer the antireflecting functionality but also enhance the local electric field, and the field concentration reduces parasitic transmittance through the PV layer of the TFSC. This effect is nonuniform over the visible range, as there are several local maxima. However, even at the minima, the PV absorption is not worsened compared to that in the absence of dents. At the wavelength of 600 nm, the collimated beam in a-Si has the characteristic width 300 nm, whereas the period of notches is 450 nm. In other words, the cross section of the wave beam illuminating one unit cell of the structure shrinks 2.25 times. This collimation was observed in simulations both in the contact layer of TCO and in the PV layer. Therefore, an array of dents promises an alternative technical solution to an array of spheres [115].

Later, it was realized that the focusing function of a truncated conical dent is weaker than that of a hemispherical dent of the same size. Also, it was found that the maximal increase of the PV absorption takes place when the dent depth (radius) is larger, namely for $r \approx 700$ nm (for air notches in silica). Larger radii than $r \approx 700$ nm are favorable for the light-trapping effect but imply an optically large period of the array and result in an increase of reflection loss (as was discussed earlier). For the array of optimized notches, the formation of the collimated beam is observed in the whole visible range, as we can see in Figure 10.16. It is analogous to the formation of a photonic nanojet; a hemispherical void cavity of micron diameter produces a photonic nanojet similarly to a dielectric microsphere.

Since the optimal dielectric antireflector is much thinner than the optimal notch depth, combining the light-trapping and antireflective properties of such an array is problematic. A possible solution is using a densely packed array of notches in the top electrode. In paper [116], it was a layer of AZO with the thickness $h = r = 600$ nm. The radius $r = 600$ nm is optimal for densely packed hemispherical voids in AZO (the period 1.2 μm). In this LTS, each notch contains a semi-diamond particle of silica (i.e., a particle whose bottom half is a hemisphere and top half is a cone). The optical contrast of AZO ($n_{AZO} \approx 2$)

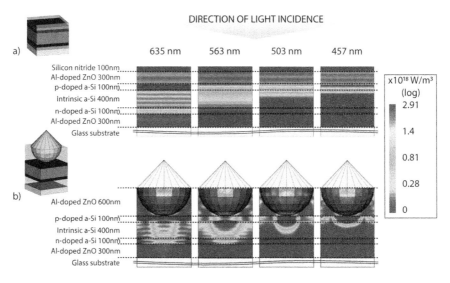

Figure 10.16 (a) Electric field intensity simulated for the reference structure with an ARC of SiN. (b) The same simulated for the structure with the LTS at four randomly chosen wavelengths. Simulations are done by M. Omelyanovich.

with silica ($n_{\text{sil.}} \approx 1.4$) is sufficient to realize the nanojet effect (for air notches, the optimal radius is smaller). Meanwhile, the array of silica cones operates like a moth-eye ARC. In Figure 10.16(a), we show the intensity map of the electric field for the reference structure – the same TFSC enhanced by an antireflector of silicon nitride – and in Figure 10.16(b), the map corresponds to the LTS described in paper [116]. Strong field concentration indicates the light-trapping effect in the whole visible range.

At first glance, it seems that the fabrication of this light-trapping and antire-flecting top electrode is difficult and expensive. However, it is not so. In the first step, a monolayer of colloidal silica microspheres is prepared by using the self-assembly. The substrate to which the colloidal silica spheres attach from the colloidal suspension is p-doped a-Si. As a result, a monolayer of silica spheres is obtained on top of the PV layer of our SC before we have started the fabrication of the top electrode. In the next step, the textured electrode of AZO is fabricated by using a machine for large-area atomic layer deposition. It produces a film centered by spherical voids of silica of radius r and thickness $h = 2r$ inside AZO. This process for the area $10\,\text{cm}^2$ is shorter than an hour.

In the last stage, the top half of this composite film is removed. This removal is done with the nanometer quality that is offered by the ion-beam etching (IBE) technique. This technique is much faster than the ion beam lithography utilizing nanofocused ion beams. Basically, the beam of Ar ions of a few nm thickness and macroscopic length (millimeter scale!) is an effective blade cutting solid dielectric or semiconductor surfaces at any angle up to the grazing one (as in our case). This IBE process takes about 11 minutes for the area of $10\,\text{cm}^2$ and results in a regular arrangement of high-quality hemispherical cavities filled with semi-diamond silica microparticles. Surprisingly, the top half of this silica sphere is not removed but is transformed into a conical protrusion. These conical protrusions serve as an ARC of the SC.

Of course, self-assembly inevitably results in defect regions of nearly one mil-limeter size where the needed monolayer of spheres is not formed. In test samples of the size about $10\,\text{cm}^2$, these regions cover about 20% of the whole area. In these defect areas, we observe isolated nanocones and clusters of nanocones (few hundreds per square millimeter). However, these defect areas scatter much more weakly than those in the earlier LTS. This is because nanocones are not as efficient backscatterers as microspheres. This experimental fact agrees with optical simulations. The angle-averaged integral reflectance of a defect region is nearly equal to that for a bare SC. Meanwhile, the areas with the regular arrangement of insertions offer a triple reduction of the averaged reflectance. Finally, in spite of the defect regions, the mean PV absorption of such TFSC growth twice as compared to the same TFSC with a simple antireflector on top of a flat AZO layer. It is important to stress that this result was obtained by illuminating the whole area of the SC [116]. The measured daytime-averaged optical efficiency of the SC is 60%. It is the best known result for flexible TFSCs up to now.

However, for commercialization of a new LTS, a demonstration of its high *optical* efficiency is not enough. The researchers have to experimentally demonstrate a record *overall* efficiency, obtained due to the new LTS. For the moment, in the literature there are data on several very efficient LTSs demonstrating enhancements close to that claimed in paper [116] (see, e.g., [117–119]). However, record efficiencies of flexible TFSCs were not obtained in any of these works. It is not surprising, because experts in experimental nanophotonics are rarely capable of fabricating high-quality SCs. Likewise, experts in experimental photovoltaics can rarely fabricate high-quality nanostructured LTSs and ARCs. Successful collaborations between these two scientific communities seldom occur. When it occurs, results can be exciting: in work [120], a collaborative team claimed a record for the mean overall efficiency of a flexible a-Si TFSC $\eta = 8.2\%$. The collaboration of nanophotonic and photovoltaic teams is a prerequisite for bringing the overall efficiency of flexible TFSCs to the Shockley–Queisser limit.

10.6 Conclusions

In this chapter, we have discussed SCs of the so-called diode type and analyzed typical reasons for losses of sunlight power in them. We have analyzed the role of the optical loss for the overall efficiency of the SCs and have shown that optical nanostructures integrated with SCs can significantly decrease their effective (integral over the frequencies and averaged over the incidence angle) optical loss factor compared to the conventional flat structures. In our analysis, we kept in mind commercialization readiness of all technical solutions that usually require nanotechnologies that would be affordable for large-area samples. We have shown that for thick SCs, the best currently known technical solution is black silicon for SCs based on c-Si and moth-eye ARCs for some other types of SCs. However, for flexible TFSCs, the promising approach is the replacement of a nanostructured ARC by a nanostructured LTS. We have shown that dielectric nanostructures and nanostructures combined with a transparent (semiconducting but not photovoltaic) electrode are much more promising for industrial adaptation than plasmonic LTSs, which are more popular among researchers.

Note that, in this book, we did not consider nanostructures dedicated to reduction of intrinsic Losses 1–3. This task refers to solid-state physics. Also, we have not considered arrayed SCs consisting of PV nanoelements, such as resonant nanowires or nanospheres of PV materials that allow resonant optical absorption of sunlight in effective areas exceeding their physical size. We have not concerned many related questions on the edge between photonics and nanophotonics. For example, we have not discussed ARCs of dielectric nanospheres for GaAs SCs, which have no conventional counterparts (because dielectric layers as well as TCOs cannot attach to a flat surface of GaAs). We have not discussed the enhancement of SCs by the so-called hot electrons produced in plasmonic nanoparticles (see, e.g., [12]). So, we have omitted many issues related to nanostructures in SCs. Our first goal was to introduce the reader into the topic.

Our second goal was to show how much work remains in the field of solar photovoltaics to a researcher working on the edge between nanophotonics and conventional photonics and electronics. We have done it in examples of available technical solutions.

Problems and Control Questions

1. Consider a simple antireflector: a layer of the dielectric medium with the refractive index n_A of thickness d sandwiched between two half-spaces – free space and a photovoltaic medium with the refractive index n_{PV}. Assume that the optical losses in media are negligibly small. Prove that at the wavelength λ_0, corresponding to $d = \lambda_0/4n_A$, the power reflectance R of the normally incident plane wave can strictly vanish. Find the relation between n_A and n_{PV} that offers this regime.

2. Calculate numerically the power reflectance R versus λ in the vicinity of the antireflective wavelength λ_0. Find the band in which $R < 0.1$ if $n_{PV} = 3.6$ (a good approximation for c-Si in the range 600–900 nm).

3. Find the antireflective condition for an arbitrary incidence angle θ. Calculate numerically the power reflectance R versus λ and θ in the vicinity of the antireflective wavelength λ_0 and antireflective angle θ_0. Take, for example, $\lambda_0 = 600$ nm, $\theta_0 = 30°$, $n_{PV} = 3.6$. Find the band of wavelengths and the sheer of angles in which $R < 0.1$, and depict this region in the coordinate frame (λ, θ).

4. Why cannot the bandgap of the PV semiconductor be chosen to make the spectral efficiency equal 100%?

5. Why is the p–n junction helpful for efficient conversion of light into electricity?

6. Why is the Shockley–Queisser limit for silicon solar cells as low as $\eta_{max} = 1/3$, and, cannot, be beaten? List all the reasons you know.

7. What is the main reason why the Shockley–Queisser limit is not achieved even in advanced solar cells?

8. Multijunction (cascaded) solar cells allow one to beat the Shockley–Queisser limit for silicon solar cells because they split the solar spectrum onto several parts. Why does this spectrum splitting increase the overall efficiency? Which type of loss is reduced in every cascade?

9. Why are multilayer antireflective coatings used in the satellite solar cells and not in the commercial ones?

10. Destructive interference of two waves reflected from two interfaces is impossible for the front glass of a solar cell because the nanometer precision parallelism of two surfaces is impossible for a macroscopic layer. However, the front glass decreases the reflectance of a bare solar cell. Why?

11. Why does the presence of a front glass erase the benefits granted by a multilayer antireflective coating compared to a simple antireflector?

12. Why is a moth-eye antireflecting coating more broadband than a simple antireflector?
13. Why is a moth-eye antireflecting coating more wide-angle than a simple antireflector?
14. Why is black silicon better for silicon solar cells than a moth-eye antireflecting coating?
15. What are advantages and drawbacks of thin-film solar photovoltaics compared to silicon phovoltaics?
16. Explain why an array of silica nanospheres on top of an amorphous-silicon solar cell operates as a bilayer antireflective coating.
17. Which type of optical loss – reflection or transmission loss – is suppressed in the light-trapping structure implemented as a pyramidal texture? What do you say about the second type of optical loss – does it keep the same level after this texturing or increase? If it increases, why?
18. What is the main drawback of plasmonic light trapping in the random metal nanostructures?
19. What is the main drawback of plasmonic light trapping structures performed as regular arrays?
20. Why is all-dielectric light trapping not very efficient if based on the cavity resonaances?
21. Why is dielectric microspheres forming the nanojet operate better when they are illuminated obliquely compared to their operation for normal incidence? Is this advantage more efficient light trapping of the obliquely incident light or something else?
22. Why does the dielectric light-trapping structure integrated with the top transparent electrode operate better than the monolayer of spheres? What is the advantage of conical protrusions?

Bibliography

[1] W. Palz, *Power for the World: Emergency of Electricity from the Sun*, Pan Stanford Publishing (2010).

[2] E. Becquerel, "Mémoire sur les effets électriques produits sous l'influence des rayons solaires," *Comptes Rendus* **IX**, 561–567 (1839).

[3] W. G. Adams and R. E. Day, "The action of light on selenium," *Philosophical Transactions of the Royal Society* **167**, 313–349 (1877).

[4] C. E. Fritts, "On a new form of selenium photocell," *American Journal of Science* **26**, 465–466 (1883).

[5] A. Stoletow, "Suite des recherches actino-électriques," *Comptes Rendus* **CVII** 91–120 (1888).

[6] A. Einstein, "Über einen die Erzeugung und Verwandlung des Lichtes betreffenden heuristischen Gesichtspunkt," *Annalen der Physik* **17**, 132–148 (1905).

[7]   A. Fahrenbruch and R. Bube, *Fundamentals of Solar Cells: Photovoltaic Solar Energy Conversion,* Academic Press (1983).

[8]   A. Luque and S. Hevedus, *Handbook of Photovoltaic Science and Engineering*, Wiley (2002).

[9]   W. Franz, "Einfluss eines elektrischen Feldes auf eine optische Absorptionskante," *Zeitschrift für Naturforschung* **13a**, 484–489 (1958).

[10] L. V. Keldysh, "Behaviour of non-Metallic crystals in strong electric fields," *Soviet Physics Journal of Experimental and Theoretical Physics* **6**, 763–770 (1958).

[11] W. Shockley and H. J. Queisser, "Detailed balance limit of efficiency of p-n junction solar cells," *Journal of Applied Physics* **32**, 510–519 (1961).

[12] V. A. Milichko, A. S. Shalin, I. S. Mukhin, et al., "Solar photovoltaics: Modern state and trends," *Physics Uspekhi* **186**, 801–852 (2016).

[13] L. Zhu, A. Raman, K. X. Wang, M. A. Anoma, and S. Fan, "Radiative cooling of solar cells," *Optica* **1**, 32–38 (2014).

[14] C. Liebert and R. R. Hibbard, *Theoretical Temperatures of Thin-Film Solar Cels in Earth Orbit*, NASA Technical Note D-4331, Nat. Aeronaut. Space Admin., WA (1968).

[15] A. Marti and A. Luque, *Next-Generation Photovoltaics*, IoP Publishing (2004).

[16] N.-G. Park, "Perovskite solar cells: An emerging photovoltaic technology," *Materials Today* **19**, 65–72 (2015).

[17] A. K. Chilvery, A. K. Batra, B. Yang, et al., Perovskites: transforming photovoltaics, a mini-review, *Journal of Photonics for Energy* **5**, 057402 (2015).

[18] M. Stolterfoht, C. M. Wolff, Y. Amir, et al., "Approaching the fill factor Shockley-Queisser limit in stable, dopant-free triple cation perovskite solar cells," *Energy & Environmental Science* **10**, 1530–1539 (2017).

[19] Y. Hamakawa, *Thin-Film Solar Cells: Next Generation Photovoltaics and Its Applications*, Springer (2004).

[20] M. A. Green, "The path to 25% silicon solar cell efficiency: history of silicon cell evolution," *Progress in Photovoltaics: Research and Applications* **17**, 183–189 (2009).

[21] M. A. Green, *Third-Generation Photovoltaics*, Springer, Berlin (2004).

[22] G. K. Dey and K. T. Ahmed, "Performance characterization of photovoltaic technology with highly efficient multi-junction solar cells for space solar power satellite systems," *Proceedings of the Third International Conference on Green Energy, Green Engineering and Technology*, Dhaka, Bangladesh, Sep. 11–12, 2015, pp. 1–6.

[23] K. Yoshikawa, H. Kawasaki, W. Yoshida, et al., "Silicon heterojunction solar cell with interdigitated back contacts for a photoconversion efficiency over 26%" *Nature Energy* **2**, 17032 (2017).

[24] C. S. Fueller and G. L. Pearson, "A new silicon p-n junction photocell for converting solar light to electricity," *Journal of Applied Physics* **25**, 676–677 (1954).

[25] J. A. Kong, *Theory of Electromagnetic Waves* (1975).

[26] A. Osipov and S. Tretyakov, *Modern Electromagnetic Scattering Theory with Applications*, John Wiley & Sons (2017).

[27] M. A. Kats, R. Blanchard, P. Genevet, and F. Capasso, "Nanometer-scale optical coatings based on strong interference effects in highly absorbing media," *Nature Materials* **12**, 20–24 (2013).

[28] H. K. Raut, V. A. Ganesh, A. S. Nair, and S. Ramakrishna, "Anti-reflective coatings: A critical, in-depth review," *Energy & Environmental Science* **4**, 3779–3804 (2011).

[29] J. Zhao and M. A. Green, "Optimized antireflection coatings for high-efficiency silicon solar cells," *IEEE Transactions on Electron Devices* **38**, 1925–1934 (1991).

[30] B. Thaidigsmann, E. Lohmuller, U. Jager, et al., "Large-area p-type HIPMWT silicon solar cells with screen printed contacts exceeding 20% efficiency," *Physica Status Solidi Rapid Research Letters* **5**, 286–288 (2011).

[31] S. H. Glick, F. J. Pern, D. Tomek, J. Raaff, and G. L. Watson, "Performance degradation of encapsulated monocrystalline Si solar cells upon accelerated weathering exposures," *Proc. NCPV Program Review Meeting*, Lakewood, Colorado, October 14–17, 2001, paper NREL/CP-520-30841. Available at www.osti.gov/bridge.

[32] H. A. Macleod, *Thin Film Optical Filters* (3rd edn.), IOP Publishing (2001).

[33] K. Nurrows and V. Fthenakis, "Glass needs for a growing photovoltaic industry," *Solar Energy Materials and Solar Cells* **132**, 455–459 (2015).

[34] P. B. Clapham and M. C. Hutley, "Reduction of lens reflection by moth-eye principle," *Nature* **244**, 281–282 (1973).

[35] A. Parretta, A. Sarno, P. Tortora, et al., "Angle-dependent reflectance measurements on photovoltaic materials and solar cells," *Optics Communications* **172**, 139–151 (1999).

[36] V. Hotar, O. Matusek, and J. Svoboda, "Laboratory detection of flat glass shapes using its reflection," *MATEC Web of Conferences* **89**, 01007 (2017).

[37] C. Ballif, J. Decker, D. Borchert, and T. Hofmann, "Solar glass with industrial porous SiO2 antireflection coating: Measurements of photo-voltaic module properties improvement and modelling of yearly energy yield gain," *Solar Energy Materials and Solar Cells* **82**, 331–344 (2004).

[38] Q. Chen, G. Hubbard, P. A. Shields, et al., "Broadband moth-eye antire-flection coatings fabricated by low-cost nanoimprinting," *Applied Physics Letters 94*, 263118 (2009).

[39] N. Yamada, T. Ijiro, E. Okamoto, K. Hayashi, and H. Masuda, "Char-acterization of antireflection moth-eye film on crystalline silicon photo-voltaic module," *Optics Express* **19**, A118–A125 (2011).

[40] S. A. Boden and D. M. Bagnall, "Optimization of moth-eye antireflection schemes for silicon solar cells," *Progress in Photovoltaics: Research and Applications* **18**, 195–203 (2010).

[41] J. Tommila, A. Aho, A. Tukiainen, et al., Moth-eye antireflection coating fabricated by nanoimprint lithography on 1 eV dilute nitride solar cell, *Progress in Photovoltaics: Research and Applications* **21**, 1158–1162 (2012).

[42] K.-S. Han, J.-H. Shin, W.-Y. Yoon, and H. Lee, "Enhanced performance of solar cells with anti-reflection layer fabricated by nano-imprint lithography," *Solar Energy Materials and Solar Cells* **95**, 288–291 (2011).

[43] L. W. Chan, D. E. Morse, and M. J. Gordon, "Moth-eye-inspired antireflective surfaces for improved IR optical systems and visible LEDs fabricated with colloidal lithography and etching," *Bioinspiration and Biomimetics* **13**, 041001 (2018).

[44] A. Scheydecker, A. Goetzinger, and V. Wittner, "Reduction of reflection losses of PV-modules by structured surfaces," *Solar Energy* **53**, 171–176 (1994).

[45] P. G. Carey, J. B. Thompson, and R. C. Aceves, "Solar cell module lamination process," US Patent No 6,340,403 B1, issued Jan. 22, 2002.

[46] X. Liu, P. Coxon, and M. Peters, "Black silicon: Fabrication methods," properties and solar energy applications, *Energy & Environmental Science* **7**, 3223–3230 (2014).

[47] P. Repo, A. Haarahiltunen, L. Sainiemi, et al., "Effective passivation of black silicon surfaces by atomic layer deposition," *IEEE Journal of Photovoltaics* **3**, 90–94 (2012).

[48] H. Savin, P. Repo, G. von Gastrow, et al., "Black silicon solar cells with interdigitated back-contacts achieve 22.1% efficiency," *Nature Nanotechnology* **10**, 624–628 (2015).

[49] Z.-Q. Zhou, F. Hu, W.-L. Zhou, et al., "An investigation on a crystalline-silicon solar cell with black silicon layer at the rear," *Nanoscale Research Letters* **12**, 623–626 (2017).

[50] Y.-J. Hung, S.-L. Lee, and L. A. Coldren, "Deep and tapered silicon photonic crystals for achieving anti-reflection and enhanced absorption," *Optics Express* **18**, 6841–6852 (2010).

[51] J. Lv, T. Zhang, P. Zhang, Y. Zhao, and S. Li, "Review application of nanostructured black silicon," *Nanoscale Research Letters* **13**, 110 (2018).

[52] T. Zhang, P. Zhang, S. Li, W. Li, Z. Wu, and Y. Jiang, "Black silicon with self-cleaning surface prepared by wetting processes," *Nanoscale Research Letters* **8**, 351 (2013).

[53] M. Osborne, "Hevel achieves heterojunction cells with 22.8% efficiency as plant ramps," PV-Tech, Jan. 29, 2018, available at www.pv-tech.org/news/hevel-achieves-heterojunction-cells-with-22.8-efficiency-as-plant-ramps.

[54] D. Andronikov, A. Abramov, S. Abolmasov, et al., "A successful conversion of silicon thin-film solar module production to high efficiency heterojunction technology," *Proc. European PV Solar Energy Conference EU PVSEC'2017*, Sep. 24–28, 2017, Brussels, Belgium, Paper 2AV.3.12.

[55] M. Petrov, K. Lovchinov, M. Mews, C. Leenderts, and D. Dimova-Malinovska, "Optical and structural properties of electrochemically deposited ZnO nanorod arrays suitable for improvement of the light harvesting in thin film solar cells," *Journal of Physics: Conference Series* **559**, 012018 (2014).

[56] S. Abolmasov, P. Roca i Cabarrocas, and P. Chatterjee, "Towards 12% stabilised efficiency in single junction polymorphous silicon solar cells:

experimental developments and model predictions," *European Physical Journal: Photovoltaics* **7**, 70302 (2016).

[57] M. R. Maghami, H. Hizam, C. Gomes, et al., "Power loss due to soiling on solar panel: A review," *Renewable and Sustainable Energy Reviews* **59**, 1307–1316 (2016).

[58] A. B. Roy, A. Dhar, M. Choudhuri, et al., "Black silicon solar cell: Analysis optimization and evolution towards a thinner and flexible future," *Nanotechnology* **27**, 305302 (2016).

[59] E. Ogorodnikov, "In the scattered light of the Russian sun," *RusNano News*, Sep. 11 (2017), available at http://en.rusnano.com/press-centre/media/20170911-expert-in-the-scattered-light-of-the-russian-sun.

[60] J. B. Heng, C. Yu, Z. Xu, and J. Fu, Solar cell with oxide tunneling junctions, Patent US2011/0272012A1, issued on Oct. 11, 2011.

[61] J. B. Heng, J. Fu, Z. Xu, and Z. Xie, Back junction solar cell with tunnel oxide, Patent US2012/0318340A1, issued on Dec. 20, 2012.

[62] J. Fu, J. B. Heng, C. Yu, Tunneling-junction solar cell with copper grid for concentrated photovoltaic application, Patent US2015/0236177A1, issued on Aug. 20, 2015.

[63] V. Vavilov, R. Aflyatonov, and A. Yakupov, "Usage of the solar radiation potential in the republic of Bashkortostan of Russian Federation," *Proc. 2nd International Conference on Industrial Engineering, Applications and Manufacturing ICIEAM'2016* Chelyabinsk, Russia, May 19–20, 2016, pp. 1–5.

[64] M. Tao, W. Zhou, H. Yang, and L. Chen, "Surface texturing by solution deposition for omnidirectional antireflection," *Applied Physics Letters* **91**, 081118-1–081118-3 (2007).

[65] W. Qarony, M. I. Hossain, M. K. Hossain, et al., "Efficient amorphous silicon solar cells: Characterization, optimization, and optical loss analysis," *Results in Physics* **7**, 4287–4293 (2017).

[66] *Ultra-Low-Cost Solar Electricity Cells, An Overview of Nanosolar's Cell Technology Platform*, Nanosolar Inc. White Paper (2009), available at www.catharinafonds.nl/wp-content/uploads/2010/03/NanosolarCellWhitePaper.pdf.

[67] A. Alkaya, R. Kaplan, H. Canbola, and S. S. Hegedus, "A comparison of fill factor and recombination losses in amorphous silicon solar cells on ZnO and SnO$_2$," *Renewable Energy* **34**, 1595–1599 (2009).

[68] *U.S. Solar Market Insight Report*, Solar Energy Industries Association Year-in-Review (2011).

[69] T.-H. Chang, P.-H. Wu, S.-H. Chen, et al., "Efficiency enhancement in GaAs solar cells using self-assembled microspheres," *Optics Express* **8**, 6519–6524 (2009).

[70] J. Grandidier, D. M. Callahan, J. N. Munday, and H. A. Atwater, "Configuration optimization of a nanosphere array on top of a thin-film solar cell," *Advanced Materials* **32**, 1272–1276 (2011).

[71] A. S. Shalin and S. A. Nikitov, "Approximate model for universal broad-band antireflection nano-structure," *Progress in Electromagnetic Research B* **47**, 127–144 (2013).

[72] S. Dias, C. Banerjee, and A. Kundu, "Silica nanoparticles on front glass for efficiency enhancement in superstrate-type amorphous silicon solar cells," *Journal of Physics D: Applied Physics* **46**, 415102 (2013).

[73] J. Grandidier, D. M. Callahan, J. N. Munday, and H. A. Atwater, "Light absorption enhancement in thin-film solar cells using whispering gallery modes in dielectric nanospheres," *Advanced Materials* **23**, 1272–1276 (2011).

[74] E. Yablonovitch, "Intensity enhancement in textured optical sheets for solar cells," *IEEE Transactions on Electron Devices* **29**, 300–308 (1982).

[75] P. Campbell, "Light trapping in textured solar cells," *Solar Energy Materials* **21**, 165–172 (1990).

[76] Z. Yu, A. Raman, and S. Fan, "Fundamental limit of light trapping in grating structures," *Optics Express* **18**, A366–A380 (2010).

[77] H. R. Stuart and D. G. Hall, "Absorption enhancement in silicon-on-insulator waveguides using metal island films," *Applied Physics Letters* **69**, 2327–2340 (1996).

[78] H. A. Atwater and A. Polman, "Plasmonics for improved photovoltaic devices," *Nature Materials* **9**, 205–209 (2010).

[79] Yu. A. Akimov, W. S. Koh, S.Y. Sian, and S. Ren, "Nanoparticle-enhanced thin-film solar cells: Metallic or dielectric nanoparticles?" *Applied Physics Letters* **96**, 073111 (2010).

[80] P. Spinelli, V. E. Ferry, J. van de Groep, et al., "Plasmonic light trapping in thin-film Si solar cells," *Journal of Optics* **14**, 024002 (2012).

[81] M. G. Deceglie, V. E. Ferry, A. P. Alivisatos, and H. A. Atwater, "Design of nanostructured solar cells using coupled optical and electrical modeling," *Nano Letters* **12**, 2894–2900 (2012).

[82] S. Pillai, K. R. Catchpole, T. Trupke, and M. A. Green, "Surface-plasmon enhanced silicon solar cells," *Journal of Applied Physics* **101**, 093105 (2007).

[83] K. R. Catchpole and A. Polman, "Plasmonic solar cells," *Optics Express* **16**, 21793–21800 (2008).

[84] V. E. Ferry, L. A. Sweatlock, D. Pacifici, and H. A. Atwater, "Plasmonic nanostructure design for efficient light coupling into solar cells," *Nano Letters* **8**, 4391–4397 (2008).

[85] Yu. A. Akimov, K. Ostrikov, and E. P. Li, "Surface plasmon enhancement of optical absorption in thin-film silicon solar cells," *Plasmonics* **4**, 107–113 (2009).

[86] H. Tan, R. Santbergen, A. H. M. Smets, and M. Zeman, "Plasmonic light trapping in thin-film silicon solar cells with improved self-assembled silver nanoparticles," *Nano Letters* **12**, 4070–4076 (2012).

[87] F. J. Beck, A. Polman, and K. R. Catchpole, "Tunable light trapping for solar cells using localized surface plasmons," *Journal of Applied Physics* **105**, 114310 (2009).

[88] H.-M. Li, G. Zhang, C. Yang, et al., "Enhancement of light absorption using high-k dielectric in localized surface plasmon resonance for silicon-based thin film solar cells," *Journal of Applied Physics* **109**, 093516 (2011).

[89] A. Tamang, H. Sai, V. Jovanov, et al., "On the interplay of cell thickness and optimum period of silicon thin-film solar cells: Light trapping and plasmonic losses," *Progress in Photovoltaics: Research and Applications* **24**, 379–388, (2016).

[90] P. H. Wang, M. Theuring, M. Vehse, et al., "Light trapping in a-Si:H thin-film solar cells using silver nanostructures," *AIP Advances* **7**, 015019 (2017).

[91] K. Domanski, J. Correa-Baena, N. Mine, et al., "Not all that glitters is gold: Metal-migration-induced degradation in perovskite solar cells," *ACS Nano* **10**, 6306–6314 (2016).

[92] R. A. Pala, J. White, E. Barnard, J. Liu, and M. L. Brongersma, "Design of plasmonic thin-film solar cells with broadband absorption enhancements," *Advanced Materials* **21**, 1–6 (2009).

[93] W. Wang, S. Wu, K. Reinhardt, Y. Lu, and S. Chen, "Broadband light absorption enhancement in thin-film silicon solar cells," *Nano Letters* **10**, 2012–2018 (2010).

[94] M. Omelyanovich, Y. Ra'di, and C. Simovski, "Perfect plasmonic absorbers for photovoltaic applications," *Journal of Optics* **17**, 125901 (2015).

[95] C. Simovski, Thin-film photovoltaic cell structure, nanoantenna and method for manufacturing. US Patent No 9252303 B2, issued on Feb. 2, 2016.

[96] C. Simovski, D. Morits, P. Voroshilov, et al., "Enhanced efficiency of light-trapping nanoantenna arrays for thin-film solar cells," *Optics Express* **21**, A714–A725 (2013).

[97] P. M. Voroshilov, C. R. Simovski, and P. A. Belov, "Nanoantennas for enhanced light trapping in transparent organic solar cells," *Journal of Modern Optics* **61**, 1743–1748 (2014).

[98] P. Voroshilov, V. Ovchinnikov, A. Papadimitratos, A. Zakhidov, and C. Simovski, "Light trapping enhancement by silver nanoantennas in organic solar cells," *ACS Photonics* **5**, 1767–1772 (2018).

[99] P. M. Voroshilov and C. R. Simovski, "Leaky domino-modes in regular arrays of substantially thick metal nanostrips," *Photonics and Nanostructures: Fundamentals and Applications* **20**, 18–30 (2016).

[100] J. N. Munday and H. A. Atwater, "Large integrated absorption enhancement in plasmonic solar cells by combining metallic gratings and antireflection coatings," *Nano Letters* **10**, 2195–2201 (2010).

[101] C. Li, L. Xia, H. Gao, et al., "Broadband absorption enhancement in a-Si:H thin-film solar cells sandwiched by pyramidal nanostructured arrays," *Optics Express* **20**, A589–A595 (2012).

[102] P. Spinelli and A. Polman, "Prospects of near-field plasmonic absorption enhancement in semiconductor materials using embedded Ag nanoparticles," *Optics Express* **20**, A641–A654 (2012).

[103] V. E. Ferry, M. A. Verschuuren, H. B. T. Li, et al., "Light trapping in ultra-thin plasmonic solar cells," *Journal of Optics* **14**, 024002 (2012).

[104] T. Matsui, H. Sai, K. Saito, and M. Kondo, "High-efficiency thin-film silicon solar cells with improved light-soaking stability," *Proc. 27-th Int. Conf. Progress in Photovoltaics*, Frankfurt, Germany, 2213–2217 (2013).

[105] J. Grandidier, R.A. Weitekamp, M.G. Deceglie, et al., "Solar cell efficiency enhancement via light trapping in printable resonant dielectric nanosphere arrays," *Physica Status Solidi A* **210**, 255–260 (2012).

[106] C. R. Simovski, A. S. Shalin, P. M. Voroshilov, and P. A. Belov, "Photovoltaic absorption enhancement in thin-film solar cells by non-resonant beam collimation by submicron dielectric particles," *Journal of Applied Physics* **114**, 103104 (2013).

[107] M. Omelyanovich, V. Ovchinnikov, and C. Simovski, "A non-resonant dielectric metamaterial for enhancement of thin-film solar cells," *Journal of Optics* **17**, 025102 (2015).

[108] A. Heifetz, S.-C. Kong, A. V. Sahakian, A. Taflove, and V. Backman, "Photonic nanojets," *Journal of Computational and Theoretical Nanoscience* **6**, 1979–1992 (2009).

[109] N. Horriuchi, "Photonic nanojets," *Nature Photonics* **6**, 138–139 (2012).

[110] D. McCloskey, K. E. Ballantine, P. R. Eastham, and J. F. Donegan, "Photonic nanojets in Fresnel zone scattering from non-spherical dielectric particles," *Optics Express* **20**, 128–140 (2012).

[111] A. Neves, "Photonic nanojets in optcal tweezers," *Journal of Quantitative* **162**, 122–132 (2015).

[112] S. Yang, F. Wang, Y. Ye, et al., "Influence of the photonic nanojet of microspheres on microsphere imaging," *Optics Express* **25**, 27551–27558 (2017).

[113] J. Y. Chen and K. W. Sun, "Enhancement of the light conversion efficiency of silicon solar cells by using nanoimprint anti-reflection layer," *Solar Energy Materials and Solar Cells* **94**, 629–639 (2010).

[114] D. A. Baranov, P. A. Dmitriev, I. S. Mukhin, et al., "Broadband antireflective coatings based on two-dimensional arrays of subwavelength nanopores," *Applied Physics Letters* **106**, 171913 (2015).

[115] P. M. Voroshilov, C. R. Simovski, P. A. Belov, and A. S. Shalin, "Light-trapping and antireflective coatings for amorphous Si-based thin-film solar cells," *Journal of Applied Physics* **117**, 203101 (2015).

[116] M. M. Omelyanovich and C. R. Simovski, "Wide-angle light-trapping electrode for photovoltaic cells," *Optics Letters* **42**, 3726–3729 (2017).

[117] Y. Wang, X. Zhang, B. Han, et al., "UV micro-imprint patterning for tunable light trapping in p-i-n thin-film silicon solar cells," *Applied Surface Science* **355**, 14–18 (2015).

[118] C. S. Schuster, S. Morawiec, M. J. Mendes, et al., "Plasmonic and diffractive nanostructures for light trapping: An experimental comparison," *Optica* **2**, 194–200 (2015).

[119] P. M. Voroshilov and C. R. Simovski, "Affordable universal light-trapping structure for third-generation photovoltaic cells," *Journal of the Optical Society of America B* **34**, D77 (2017).

[120] C. Zhang, Y. Song, M. Wang, et al., "Efficient and flexible thin film amorphous silicon solar cells on nanotextured polymer substrate using sol-gel based nanoimprinting method," *Advanced Functional Materials* **27**, 1604720 (2017).

Notes

1 Modern estimates have shown that the gold film in this experiment was 40–60 nm thick.

2 In fact, quasi-Fermi levels W_{eF} and W_{hF} have no well-defined physical meaning in the quantum theory of a heterogeneous semiconductor. These values are used in the classical model through E_d and V_{oc}.

3 One also often considers the spectral internal quantum efficiency – that referred to the unit interval of wavelengths within the solar spectrum. This value is more dispersive than the spectral photoelectric response because the spectral density of Loss 2 and Loss 3 are both frequency dependent.

4 Nowadays, it is a part of Tesla.

Nanostructures for Enhancement of Thermophotovoltaic Systems

11.1 What Are Thermophotovoltaic Systems?

Thermophotovoltaics is a direct heat-to-electricity conversion method suggested by Henry Kolm and Pierre Aigrain in 1950s [1, 2]. The thermophotovoltaic (TPV) concept is straightforward: *thermal radiation* of a hot body (emitter) is converted by a photovoltaic (PV) cell into electricity, in the same way as the solar radiation is harvested and converted by a PV cell.

Initially, the idea was to complement a solar cell (SC) by a *TPV generator*. As we know from the previous chapter, the recombination loss (Loss 3) implies near-infrared radiation produced by SCs. This radiation results in a narrow local maximum in the background of the usual very broadband spectrum of thermal radiation produced by the SC heated by sunlight. The energy corresponding to this local maximum can be captured and reused by an IR PV diode complementing the SC. This was the historically first idea of the TPV generator.

Already in 1950s, a semiconductor (GaAs) whose bandgap frequency is located at the low-frequency edge of the near-IR band was available. The corresponding PV diode operating in the narrowband regime (i.e., converting only the radiation corresponding to Loss 3 in the SC) manifested a low level (for that time) of its own recombination loss (25%–30%) and high value (for that time) of the fill factor (about 0.7). In [1], it was suggested to complement a usual silicon SC by a gallium arsenide PV diode, slightly increasing the total electric power output. Though this increase was very small, in paper [2], it was noticed that the PV conversion of thermal radiation may hold independently from the presence of a narrow peak in the IR radiation spectrum. The PV conversion of thermal radiation differs from solar photovoltaics only by the operation frequency band. Electricity can be extracted from the radiation of such sources as flame, flue gas in a stalk of a *melting furnace* or a *waste combustor*, or a specially designed *gas combustor* that may also heat cold water for heating a house.

Prior to [2], the idea of heat-to-electricity conversion was commonly thought as that related to *thermoelectric elements* (TEEs). TEEs are devices based on the well-known Peltier effect. They cannot operate at as high temperatures as that of the flame or a flue gas. Efficiency of a TEE is limited by the compromise between the thermal and electric conductivities of materials, and the corresponding limitation of the efficiency is severe. The high temperatures at which a TPV generator may work and the potentially high efficiency of the TPV generator promised a technical breakthrough in the generation of sustainable energy [3].

However, the real progress in the field of TPV electric generators in the period precedent to the era of nanophotonics was quite modest. In the next section, we will see why it was so.

Direct conversion of radiation of heat sources is hardly possible because both flame and flue gas, in addition to producing thermal radiation, also produce soot. Soot is not transparent for the IR radiation. The only way to receive a radiative heat flux by the PV cell is to use an emitter: a layer separating the PV diode from the heat source. Its front interface, illuminated by the heat source, is covered with soot. Soot has the same temperature as that of the heat source (typically, $T = 1,500$–$2,000$ K) and behaves as a *black body* practically in thermodynamic equilibrium with the heat source. However, it is only a *quasi-equilibrium*. Some part of the thermal energy of the heat source is transferred by the emitter from the front interface to the rear one from where it is radiated toward the PV diode separated from the emitter by a vacuum gap.

The gap between the emitter and the PV panel in the TPV system is called *TPV cavity*. This term in the theory of TPV systems does not imply any resonant width of the gap d; it simply allows sufficient thermal isolation of the hot and cold parts of the system. A conventional (far-field) TPV system in most practical cases is a concentric (tubular) structure introduced into a hot medium. However, the curvature radii of both the cylindrical PV diode and the emitter are much larger than the gap d, which in its turn is much larger than the radiation wavelengths λ. This means that, locally, a TPV system can be considered as a parallel-plate structure, schematically shown in Figure 11.1(a). An important component of a TPV system is a narrowband filter covering the PV cell, as shown in this figure. The crucial importance of the filter is explained in the next section. The *heat sink* shown in this scheme prevents critical heating of the PV cell (due to both thermal conduction from the heat source through the lateral walls of the tube and the radiative heating of the PV diode). The sink can be implemented as a textured metal plate called an air heat exchanger if the power received by the PV diode is sufficiently low. If the power is high, the sink is performed as a

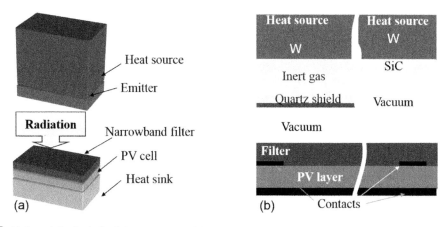

(a)

(b)

Figure 11.1   (a) General sketch of a far-field TPV generator. (b) Two variants of the practical design of far-field TPV systems: one with a metal emitter and one with a bilayer emitter.

tap-water cooling frame. Then the TPV system also serves as a source of a low-grade heat transported by the heated tap water. Both the TPV cavity and the heat sink contribute to *thermal management*, supporting suitable operation temperatures of the PV cell.

If the heat source temperature T_s exceeds 1,500 K, the front interface of the emitter must be performed of a *refractory metal* (e.g., tungsten). Refractory metals better stand the mechanical stresses caused by open-flame or *flue gases* than other solid materials. Since hot metal bodies emit electrons from interfaces with vacuum (i.e., metal elements destroy themselves and destroy the PV cell), in the case of a solid metal emitter, the TPV cavity is partially filled with an inert gas suppressing electron emission [3–7]. In this case, one inserts a millimeter-thick quartz shield into the TPV cavity, as shown in Figure 11.1(b). There is a vacuum gap between the shield and the PV cell.

If $T_s < 1,500$ K, the emitter can be made of refractory semiconductors such as silicon carbide. In the case of higher temperatures, it may be made as a bilayer of a refractory metal (the front interface) and a semiconductor (the rear interface). Open surfaces of both SiC and W possess high emissivity (close to 0.9 in the mid-IR range) due to submicron roughness of their surfaces. This roughness is achieved automatically during the fabrication process and makes these surfaces similar to that of black silicon [7]. In other words, both SiC and W are not only thermally stable, but in their electromagnetic properties, they are close to the black body. These materials are, therefore, very good thermal emitters. Both variants of the design of a TPV system – with a solid metal emitter and with a bilayer emitter – are shown in Figure 11.1(b). Both of these design solutions are equally practical for high-temperature TPV systems. A quartz shield and an inert gas imply a technical complication and higher price, but the working life of this system is comparatively long. In a bilayer emitter, the internal interface is either specially polished or contains thermal grease laminating the natural roughness [5–7]. Both polishing and the presence of grease reduce the mechanical robustness of the structure to the stress caused by heat, and the life of the bilayer emitter in a high-temperature flame or flue gas is comparatively short. However, recently, a technology of a bilayer emitter without polishing or a thermal grease was developed. Ytterbium oxide turned out to be not worse than tungsten in what concerns the thermal stress at realistic temperatures of heat sources, and a crystalline SiC microfilm can be chemically grown on it [8]. However, TPV systems using such bilayer emitters will probably be more expensive and will hardly supersede solid W and bilayer W-SiC emitters in the near future.

11.2 Conventional TPV Systems

11.2.1 More about the TPV Cavity and the TPV Filter

In the first TPV systems, the TPV cavity was simply an open-air gap. This system had low efficiency, and a noticeable part of thermal radiation of the emitter was lost in the ambient [4, 5]. In confined (e.g., tubular) systems, the air in the cavity conducts heat, and the temperature of the PV diode grows, approaching that of the emitter. Therefore, a vacuum gap is needed [3–7].

Let us discuss the importance of thermal decoupling of the hot and cold parts of the system. The simplified equivalent scheme of a PV diode was introduced in the previous chapter. In accordance with that, the total current I of the illuminated PV diode results from the short-circuit current I_{sc} and the dark diode current I_{diode} subtracted from it. Therefore, $V = (I_{sc} - I_{\text{diode}})R_L$, where R_L is the resistance of the external load. The dark current grows with the temperature due to the Peltier thermoelectric effect. The higher the temperature of the PV cell is, the lower V is and, respectively, the lower the output power is. Thus, poor thermal decoupling between the heat source and the PV cell results in "destructive interference" of the PV and TE effects.

This contest is not the only factor restricting the maximal allowed temperature of the PV cell. There is also the voltage factor representing an effective reduction of the semiconductor bandgap with the temperature. As a result, if the temperature of the PV cell top exceeds $T = 400$ K, the heat-to-electricity conversion efficiency becomes thermoelectric, and since the PV diode is a poor TEE, this efficiency does not exceed 1%. Therefore, the TPV cavity is considered to be a necessary part of a TPV electric generator. This cavity is a challenging component of the system because it requires parallelism over a large area to be kept in spite of the permanent thermal stress experienced by the emitter, and this results in its accumulating deformation. The maximal absolute deformation of the emitter due to this stress determines the minimal allowed value of d that is practically is as large as few mm. There were attempts to ensure the parallelism of the hot and cold surfaces filling the gap by a dielectric (e.g., see an overview in [7]). The use of a thick dielectric spacer instead of a vacuum gap theoretically allows thermal decoupling (if d is large enough) and simultaneously allows us to increase the *radiative heat transfer* because the power flux of thermal radiation from the black body into a dielectric is proportional to the permittivity of the latter [9]. In view of a dramatic simplification of fabrication, this direction promised a technical breakthrough. However, those theoretical studies (united under the common name of *dielectric photon concentration*) have not resulted in anything practical. The TPV generators with a dielectric spacer instead of a cavity could not efficiently work due to two reasons. The first one was the presence of optical absorption band (or several bands) within the spectrum of thermal radiation for all refractory dielectric materials (the refractory property is necessary because the dielectric spacer touches the emitter). The absorption of thermal radiation in the dielectric spacer results in its strong heating and hinders the thermal management of the PV diode. The second reason was the electromagnetic impedance mismatch of the emitter and the dielectric medium. A hot surface of silicon carbide (graphite, tungsten, etc.) mimics the black body well when this surface radiates the heat into free space. However, when a dielectric spacer is introduced between such the emitter and the PV cell, a lot of radiated power is reflected from the dielectric interface. This reflection is quite uniform over the whole spectrum and covers the operation band. Therefore, the expected enhancement is not achieved whereas the cooling of the PV cell becomes difficult.

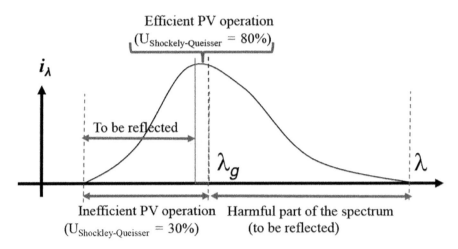

Efficient PV operation
$(U_{\text{Shockely-Queisser}} = 80\%)$

i_λ

To be reflected

λ_g

λ

Inefficient PV operation   Harmful part of the spectrum
$(U_{\text{Shockley-Queisser}} = 30\%)$ (to be reflected)

Figure 11.2 Conventional (black-body) thermal radiation spectrum with its harmful low-frequency part ($\lambda > \lambda_g$), its high-frequency part corresponding to low-efficiency PV conversion, and its central part to be filtered, which corresponds to the highest Shockley–Queisser limit.

As a result, the TPV cavity is still considered to be a necessary component of practical TPV generators [7, 8, 10].

Now let us clarify the role of the *TPV filter*. Thermal radiation of a conventional emitter mimicking a black body obeys to the *Planck and Wien laws*. The typical Planckian spectrum $i(\lambda)$ of thermal radiation is depicted in Figure 11.2. In accordance with the Wien law, the black-body emitter with radiative temperature $T_e = 1,000$ K has the maximum of radiation at $\lambda_m = 2.3\,\mu$m. Assume that the bandgap wavelength λ_g of the PV material is equal to $\lambda_m = 2.3\,\mu$m. Then, nearly 60% of the spectrum corresponds to $\lambda > \lambda_g$. This is the harmful part of the thermal emission spectrum because the corresponding radiation is sub-band (Loss 1). Thus, in the absence of a frequency filter, 60% of the power absorbed by the PV diode is dissipated into heat, eventually increasing the temperature of the PV diode to prohibited values. If $\lambda_g > \lambda_m$, the percentage of the harmful radiation increases. If $\lambda_g \ll \lambda_m$ so that the whole spectrum of thermal radiation becomes useful – that is, all thermal radiation power can be converted to electricity – it also does not solve the thermal management problem. In this case, the relaxation loss (Loss 2 in the theory of SCs) makes the PV conversion very inefficient. Recall that even for the Sun, whose thermal radiation corresponds to $T_e = 6,000$ K, Loss 2 for silicon SCs exceeds 40%. For an emitter with $T_e = 1,000$ K, the spectrum is about twice broader, and Loss 2 for such an ultra-broadband PV diode would be much higher. Since Loss 2 means dissipation, that PV diode would be heated by thermal radiation to prohibited values of temperature. Therefore, the optimal solution is to choose the PV material so that $\lambda_g \approx \lambda_m$ and to apply an optical filter reflecting the low-frequency radiation ($\lambda > \lambda_m$) back to the emitter.

In [11], it was noticed that this reflection not only mitigates strong heating of the PV cell but converts harm to advantage. Unlike a SC, where the reflected solar power is lost, in a TPV system, the thermal radiation of the emitter is confined.

Therefore, the radiative heat reflected by the filter heats the emitter, helping to maintain the high temperature of the emitting surface T_e rather close to T_s – the temperature of the heat source. In work [12], it was experimentally shown that the presence of the filter with the functionality of antireflecting coating for high frequencies ($\lambda < \lambda_m = \lambda_g$) reflecting low frequencies ($\lambda > \lambda_m = \lambda_g$) results in an increase of the electric output for modest temperatures and of the heat source and enables TPV systems operating at $T_e > 1,500$ K.

The next step in the development of TPV systems was done when it was understood that the same reasoning was adequate for a narrowband optical filter. Really, the high-frequency part of the radiation spectrum is still very broad and practically corresponds to the Shockley–Queisser limit of a single-junction PV diode of the order of 30%. A minimum of 70% of the power absorbed by the PV diode is converted into heat, and in early TP generators, this value was close to 90% [3, 5, 6]. Since the reflected power is not lost and the backward thermal flux supports T_e nearly equal to T_s, it is reasonable to transmit only a narrowband part of the thermal radiation spectrum to the PV cell. In work [13], it was proven that a bandpass filter that reflects 90% of the emitted radiative heat and transmits only 10% in the band covering $\lambda_g = \lambda_m$ offers higher electric power output than the previously used high-frequency filter. This idea is illustrated in Figure 11.2. The optimal relative bandwidth of the transmitted thermal radiation is usually 10%–15% depending on T_e, and the location of this passband on the axis of wavelengths (frequencies) is determined by the PV cell. This point will be considered in the next subsection in more detail.

However, narrowband frequency filtering is not the best solution for TPV systems [6, 7, 14]. The claim that the reflected power in the TPV system is not lost at all is too simplistic. First, this filtering results in some losses inside the filter. A narrowband optical filter maximally close to an ideal one – completely reflecting the frequencies beyond the operation band of the PV diode and completely transmitting the frequencies inside the band – is a multilayer with a high optical contrast of the constitutive layers. Low-permittivity layers should alternate with high-permittivity ones in this filter [15]. High permittivity in the IR range implies a semiconductor material. Most part of semiconductors in the IR range have noticeable optical losses (the only exception is GaAs; however, this material is technologically not suitable for optical filters). Thus, narrowband filtering results in some parasitic absorption of radiation. Besides this straightforward loss of efficiency, there is also an indirect loss due to the heating of the PV diode. As it is shown in Figure 11.1, the diode is thermally connected to the filter, because one more cavity between them would be too expensive of a solution. Therefore, an increase of the temperature of the filter implies an increase of the dark currents in the PV cell – i.e., a decrease of its efficiency. Second, a tubular TPV system is not an infinite cylinder. It has two bases – lateral walls – made of a thermoinsulating material. Thermal radiation – that produced by the emitter and that reflected by the filter – impinges these walls, and a certain portion of it is absorbed there.

Now we approach understanding of a modest progress in the TPV industry and a rapid progress in the TPV research in recent times. The early progress was modest because a lot of effort was invested into an improvement of parameters that could be improved only slightly. For example, the characteristics of narrowband filters could be improved, but the restrictions relating the needed frequency dispersion and the optical loss could not be overcome. When nanophotonics research actively started, it became clear that one may reject the whole idea of narrowband filtering, keeping the advantageous narrowband regime for the PV diode. For it, instead of a lossy narrowband filter, one may use a lossy narrowband emitter, made of a nanostructured metamaterial. Frequency selectivity of thermal emission is achievable due to the resonant nature of the metamaterial. Of course, nanostructuring of TPV emitters implies higher costs. However, we believe that development of nanotechnologies will make TPV systems with narrowband emitters sufficiently competitive with conventional TPV systems comprising narrowband optical filters.

11.2.2 Radiative Heat Transfer in TPV Systems

Here we recall basics of thermal radiation and define the radiative heat transfer parameter (RHT) as a basic characteristic of a TPV system. The notations and definitions correspond to those of [7]. In the theory of thermal radiation, one usually prefers to use either linear frequencies $\nu = \omega/2\pi$ or wavelengths $\lambda = c/\nu$. The power emitted in a unit interval of frequencies (or in a unit interval of wavelengths) per unit area of a thermal emitter of area S is called spectral radiance:

$$i_\nu = \frac{d^2 P_{\text{rad}}}{d\nu \, dS}, \qquad i_\lambda = \frac{d^2 P_{\text{rad}}}{d\lambda \, dS}. \tag{11.1}$$

Spectral radiance of a black body at the temperature T_e into a lossless dielectric medium with refractive index n obeys the Planck law [9]:

$$i_{B\nu} = \frac{2n^2 h\nu^2}{c^2} \frac{1}{e^{\frac{h\nu}{k_B T_e}} - 1}, \qquad i_{B\lambda} = \frac{2hc^2}{n^2 \lambda^5} \frac{1}{e^{\frac{hc}{k_B T_e n\lambda}} - 1}. \tag{11.2}$$

The qualitative difference of the two formulas in (11.2) results from the relationship for differentials $d\lambda = c \, d\nu/\nu^2$, approximately applicable to narrow bands of wavelength and frequencies. The exponentials in the so-called Planck's oscillator term, expressed either in wavelengths or in frequencies, are also different. The integration over all frequencies or all wavelengths gives, naturally, the same result for the total black-body radiance:

$$i_B = \int_0^\infty d\lambda \, i_{B\lambda} = \int_0^\infty d\nu \, i_{B\nu} = \frac{n^2}{\pi} \sigma T_e^4, \tag{11.3}$$

where $\sigma = 2\pi^5 k_B^4/15h^3 c^2$ is called the Stefan–Boltzmann constant. Formula (11.3) is the Stefan–Boltzmann law. The factor $n^2 = \varepsilon$ in this formula has already been discussed regarding the direction of dielectric photon concentration in TPV researches.

If the emitter is not a black body, its spectral radiance contains a factor called the spectral emissivity, usually denoted as $e(\nu)$ (or $e(\lambda)$):

$$i_\nu = i_{B\nu} e(\nu), \quad i_\lambda = i_{B\lambda} e(\lambda). \tag{11.4}$$

In accordance with the Kirchhoff law, $e(\nu)$ (or $e(\lambda)$) is simply equal to the spectral absorptivity $A(\nu)$ (or $A(\lambda)$) of the emitting surface. Absorptivity of a surface is simply the plane-wave absorption coefficient $a(\theta, \phi)$ averaged over all incidence angles:

$$A = \frac{1}{2\pi^2} \int_0^{\pi/2} d\theta \int_0^{2\pi} d\phi \, a(\theta, \phi), \quad a(\theta, \phi) = 1 - R(\theta, \phi). \tag{11.5}$$

Here, $R(\theta, \phi)$ is the power reflection coefficient at a given frequency or wavelength for a given polar and azimuthal incidence angles (θ, ϕ). The Kirchhoff law asserts that

$$e(\nu) = A(\nu), \quad e(\lambda) = A(\lambda). \tag{11.6}$$

Therefore, in order to calculate the thermal radiation from a body of a given area at a given temperature, it is enough to solve an auxiliary problem of the local plane-wave reflection by this area for all incidence angles. The problem of the plane-wave reflection is relatively simple. For an emitter formed by a stack of flat homogeneous layers (isotropic or anisotropic), this problem solution is known in closed form (see, e.g., [16, 17]).

Kirchhoff's law (11.6) remains valid also for periodic nanostructured or nanotextured surfaces if they are effectively homogeneous in the operation frequency range of the emitter [18]. In our terminology, it means that the Kirchhoff law is also applicable to the surfaces of effectively homogeneous metamaterials and metasurfaces. In these cases, the plane-wave reflection problem can be solved numerically for a set of incidence angles and the integration in (11.5) is replaced by summation.

The directionality of thermal radiation produced by a unit surface of a homogeneous emitter is characterized by the Lambertian pattern $\cos \theta$ [9]. It means that the maximum power flux is in the normal direction. However, this directionality plays no role in the radiative heat transfer between two parallel surfaces. The spectral RHT$_\lambda$ is defined as the spectral power flux from the unit area of a hot surface to the unit area of a cold surface located in front of it. The sketch in Figure 11.3(a) explains why this is so. This is the result of the translation invariance: radiation of the hot unit area in the lateral directions is reimbursed by the same amount of radiation coming to the cold unit area from other areas of the hot surface. In the wavelength formalism, we have

$$RHT_\lambda \equiv \frac{d^2 P_{HC}}{d\lambda \, dS}, \tag{11.7}$$

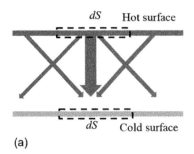
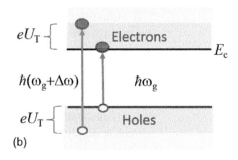

Figure 11.3 (a) The radiative heat transfer between a hot and cold unit surfaces dS is equal to the thermal radiance (irradiance) of the hot surface. Thermal radiation from the top unit area dS to lateral directions is reimbursed by the radiation incoming from adjacent areas. (b) The ideal band of photons corresponding to the minimum of the relaxation loss factor (Loss 2) corresponds to the double thickness eU_T of the edge state determined by the temperature T of the PV layer.

where P_{HC} is the thermal power transmitted from the hot surface to the cold one when their areas are large enough and equivalent. As it is clear from the figure, RHT_λ is equal to the spectral radiance i_λ if there is no reflection of the radiation from the cold surface and if the cold surface does not radiate itself. Notice that in the theory of TPV systems, the spectral thermal radiance (as well as its integral – total radiance) is often called *irradiance* [7, 9].

In a realistic TPV system, RHT from the emitter to the PV layer is not equal to i_λ due to two factors. The first factor is the presence of a filter and a top contact mesh. The second factor is thermal radiation of the cold surface. For simplicity, we can neglect the impact of the mesh, which is relatively small compared to that of the filter. We also can neglect the backward RHT from the cold surface, because its temperature T is of the order of 300–400 K. Since formula (11.3) establishes the proportionality of irradiance to the fourth power of T, the backward flux of thermal radiation is really negligibly small compared to the forward one, which is created by the emitter. Then the TPV RHT – the power transmitted between two unit areas (of the emitter and of the PV semiconductor) per unit interval of wavelengths – is equal to

$$RHT_\lambda = i_{e\lambda} TR(\lambda), \quad TR(\lambda) \approx 1 - R(\lambda). \tag{11.8}$$

Here, $i_{e\lambda}$ is the spectral irradiance of the emitter, equal to the black-body irradiance multiplied by the emitter emissivity; $TR(\lambda)$ and $R(\lambda)$ are, respectively, power transmittance of the filter and power reflectance of its top surface. The approximate equivalence $TR(\lambda) \approx 1 - R(\lambda)$ implies that a negligibly small amount of power is absorbed in the filter. In the absence of this parasitic absorption, the power balance requires exact equivalence.

In the ideal case, $R(\lambda) = 0$ for the operation band of the PV diode $\lambda_g - \Delta\lambda < \lambda < \lambda_g$, and $R(\lambda) = 1$ beyond this interval. The reflected thermal radiation creates some backward radiative heat flux even if the irradiance of the PV cell is negligible.

11.2.3 The Optimal Operation Band and the Shockley–Queisser Limit for TPV Systems

Here, we clarify the question of the optimal operation band $\Delta\lambda$ of a TPV generator. The optimum results from the consideration of two factors – efficiency and electric output. The narrower is the operation band; the lower is Loss 2 of the PV diode. However, this is true until a certain limit. This limit is dictated by the uncertainty principle of quantum mechanics. In accordance with this principle, the energy levels of any quantum system have finite thickness ΔW (inversely proportional to lifetime of the level). Photoinduced charges in the PV diode will have this uncertainty in their energy even if induced by a monochromatic light. This implies a certain irreducible value of Loss 2. Further reduction of the incident light bandwidth is useless for the increase of the efficiency. Moreover, in the case when the frequency selectivity is achieved with the use of a narrowband filter, this reduction would be harmful for the electric output. Really, transmitting the narrower spectrum of thermal radiation into the PV diode, we decrease the incident power, and, since the efficiency does not grow as this spectrum shrinks, we decrease the electric output. Therefore, the optimal bandwidth of the TPV generator in terms of the energy of photons is equal to $2\Delta W$, where ΔW is the thickness of the energy levels at the edges of the semiconductor bandgap. In terms of frequencies, it is $2\Delta W/\hbar$, where factor 2 implies that both electrons and holes are photogenerated carriers. The central frequency of this optimal band is, of course, ω_g because in the approximation of ideally discrete energy levels, Loss 2 is absent for the monochromatic light with this frequency.

The thickness ΔW of the edge states of a semiconductor in the interval of temperatures 0°C–100°C is approximately equal to $eU_T \equiv k_BT$ – this value is sometimes called thermal potential energy [9]. Here, k_B is the Boltzmann constant. Therefore, the irreducible value of Loss 2 is equal to $2eU_T/\hbar\omega_g$. Figure 11.3(b) illustrates this fact. In terms of the wavelength, the band $\Delta\omega = \Delta W/\hbar$ corresponds to $\Delta\lambda = 2\pi c\Delta\omega/\omega_g^2$. It is an approximate formula that takes into account that the operation band is very narrow and replaces the finite variation by differentials (as in the preceding formulas for thermal radiance).

For InAs having the bandgap in the mid-IR range ($\omega_g = 2\pi \cdot 83$ THz) at room temperature (300 K), irreducible Loss 2 is equal to 12% [7]. The corresponding term $1 - \text{Loss 2}$ for InAs at room temperature is, therefore, equal to 88%. It corresponds to the maximally achievable overall efficiency (Shockley–Queisser limit) exceeding 40% [7].

For crystalline silicon ($\omega_g = 2\pi \cdot 357$ THz), the possible improvement is even more spectacular. The optimization of the operation band at room temperature gives, in this case, Loss 2 of the order of 1%. Recall that, for Si solar cells, this factor was equal to 30%–40%, whereas the maximally achievable overall efficiency was estimated as 30%. The increase of the factor $1 - \text{Loss 2}$ for silicon PV cell from 60% to 99% implies the corresponding increase of the Shockley–Queisser limit from 30% to 48%. However, silicon is not the best material for

IR photovoltaics. Since the optimal operation band is centered at the bandgap frequency (for c-Si, nearly equal to 360 THz), this band, called the *interband range*, would require an emitter with an unrealistic temperature close to that of Sun. For a PV layer of a conventional TPV system operating at $T_e = 1{,}500$–$2{,}000$ K suitable materials are, for example, InGaAs or GaSb, for which the ultimate efficiency U_{SQ} is estimated as 75%–80% [7]. It is also possible for chemists to create new multicomponent PV materials with even higher ultimate efficiencies.

11.2.4 TPV Generators: Domestic, Industrial and Solar Stations

Practical TPV systems operating as electric generators for domestic power supply or for industrial waste-heat harvesting were developed in 2002–2003 by the American company JX Crystals [10, 19–21]. It was the first company who made sufficient investments in high-power thermophotovoltaic generators. A TPV system for domestic applications, whose cross section is shown in Figure 11.4(a), represents a concentric tube whose tubular PV panel of GaSb PV cells surrounds a concentric gas burner. The purpose of the TPV tube is twofold: generating electricity and utilization of low-grade heat. The PV panel operates at the mean temperature close to 100°C, which is possible to maintain with air cooling. The concentric bilayer[1] emitter comprises an interior layer of W impinged by the flame of the tubular gas burner, whose temperature is close to 1,700 K. The narrowband optical filter applied to the PV panel consists of nine nanolayers of quartz and silicon and transmits wavelengths in the band 1.5–1.8 μm, reflecting the out-of-band radiation. The operation band is optimal for GaSb. Though its bandgap wavelength is $\lambda_g = 1.7$ μm, the uncertainty of the edge state also implies a noticeable PV response at 1.7–1.8 μm. The overall efficiency of the PV cell of

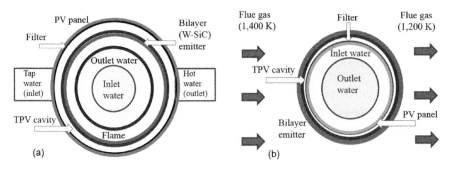

Figure 11.4 High-power TPV tubes combining water heating with generation of electricity. (a) Sketch of the cross section of a TPV generator tube for domestic applications, suggested in [19, 20] (author's drawing, after those works; some technical details are omitted). Cooling fins of the PV panel, outer shell, and some other irrelevant details are omitted. (b) A similar cross section of a TPV tube for application in the waste-combusting plant; see [10, 21], (Author's drawing, after those works; some technical details are omitted).

this TPV system is nearly equal to 29%, and its electric output is 1.5 W per cm^2 and 703 W in total.

An experimentally confirmed PV efficiency this high became possible because of two main advantages of TPV systems compared to solar cells: a high Shockley–Queisser limit (75%) due to the narrow operation band of the PV diode and high optical efficiency due to low reflection loss. Of course, the optical efficiency is not very close to unity. First, thermal radiation impinges the lateral thermo-insulating walls of the TPV cavity made of porous quartz – a material whose thermal conductivity is close to 0.3 W/m°K. About 7% of the heat produced by the burner is absorbed in these walls, which also absorb the IR radiation of the emitter. Corresponding optical loss is of the order of 10% [22]. Next, the TPV efficiency – that taking into account the loss in the electric network of the PV panel – is lower than the overall efficiency of the PV cell. Finally, the efficiency of the gas burner is also not equal to 100%. The overall fuel-to-electricity conversion efficiency turns out to be about 16% [20]. A similar result was obtained for a TPV tube mounted in the transition from the combusting camera of a waste-combusting plant to the stalk. The sketch of this tube described in works [10, 21] is presented in Figure 11.4(b). The main difference of this design solution is the outer location of the heat source. Again, the same cold water cools the PV panel and is heated for further use of its low-grade heat. Flue gas in the exhausting part of the combusting camera is slightly cooled. The drop of the temperature and the gas output per unit time show how efficiently the thermal energy is converted.

In fact, fuel-to-electricity conversion efficiency equal to 16% is not as good as the same parameter of the so-called *Stirling generator* – an electric generator where heat is converted to electricity via mechanical work of a vapor machine. In accordance with [23], the overall fuel-to-electricity conversion efficiency of the best available Stirling machine is nowadays about 41%. However, unlike this machine, the TPV generator has no moving parts, does not require technical service, and is very compact. Therefore, the appropriate prototype for comparison of the TPV tube efficiency is a thermoelectric generator. The heat-to-electricity conversion efficiency even for a single TEE in the best-known cases is about 1.5%. It appears that there is no potential for a significant breakthrough in thermoelectricity, and there is no reason to expect that the overall efficiency of thermoelectric generators will overcome the threshold of 4% [24]. Thus, the fuel-to-electricity conversion efficiency of the order of 16% is an excellent result.

The main drawback of high-power TPV tubes is their short lifetime. The permanent impact of the flame/flue gas creates mechanical stresses in the emitter. These stresses accumulate and finally result in distortions destroying the TPV cavity. Since the lifetime of both domestic and industrial TPV tubes is about one year – 10 times shorter than that of a typical solar panel – the cost of electricity produced by these generators is high, and the demand for these devices is very restricted.

Another type of a TPV generator that theoretically promises a long lifetime is a solar TPV station. In accordance with [7], this idea is rather old, and we

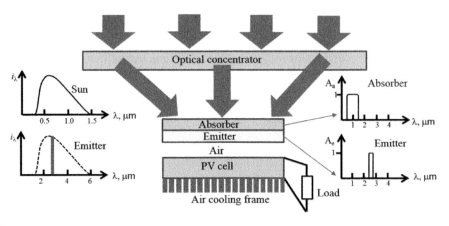

Solar TPV station and its operation illustrated by four relevant spectra. Author's drawing, after [25, 26].

may refer to those solar TPV stations that do not comprise nanostructures as conventional TPV systems.

The conceptual scheme of a solar TPV station station is shown in Figure 11.5. Solar light harvested from a large area by an array of lenses (this array is called an *optical concentrator*) is focused on a plate that absorbs the sunlight (whose spectrum called insolation is shown on the top-left inset) like a black body. This way, the sunlight is efficiently converted into heat. Such absorbing media do not exist in nature (black dyes rapidly degrade under solar exposure; the soot scatters the visible light). Stable absorbers of sunlight were created in 1980s and 1990s – they are either plasmonic composites comprising Au nanoparticle with different dimensions (from 2–3 to 50–100 nm [27]) or metal-ceramic absorbers [28, 29] composed by non-plasmonic metal nanoparticles and microparticles in a semitransparent ceramic matrix. These particles have resonant sizes at different wavelengths of the solar spectrum, and the omnidirectional resonant scattering increases the absorption in ceramics. Both plasmonic and non-plasmonic solar absorbers look black and do not lose this color under solar exposure.

An absorber having a temperature of the order of $1,000 - 1,500$ K radiates very weakly. The band of its thermal radiation corresponds to the mid-IR range, whereas the absorptivity/emissivity is high only in the visible and near-IR ranges. In the ideal case illustrated by the top-right inset, $e = 0$ in the range of λ, where i_λ is non-zero and vice versa. Then the absorber does not radiate, and all its heat is transferred to the emitter. In this situation, the temperature of the emitter is the same as that of the absorber. Here, the TPV vacuum cavity is absent; the structure is open to the air. Natural thermal management was expected to be one of the key advantages of solar TPV stations. It should be achieved due to evacuation of heat via thermal convection. Moreover, a significant portion of the emitted radiative heat is also lost in this geometry. However, the authors of this concept hoped that, in a comparison of the solar TPV station and a PV solar panel, the impact of these losses will be compensated for by the high Shockley–Queisser limit.

An important point is the frequency selectivity of the emitter necessary for efficient operation, as illustrated by the bottom-right inset. There are three reasons why it is so. First, in this open design, reflections from the narrowband filter, if applied on the PV cell, would result in one more loss. Second, if the thermal radiation of the emitter were equal to that of black-body (dashed line on the bottom-left inset) power, though mostly useless for the PV conversion, would be too high for maintaining the needed temperature regime in both the absorber and the emitter. This is so because the emitter of the solar TPV station is not connected to the thermal reservoir (a heat source), and its thermal regime is especially sensitive to radiative losses [7]. The radiation in the whole black-body spectrum would result in radiative cooling of the emitter, accompanied by a high temperature gradient, resulting in the failure of the quasi-equilibrium approximation. This failure disables the prerequisite of the theoretical model, and predicting the operational characteristics of the solar TPV station becomes very difficult. The narrowband absorptivity/emissivity of the emitter $A_e(\lambda)$ shares a narrow band from the black-body spectrum, and only this range of wavelengths is radiated toward the PV cell. Most of the thermal energy stays confined inside the emitter, which prevents noticeable violations of its thermal equilibrium. The third reason why the frequency selectivity should be achieved in the emitter is the aforementioned optical loss inevitable for narrowband optical filters [15]. This optical loss results in additional heating, which makes the air thermal management of the PV cell more difficult.

The operational band of the TPV system operating at $T_e = 1,500$ K is $\lambda = 2.6$–3.1 μm, which is the optimal band for a PV diode based on one of the types of InGaAs [7]. This material has numerous modifications since its chemical formula is $In_xGa_{1-x}As$, where x can vary from 0 to 1. These modifications correspond to different bandgap wavelengths, varying in the range $\lambda_g = 2.1$–2.9 μm. PV diodes based on the material with $\lambda_g = 2.6$ μm were created, and the main issue became the creation of an emitter with the needed frequency selectivity. The simplest version of such an emitter would be a solid refractory material covered with a multilayer refractory structure with the needed filtering properties. However, the realization of narrowband filters of refractory materials is very difficult, and such filters were not created until recently. Here, nanophotonics entered the game – several years ago, a nanostructured frequency-selective emitter made of carbon nanotubes was created, and the first working *TPV solar station* was reported in experimental paper [26]. The overall efficiency of this system was 3.2%. Such a low efficiency is not surprising because of the layered geometry of that station – similar to that depicted in Figure 11.5. It implies high optical and convective losses of solar power. To achieve much higher efficiencies, a solar TPV station must be implemented as a confined system.

11.3 Advanced TPV Systems

11.3.1 Solar TPV Station: Confined Design and Metasurface-Based Emitter

Theoretically, a solar TPV station may manifest surprisingly high values of efficiency for conversion of solar light into electricity (e.g., 45% in [25]). However, to realize such performance, one has to eliminate significant losses inherent to

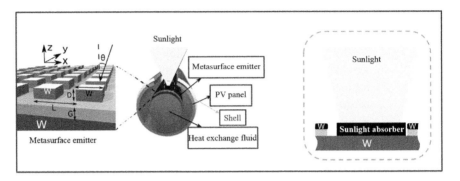

Figure 11.6 Advanced solar TPV station with its key components: on the left panel, the metasurface-based emitter is shown; on the right panel, the solar absorber is depicted. Author's drawing, after [30, 31].

open TPV systems, discussed earlier. The theory of such advanced TPV solar stations has been developed in works [30–32], where a long tubular structure was considered (the impact of its finite length has not been studied yet). The tube is oriented from east to west and comprises a metal shell with an axial slit in its top part. This slit is covered with glass in order to reflect (at least partially) thermal radiation while transmitting the focused sunlight. The slit is located in the plane of the Sun movement; i.e., on sunny days, it is illuminated for the whole day. The cross section of the structure is shown in Figure 11.6, together with the key components enlarged in two insets. One component is the solar power absorber. During the daytime, the same amount of concentrated sunlight (focused by a cylindrical mirror, which is located separately and is not shown) illuminates the solar absorber (through the slit in the metal shell) – a black plate, shown in the right inset. Sunlight has the spectrum mainly concentrated in the band 320–1350 nm, whereas the absorber heated by sunlight in the steady regime to the temperature $T_e \approx 1,400°$ K radiates in the range 1.5–10 µm, where its emissivity/absorptivity is very low. Therefore, the radiation of the absorber is weak, and only its small part penetrates outside through the slit. To reduce this radiative loss, the slit may be performed as a high-frequency optical filter with a sharp slope of the transmittance from almost unity (higher frequencies) to almost zero (low frequencies) at the effective bound of the solar spectrum – the wavelength of 1,320 nm. Such filters have been developed recently using the achievements of nanophotonics. This structure is called *photonic thermos* [33]. A photonic thermos in the slit is not shown in Figure 11.6 for simplicity.

Absorbers of sunlight have been already discussed. A strip of this material parallel to the slit is located on the surface of a hollow tungsten tube whose diameter is equal to 15 cm. Tungsten is a very good heat exchanger – its thermal conductivity is as high as that of copper. Therefore, the *frequency-selective emitter* located aside of the light-absorbing strip has the same temperature, $T_e \approx 1,000°$ C, as the absorber.

This emitter is the second key component of the system shown in Figure 11.6. It is a metasurface implemented as an array of submicron patches of tungsten raised due a tiny dielectric gap (refractory dielectric withstanding high temperatures is, for example, SiC) over the cylindrical surface of tungsten. The gap G is very small compared to the wavelengths of the emitted radiation.

If we consider this metasurface as a planar array (that is reasonable because the radius of the cylinder is incomparably larger than λ), we can numerically calculate the absorption coefficient $a(\lambda, \theta)$ of this metasurface quite easily and see that it exceeds 0.5 in the interval $\lambda = 1.7$–$2.3\,\mu$m. In other words, the emissivity $e(\lambda) = A(\lambda)$ exceeds 0.5 in the operation band of the emitter. The band 1.7–$2.3\,\mu$m is the optimal operation band for a PV diode based on InGaAs (see the end of Section 11.2.3). These PV diodes form a cylindrical panel (on the internal surface of the metal shell). The TPV cavity between this panel and the emitter grants thermal isolation. The working temperature of the PV panel is sufficiently low (theoretically, about $100°$C). The PV conversion efficiency of the PV panel is high due to the narrowband nature of filtered thermal radiation and theoretically equals 41%. The overall TPV efficiency of such solar station presumably exceeds 20% [31, 32].

Such TPV solar stations should be combined either with a boiler or with a Stirling machine [23] that can enhance the light-to-electricity conversion efficiency up to 60% [32]. The optimal temperature of the emitter corresponds to the thermal regime in which the excessive heat is evacuated by a heat exchange fluid circulating in the central tube thermally connected to the absorber and the emitter. Its shell is sufficiently thick, and the thermal gradient across it allows the fluid to have a reasonable temperature estimated as $400°$C–$600°$C. Fluids that can stand this temperature are available.

The main point of interest for us is the metasurface thermal emitter whose operation characteristics are better than those of the emitter employed in work [26]. In the follow section, we study the *frequency-selective absorptivity/emissivity* of this metasurface.

11.3.2 Emissivity of Resonant Metasurfaces

The considered metasurface is depicted in Figure 11.7(a). Its main property is resonant absorptivity – the absorption coefficient $a(\lambda, \theta, \phi)$ in a broad range of incident angles is close to 100% within the operation band and close to zero beyond it. This absorber has the design parameters $D = 10$, $w = 230$, $L = 300$, and $G = 17$ nm (see Figure 11.7). Both the nanopatches and the ground plane are made of tungsten, and the dielectric spacer is made of aluminum nitride. Note that tungsten in the whole IR range is a highly conducting material. The skin depth in the operation band is close to 30 nm.

The operation of this metasurface is similar to that of its microwave analog with absorbing patches (see, e.g., in [34]). Let us consider the operation of the last one in the simplest case of the normal incidence, as shown in Figure 11.7(a). Let the grid of very thin patches be made of a conductive material with loss. Then it is a lossy capacitive grid due to the capacitive coupling between the adjacent patches and resistive losses in them.

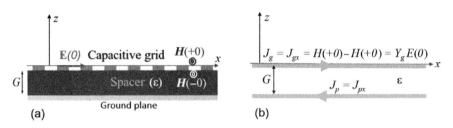

Figure 11.7 (a) Normal incidence of a plane wave on a microwave absorber or to a frequency-selective emitter based on a capacitive grid. The vertical cross section. (b) Homogenized capacitive grid is characterized by the induced surface current density $\mathbf{J}_g$ related to the tangential electric field via the grid admittance. If $G \ll \lambda/\sqrt{\varepsilon}$, the surface current $\mathbf{J}_p$ induced on the ground plane flows in the opposite direction.

The surface current density J_g induced by a normally incident plane wave in this grid is the surface current density found obtained by the spatial averaging of the microscopic currents induced in the patches over the grid plane $z = 0$; see Figure 11.7(b). In accordance with Maxwell's boundary condition, $J_g = H(+0) - H(-0)$, where $H(\pm0) \equiv H_y(\pm0)$ is the surface-averaged magnetic field on the top and bottom sides of the grid. The grid admittance Y_g relates J_g to the surface-averaged tangential electric field in the grid plane $E(0) \equiv E_x(0)$, which is continuous across the grid $E(0) = E(+0) = E(-0)$ because the grid thickness is negligibly small compared to the wavelength. Thus, we can write

$$J_g = H(+0) - H(-0) = Y_g E(0). \qquad (11.9)$$

Next, the surface-averaged magnetic field at the bottom side of the grid is related to the averaged electric field $E(-0) = E(0)$ via the surface admittance Y_s of the metal-backed dielectric spacer of thickness G:

$$H(-0) = Y_s E(0), \qquad Y_s = -j/(\eta \tan kG), \qquad (11.10)$$

where $\eta = \sqrt{\mu_0/\epsilon_0 \varepsilon}$ and $k = (\omega/c)\sqrt{\varepsilon} \equiv (\omega/c)n$ are, respectively, the wavenumber and the wave number of the dielectric medium ($n = \sqrt{\varepsilon}$ is the refractive index). The second relation in (11.10) represents the input admittance of a short-ended piece of a transmission line. Really, for a plane wave propagating in the arbitrary layered structure, the transmission-line model is strictly correct. The dielectric medium of relative permittivity ε is an effective TEM transmission line, and the ground plane with its zero surface impedance is the short termination. Therefore, the surface admittance of the metal-backed dielectric layer in the approximation of perfect conductivity of the metal ground is really the well-known input admittance of a short-circuited transmission line of length G. The perfect-conductor approximation for the ground plane is adequate because the ground plane is much thicker than the skin depth and the absorption of the wave in it is negligibly small.

From formulas (11.9) and (11.10), it is easy to find that

$$H(+0) = (Y_s + Y_g)E(0) = (Y_s + Y_g)E(+0). \qquad (11.11)$$

If $Y_s + Y_g = 1/\eta_0 \equiv \sqrt{\epsilon_0/\mu_0}$ – i.e., the surface admittance of the whole structure in the plane $z = 0$ is equal to that of free space, there is no reflection. There is also no transmission because the structure is impenetrable. This means that the last equation is the condition of perfect absorption. This condition can be written, taking into account relations (11.10) and (11.11), as follows:

$$Y_g = \frac{1}{\eta_0} + j\frac{1}{\eta \tan kG}. \tag{11.12}$$

Recall that our grid possesses both capacitive and resistive properties. The grid resistance R_g and the grid capacitance C_g in the equivalent scheme of the grid are connected in series because the conductivity currents in the patches are continued by the displacement currents in the gaps between them. Therefore, we can express the grid impedance $Z_g \equiv 1/Y_g$ as

$$Z_g = R_g + \frac{1}{j\omega C_g} = \frac{1}{Y_g}. \tag{11.13}$$

Equating Y_g given by (11.12)–(11.13), we obtain the needed grid capacitance C_g and resistance R_g:

$$C_g = \frac{1}{\omega\eta \tan kG} = \frac{n}{\omega\eta_0 \tan k_0 nG}, \quad R_g = \frac{\eta_0}{1 + \omega^2\eta_0^2 C_g^2}. \tag{11.14}$$

Of course, both these conditions can be strictly satisfied only at one frequency, $\omega = \omega_0$. As a rule, the dielectric spacer is optically thin – i.e., $kG \ll \pi$. Then we have

$$C_g = \frac{1}{\omega_0^2\mu_0 G}, \quad R_g = \frac{\eta_0}{1 + (k_0 G)^{-2}}. \tag{11.15}$$

Conditions (11.15) are feasible in both microwave and infrared ranges. In the microwave range, it is possible to implement the absorber using nonmagnetized iron or graphite as the material of patches [34], in the IR range, using tungsten. A similar analysis can be done for oblique incidence. This analysis shows that the values satisfying conditions (11.15) grant nearly the same resonant absorption also for a rather broad angular spectrum of TM-polarized waves. For TE-waves, the angular stability of the absorption is worse; however, the resulting absorption for unpolarized light averaged over all possible incidence angles is acceptable.

Frequency dependencies of the absorption coefficient of a microwave absorber with iron patches are presented in Figure 11.8 for three angles of incidence [35]. Similar frequency selectivity and similar angular stability of the resonance band were obtained in work [35] for a microwave absorber with graphite patches and for an IR absorber/emitter with tungsten patches in paper [36]. The last design (with aforementioned parameters) was selected as an emitter for the prospective solar TPV station discussed previously.

The frequency-selective emissivity/absorptivity $A(\lambda)$ that results from the averaging of the absorption coefficient a over all incidence angles and both TE- and TM-polarizations is multiplied in accordance with (11.4) by the blackbody irradiance (11.2) and gives the irradiance of the emitter. In the present case, when the filter is absent, it is equal to RHT.

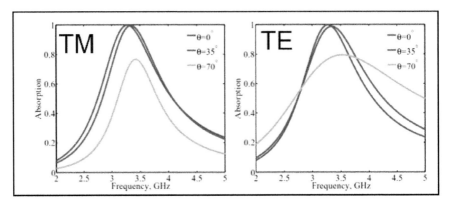

Figure 11.8 Frequency dependencies of absorption coefficients for TM-polarized (a) and TE-polarized (b) waves illuminating a selective microwave absorber (iron patches). Reprinted from [34] with permission. Copyright 2015 APS

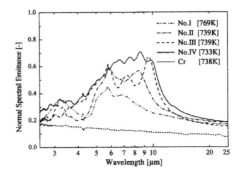

Figure 11.9 Emissivity (spectrum of thermal radiation normalized to the black-body radiation spectrum) measured for a metasurface emitter of Cr nanopatches over a Cr plane. Frequency selectivity is seen even for rather low temperatures. The dotted curve corresponds to the emissivity of a flat surface of chromium. Reprinted from [37] with permission of AIP Publishing

Frequency-selective emissivity has been obtained not only for emitters dedicated to solar TPV stations. For example, in [37], measurements of highly frequency-selective thermal radiation were reported for a similar emitter as discussed earlier with a little difference – nanopatches were made of chromium and located over a chromium plate. Also, geometric design parameters of the chromium metasurface were slightly different from those of the tungsten one. Four curves of the emissivity/absorptivity dispersion are presented in Figure 11.9 for four values of the emitter temperature, from 460° to 491°. For comparison, the emissivity of a flat Cr plate at a similar temperature is also shown. We see that the emissivity of chromium slightly varies around 0.15, whereas for the metasurface of chromium it achieves 0.7 at the maximum and drops to 0.2 at the edges of the radiation spectrum. For such a low temperature, this frequency selectivity is quite high.

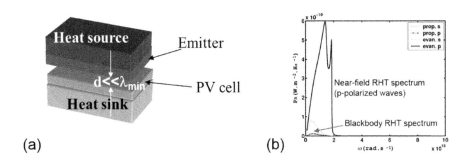

Figure 11.10 (a) Near-field TPV system. (b) RHT spectra typical for a near-field TPV system with $d < 100$ nm. Contributions of evanescent waves of TE (s) and TM (p) polarizations are shown separately. The spectrum of RHT for propagating waves is the same for two polarizations, and their sum is the RHT spectrum of two parallel black bodies. Reprinted from [38] with permission of AIP Publishing.

11.3.3 On Near-Field TPV Systems

Besides the advanced solar TPV stations discussed earlier, several research groups have developed other advanced systems, called *near-field TPV systems*, illustrated by Figure 11.10(a). In these systems, the thickness d of the TPV cavity is much smaller than the minimal wavelength of the radiative heat spectrum transferred across this TPV cavity of a tiny thickness. In a near-field TPV system, the radiative heat is carried mostly by evanescent waves, in contrast to conventional solutions, where the heat is carried by propagating waves. In the absence of a PV cell, evanescent waves produced by a heated emitter decay exponentially outside of its volume and are concentrated in the vicinity of the emitter surface. In the presence of a PV cell, the exponentially decaying evanescent waves reach the PV cell surface and are partially reflected back. Whereas a single evanescent wave cannot transfer energy along its decay direction, a pair of mutually coherent evanescent waves may transfer energy. The energy transport occurs if the reflection coefficient is complex [18]. Moreover, in this case, the power transmission coefficient can have "thickness resonances," in analogy with Fabry-Perot resonances for propagating waves. This effect is called either *frustrated total internal reflection* or photon tunneling. Peculiarities of this effect in near-field TPV systems are studied in [18, 39] and in many other papers and even books.

In this introductory book, we have no opportunity to consider near-field TPV systems in such detail. We only state that the RHT in these systems is much higher than in the far-field TPV systems with the same temperatures of the emitter and PV cell. It drastically exceeds the RHT between two parallel black-body surfaces. Since the black-body irradiance is described by the PLanck law, RHT in near-field TPV systems is called *super-Planckian RHT*.

Notice that photon tunneling through vacuum nanogap is very sensitive to material properties. It vanishes when the bottom side of the TPV cavity is not a doped semiconductor but a dielectric (for example, when it is the top surface of a multilayer optical filter). Thus, the use of narrowband filters is impossible in near-field TPV systems. Radiative heat is transported directly from the emitting medium to the PV medium, and both media are doped semiconductors with close electromagnetic properties in the IR range [38, 40, 41].

Figure 11.10(b) shows a typical RHT spectrum in such a structure for $d = 30$ nm [38]. The RHT is carried almost completely by evanescent waves, and the contributions of TE-polarized (s) and TM-polarized (p) evanescent waves are different. Both of these contributions are depicted separately in comparison with the RHT carried by propagating waves. The contributions of propagating waves with TE-polarization and TM-polarization are equivalent. Their sum is practically equal to the RHT between two parallel black bodies because the emissivity/absorptivity of both materials (the tungsten emitter and the GaSb PV layer) is close to unity.

Such broadband RHT would imply strong heating of the PV cell by sub-bandgap radiation if the temperature of the emitter were the same as the temperature in typical far-field TPV generators. Therefore, near-field TPV systems, as a rule, imply low-temperature emitters and cannot work as electric generators. They are used mainly as sensors of tiny temperature deviations [40, 41].

However, recently, it was theoretically revealed that a near-field TPV system with a gap $d = 100$ nm whose emitter and PV cell are both made of so-called hyperbolic metamaterials enables frequency-selective RHT [42]. In this case, the RHT spectrum in the operation band has peak values about one order of magnitude higher than RHT between two parallel black bodies, and the same hyperbolic metamaterial used on the cold and hot sides (used as the PV cell and the emitter, respectively) allows narrow-band RHT that grants a possibility to this near-field TPV system to work as a electricity generator. The suggested PV cell is a series connection of GaSb layers alternating with GaN layers, possessing metal properties in the operation band. The impact of contact electrodes was not studied yet. The same multilayer of GaSb and GaN can operate at high temperatures as a frequency-selective emitter with the operation band in the near-IR range corresponding to a GaSb diode. However, before an experimental implementation of such structures (which is challenging), additional theoretical studies are needed because the preliminary study of [42] does not take into account the temperature variation of the electromagnetic properties of GaSb and GaN and the significant temperature gradient across the emitter thickness.

Further, in works [43, 44], it was claimed that the role of a narrowband filter compatible with the super-Planckian RHT in a near-field TPV system can be played by a heavily doped monolayer graphene, as shown in Figure 11.11(b). The needed doping level of graphene is, in principle, achievable on Si substrates. However, additional studies are also needed for this structure because the silicon of a PV cell must also be strongly doped, and this doping may drastically change the properties of graphene compared to the model adopted in [43, 44]. Finally, there is also a technical issue that restricts creation of near-field TPV generators based on the exciting ideas of frequency-selective super-Planckian RHT achieved with the use of hyperbolic matematreials or graphene. This technical issue is a short lifetime of the vacuum gap $d < 100$ nm between the two surfaces if one

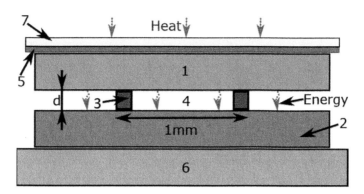

Figure 11.11 The sketch of a micro-TPV system with a submicron TPV cavity, showing the main components of the system. Here, 1 is the SiC emitter, 2 is the narrowband filter, 3 are quartz spacers, 4 is the flat TPV cavity, 5 is the heat exchanger (thermal connector), 6 is the PV cell, and 7 is the refractory metal layer. Author's drawing, after [45].

of them is in contact with a heat source at high temperatures. Temperatures of the order of 1,000 °C produce thermal stress, which results in mechanical distortion of the emitting surface. Experiments of the group of R. DiMatteo have shown that in the area exceeding 1 mm^2, one may keep the hot and cold surface untouched if $d > 500$ nm [45, 46]. However, the 1 mm^2 area is not sufficient for collecting practically significant photocurrent [47], and the gap $d = 500$ nm is not tiny enough for truly super-Planckian RHT. Therefore, the market perspective of near-field TPV generators appears, for the moment, rather doubtful.

11.3.4 Micro-TPV Systems

Portable TPV generators called *micro-TPV systems* or *TPV microgenerators* are dedicated for low power output – dozens of W [7]. Their heat source is the flame in a mini-combustor with the maximal volume of the order of 1–5 cm^3. Some of these generators have a flat TPV cavity (see, e.g., in [22]). Even in the tubular geometry, the length of the tube and its diameter are of the same order; otherwise, compact implementation of the burner is impossible. Therefore, two lateral thermo-insulating walls have the area comparable with that of the working cylindrical surface. In this case, a narrowband filter positioned at a cylindrical PV panel with corresponding reflection of nearly 90% of thermal emission would imply high losses of radiation in the thermo-insulating walls. The reflection is, therefore, undesirable. The commercialized TPV generators comprise nanostructured frequency-selective emitters. Emitters used in known micro-TPV systems are not based on capacitive grids.

In [48], the frequency-selective emitter of a flat micro-TPV generator is a multilayer nanotextured plate. The radiating layer is c-Si textured by an array of parallel cylindrical nanocavities (notches). The whole radiating surface is covered with Ti of a few nm thickness that is necessary for preserving the hot c-Si from destruction (its crystalline structure withstands high temperatures, but the open surface is destroyed rapidly). This notched bilayer nanotexture is prepared by

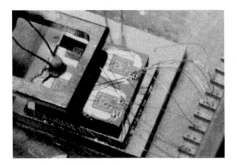

Figure 11.12   A micro-TPV system with a submicron TPV cavity. Left panel: an experimental setup (magnified scale). Reprinted from [46] with permission of the authors. Right panel: a working prototype of a micro-TPV in a real scale. Reprinted from [49] with permission of AIP Publishing.

using electron-beam lithography and other expensive techniques (such as deep reactive-ion etching). The refractory layer contacting with the flame is that of W. In work [50], a more affordable technical solution was found for a micro-TPV system with similarly flat geometry. Here, the submicron notches are made on the radiating side of the solid tungsten layer. These technical solutions evoked interest to the flat geometry of the TPV cavity. This geometry allows one to obtain a better parallelism of a very tiny TPV cavity. The authors of [45, 51] targeted the parallelism of a flat TPV cavity of thicknesses within $d = 500–1,000$ nm over a macroscopic area. This parallelism has been achieved for a long time at the temperatures $T_e = 1,000–1,100$ K of the emitter performed as an electric heater [46]. These applied studies have been reviewed in [49].

In [45], the parallelism of the emitter and the PV cell is ensured by spacers of nearly cubic shape (the cube side is 500 nm) made of porous quartz. These spacers are arranged into several chains distanced by 1 mm from one another. The distance between the spacers in a chain is of the order of 1 μm. The vertical cross section of the TPV structure perpendicular to the chains is sketched in Figure 11.11 where two chains are seen, i.e., the horizontal size of the whole TPV generator is between 2 and 3 mm. Compared to the original sketch from [45] a few details having no principal importance are omitted.

In the experimental demonstrators described in [46, 49] and depicted in Figure 11.12, the TPV cavity was adjusted by a piezoelectric system. The horizontal size of this TPV generator is enlarged up to 2 cm and $T_e \approx 1,200$ K. In the left panel, the first experimental setup is shown, where one used electric heating element as the heat source. In the right panel, we see a working prototype with an incorporated mini-combustor. This experimental device manifested the electric output close to 2 W/cm^2 [49]. This output is nearly the same as that manifested by conventional (commercially available) TPV devices [10, 21]. However, it was obtained for a much lower temperature: 1,200 K against $T_e = 1,600–1,700$ K required for its conventional prototype. Adopting the Stefan–Boltzmann law for the emitted power, we see that the efficiency in the novel micro-TPV system is about four times higher. And this gain in efficiency is achieved in spite of the flat geometry of the generator, which is not so advantageous as the tubular one

[25, 47, 52]. We do not know exactly why this design has not been commercialized since 2004, but we assume that the reason is namely the intrinsic drawbacks of a flat TPV cavity: parasitic heat conductance through lateral thermo-insulating walls and higher loss of useful radiative heat compared to the tubular TPV systems. However, theoretical investigations targeted to enhancements of such micro-TPV generators by nanostructured metamaterials are known (see, e.g., in [53–55]).

11.3.5 TPV Generator without a TPV Cavity

The aforementioned idea of *dielectric photon concentration*, which would allow TPV generators without a cavity with a simultaneous increase of thermal radiation granted by the substitution of vacuum with a dielectric, has been rejected after several theoretical studies. The main reason for that was critical heating of the dielectric spacer. If quartz and other similar dielectric materials used to replace vacuum were lossless, one could really achieve both matching of the dielectric spacer to the emitter and good thermal management of the PV cell. To do that, one would only need to properly choose the spacer thickness, which should be much larger than that of a usual TPV cavity. The flux of conductive heat F_{ch} is proportional to temperature gradient, and this gradient for a given emitter temperature T_e and the maximal allowed PV cell temperature T depends on the thickness d of the spacer: $F_{ch} \sim (T_e - T)/d$. If F_{ch} is reduced to such levels of heat power that can be evacuated by a cooling frame, the vacuum gap would not be necessary. Simultaneously, the permittivity ε of the spacer (if it is matched to the emitter) would offer enhancement of thermal radiation ε times. Matching can be obtained using the Fabry-Perot resonance of the spacer that grants an additional frequency selectivity of the RHT: in the resonance band, RHT will be ε-fold compared to the system with a vacuum TPV cavity, and beyond the resonance, it will be reduced.

However, this approach does not work straightforwardly because all thermally robust dielectrics have some resonant absorption bands in the IR range, which is a very broad range of frequencies from millimeter waves to visible light. Despite the frequency selectivity of the dielectric spacer with proper thickness, some spectral components experience strong dissipation in the dielectric spacer, resulting in its radiative heating. This heating makes the radiative and conductive heat fluxes coupled, and thermal management of the system becomes practically impossible. This is why all dielectric media are unsuitable for cavity-free TPV generators.

In work [56], the concept of a TPV system, where the dielectric spacer is a slab of monocrystalline gallium arsenide, was introduced. This is a material with unique infrared transparency that has no absorption bands in an ultra-broad range covering both far-IR and mid-IR bands. As to the near-IR range, this part of the spectrum is theoretically absent in the RHT spectrum due to an amazing frequency selectivity of the multilayer thermal emitter introduced in [56]. The emitter separating the heat source from the GaAs slab is a multilayer structure of refractory metal and semiconductor nanolayers. The structure is effectively a micron-thick slab of a metamaterial and is made so that the transversal component of the effective permittivity tensor of this metamaterial is smaller

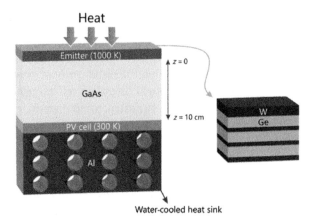

Figure 11.13 Proposed TPV generator without a vacuum cavity theoretically operating at $T_e = 1{,}000$ K with the efficiency 51%. Drawing courtesy of S. Mirmoosa.

than unity in the range 70–130 THz. This emitter is shown in the inset of Figure 11.13, and the metamaterial is simultaneously *epsilon-near-zero* (ENZ) metamaterial and *hyperbolic metamaterial* because the normal component of its effective permittivity tensor is negative in the operation band. Beyond the operation band, the emission can be significant; however, in the near-IR band, where GaAs is absorptive, its level is negligibly low.

The operation band 70–130 THz should be optimal for operation of a PV cell. Such PV cells are not known, and it was assumed that a hypothetic (but feasible!) multicomponent semiconductor could be created for this purpose. For this TPV generator (with a hypothetical optimal PV cell), a very high frequency-selective RHT was theoretically predicted, significantly exceeding the conductive heat transfer and granting a high value of TPV efficiency assuming that the efficiency of the PV cell is its Shockley–Queisser limit. The thickness of the monocrystal GaAs of 10 cm is theoretically sufficient to ensure thermal management for $T_e = 1{,}000$ K, granting an amazing estimate of 20 W/cm^2 for the electric output of this generator [56]. The thickness 5 cm of the GaAs spacer corresponds to the still amazing (for such modest temperature) electric output of 7 W/cm^2. Assuming that the overall efficiency of the prospective PV cell will be twice as low as the Shockley–Queisser limit, the expected power output is still at the record level, 3.5 W/cm^2 for a TPV generator comprising such metamaterial emitter operating at $T_e = 727°$C, and a monocrystal GaAs spacer of realistic (5 cm) thickness between the emitter and the PV cell. It appears that such high electric output has not been even theoretically expected for any other TPV generator working at the emitter temperatures $T_e < 1{,}000°$C.

11.4 Conclusions

In this chapter, we have explained the basics of TPV systems operating as electric power generators. We have presented conventional and advanced TPV generators. For advanced far-field TPV systems such as solar TPV stations, the key component is the frequency-selective emitter. This emitter can be a metasurface with all-angle resonant emissivity/absorptivity in the near-IR range. The emissivity can be estimated by using an analogy with similarly designed microwave absorbers. We have studied the operation of this absorber and analytically derived the thermal irradiance of the emitter developed for advanced solar stations.

Further, we have reviewed two other types of advanced TPV systems: near-field TPV systems and micro-TPV systems. We have analyzed in more detail the TPV systems promising for electric power generation. We have seen that a multilayer metamaterial in a cavity-free TPV system is a promising technical solution offering drastic enhancement of narrowband radiative heat transfer, resulting in very high values for the electric power output of such TPV generators.

In summary, humans need more renewable sources of electricity. However, the market demand for thermophotovoltaics is not yet sufficient because the known technical solutions are, in spite of their high efficiency, not attractive for everyday use: they are very expensive, their electric output is restricted, and their lifetime is short, making the leasing cost of electric energy very high. New solutions increasing the exploitation period of TPV generators and enhancing their electric output may change the situation dramatically. A real breakthrough is, in our opinion, the possible involvement of metamaterials as artificial materials with the optimal properties. Growing interest in metamaterials for TPV applications is confirmed by the increasing number of published papers on this topic. It has been exponentially growing since 2013, and it allows us to hope that advanced TPV electric generators will be experimentally tested and commercialized in the near future.

Problems and Control Questions

1. Consider a metasurface absorber with a lossy capacitive grid over a ground plane impinged by a normally incident plane wave. Assume that the patches are lossy and the spacer of thickness d between the ground plane and patches is free space. Derive the perfect absorption condition for the grid admittance of the capacitive grid

$$Y_{cg} = \frac{1}{\eta_0} + \frac{j}{\eta_0 \tan k_0 d} \tag{11.16}$$

without applying the transmission-line model. Use the formula for the electric field $E = E_x$ produced by an in-phase sheet of x-directed surface current J at the distance z from the sheet:

$$E_J(z) = -\frac{\eta_0}{2} J e^{-jk_0 z}. \tag{11.17}$$

Recall the definition of the grid impedance and the PEC boundary condition on the ground plane and require the absence of reflection.

2. Applying either the transmission-line model for the TE-incidence and for the TM-incidence or the same approach as in Problem 1, derive the condition of complete absorption if the absorber of Problem 1 is impinged by a plane wave under a certain angle θ and frequency ω is low enough so that the spacer d is optically thin.

3. Why TPV systems need a vacuum cavity?

4. What is the role of the narrowband filter in a TPV system? Why does reflection of a significant portion of the radiative heat not result in a critical drop of efficiency?

5. Thermal radiation has a nonuniform angular pattern described by the Lambert law. Thermal irradiance is the integral of this pattern. Why is radiative heat transfer between two parallel black-body surfaces equal to thermal irradiance?

6. Why is the Shockley–Queisser limit of the ultimate efficiency higher for TPV systems than for solar cells?

7. Which factors determine the minimal bandwidth corresponding to the optimal operation of the PV cell in the TPV system? Why is further shrinking of this band useless and broadening harmful?

8. Why do solar TPV stations need a frequency-selective emitter instead of a filter applied to the PV panel?

9. What is the physical mechanism of the frequency selectivity of the IR metasurface emitter with a lossy capacitive grid? Is it the resonance of the Fabry-Perot type or something else?

10. Under what condition can the resonance of the IR metasurface emitter with a lossy capacitive grid be treated as the resonance of an effective *RLC*-circuit?

11. A TPV cavity appears to be unnecessary for a near-field TPV system. Really, this system is not an electric generator but a temperature sensor measuring slight differences of the heat source from the temperature of the PV cell. The heat source that can be, e.g., a microchip or a tip of a near-field optical microscope thermally connected to the emitter. The conductive heat flux from such heat source to the PV cell cannot destroy its PV operation. Then why are huge efforts done in order to ensure the parallelism of the emitter and the PV cell separated by a deeply submicron vacuum cavity? Why cannot a near-field TPV system be monolithic with a thermally connected emitter and PV cell?

12. Why does a layer of wire medium inserted into a micron or slightly sub-micron gap of a micro-TPV system allows super-Planckian radiative heat transfer?

13. What is the reason of high frequency selectivity of this transfer?

14. What is the role of the metamaterial emitter in the cavity-free TPV system with a thick GaAs slab replacing the cavity?

Bibliography

[1] H. H. Kolm, *Solar-Battery Power Source: Quarterly Progress Report of Magnetic Laboratory*, Massachusetts Institute of Technology (1956).

[2] P. Aigrain, *Thermophotovoltaic Conversion of Radiant Energy: Lecture Notes*, Massachusetts Institute of Technology (1956).

[3] D. Chubb, *Fundamentals of Thermophotovoltaic Energy Conversion*, Elsevier Science (2007).

[4] J. P. Benner and T. J. Coutts, "Thermophotovoltaics," in *The Electrical Engineering Handbook*, R.C. Dorf, ed., CRC Press (2000).

[5] T. J. Coutts, "Thermophotovoltaic generation of electricity," in *Clean Electricity from Photovoltaics*," Chapter 11, vol 1. Series on photoconversion of solar energy, M. D. Archer and R. Hill, eds., Imperial College Press (2001).

[6] M. G. Mauk, *Survey of Thermophotovoltaic (TPV) Devices*, Springer (2007).

[7] T. Bauer, *Thermophotovoltaics: Basic Principles and Critical Aspects of System Design*, Springer-Verlag (2011).

[8] B. Bitnar, W. Durisch, and R. Holzner, "Thermophotovoltaics on the move to applications," *Applied Energy* **105**, 430–438 (2013).

[9] R. Siegel and J. Howell, *Thermal Radiation and Heat Transfer*, 4th edn., Taylor and Francis (2001).

[10] C. Ferrari, F. Melino, M. Pinelli, and P. Ruggero-Spina, "Thermophotovoltaic energy conversion: Analytical aspects, prototypes and experiences," *Applied Energy* **113**, 1717–1730 (2014).

[11] B. D. Wedlock, "Thermal photovoltaic effect," *Proceedings of the Third IEEE Photovoltaic Specialists Conference*, April 10–14, 1963, Washington, DC, pp. A410–A413.

[12] J. J. Werth, Thermo-photovoltaic converter with radiant energy reflective means, US Patent No 3331707 (1963).

[13] B. Wernsman, R. R. Siergiej, S. D. Link, et al., "Greater than 20% radiant heat conversion efficiency of a thermophotovoltaic radiator/module system using reflective spectral control," *IEEE Transactions on Electron Devices* **51**, 512–515 (2004)

[14] B. S. Good, D. L. Chubb, and R. A. Lowe, "Comparison of selective emitter and filter thermophotovoltaic systems," *Proceedings of the Second NREL Conference Thermophotovoltaic Generation of Electricity*, Colorado Springs, July 16–20, 1995, pp. 16–34.

[15] F. O'Sullivan, I. Celanovic, N. Jovanovic, et al., "Optical characteristics of 1D Si/SiO2 photonic crystals for thermophotovoltaic applications," *Journal of Applied Physics* **97**, 033529 (2005).

[16] J. A. Kong, *Electromagnetic Wave Theory*, Wiley (1986).

[17] A. Osipov and S. Tretyakov, *Modern Electromagnetic Scattering Theory with Applications,* John Wiley and Sons (2017).

[18] Zh. Zhang, *Nano/Microscale Heat Transfer*, McGraw-Hill (2007).

[19] L. M. Fraas, J. E. Avery, and H.-X. Huang, "Thermophotovoltaics: Heat and electric power from low-bandgap 'solar' cells around gas fired radiant tube burners," *Proceedings of the 29th IEEE Photovoltaic Specialists Conference*, May 19–24, 2002, New Orleans, Louisiana, pp. 1553–1556.

[20] L. M. Fraas, J. E. Avery, and H.-X. Huang, "Thermophotovoltaic furnace generator for home using low-bandgap GaSb cells," *Semiconductor Science and Technology* 18, S247–S253 (2003).

[21] L. M. Fraas, Thermophotovoltaic generator in high temperature industrial process, US Patent No 6538193 B1 (2003).

[22] G. Mattarolo, "Development and modelling of a thermophotovolatic system," Dr.-Eng. Dissertation, University of Kassel, Germany (2007). Available at www.uni-kassel.de.

[23] G. Walker, *Stirling Engines*, Clarendon Press (1980).

[24] C. B. Vining, "An inconvenient truth about thermoelectrics," *Nature Materials* 8, 83–87 (2009).

[25] D. Chester, P. Bermel, J. D. Joannopoulos, M. Soljacic, and I. Celanovic, Design and global optimization of high-efficiency solar TPV systems with tungsten cermets," *Optics Express* 19(S3), A245–A257 (2011).

[26] A. Lenert, D. M. Bierman, Y. Nam, et al., "A nanophotonic solar thermophotovoltaic device," *Nature Nanotechnology* 9, 126–130 (2014).

[27] M. K. Hedayati, F. Faupel, and M. Elbahri, "Review of plasmonic nanocomposite metamaterial absorbers," *Materials* 7, 1221–1248 (2014).

[28] M. Kohl, M. Heck, S. Brunold, U. Frei, and B. Carlsson, "Selective solar absorber coatings on receiver tubes for CSP stations," *Solar Energy Materials and Solar Cells* 84, 275–280 (2004).

[29] C. E. Kennedy, "Review of mid- to high-temperature solar selective absorber materials," National Renewable Energy Laboratory Technical Report (2002). Available at http://large.stanford.edu/publications/power/references/troughnet/solarfield/docs/31267.pdf.

[30] E. Rephaeli and S. Fan, "Absorber and emitter for solar thermophotovoltaic systems to achieve efficiency exceeding the Shockley-Queisser limit," *Optics Express* 17, 15145–15159 (2009).

[31] C. Wu, B. Neuner III, J. John, et al., "Metamaterial-based integrated plasmonic absorber/emitter for solar thermo-photovoltaic systems," *Journal of Optics* 14, 024005 (2012).

[32] M. De Zoysa, T. Asano, K. Mochizuki, et al., "Conversion of broadband to narrowband thermal emission through energy recycling," *Nature Photonics* 6, 535–539 (2012).

[33] W. T. Lau, J.-R. Shen, and S. Fan, "Universal features of coherent photonic thermal conductance in multilayer photonic band gap structures," *Physical Review B* 80, 155135 (2009).

[34] Y. Ra'di, C. R. Simovski, and S. A. Tretyakov, "Thin perfect absorbers for electromagnetic waves: theory, design, and realizations," *Physical Review Applied* 3, 037001 (2015).

[35] M. Li, S. Q. Xiao, Y.-Y. Bai, and B.-Z. Wang, "An ultrathin and broadband radar absorber using resistive FSS," *IEEE Antennas and Propagation Letters* **11**, 748–751 (2012).

[36] C. Wu , B. Neuner III, J. John, et al., "Large-area wide-angle spectrally selective plasmonic absorber," *Physical Review B* **84**, 075102 (2011).

[37] S. Maruyama, T. Kashiwa, H. Yugami, and M. Esashi, "Thermal radiation from two-dimensionally confined modes in microcavities," *Applied Physics Letters* **79**, 1393 (2001).

[38] K. Park, S. Basu, W. P. King, and Z. M. Zhang, "Performance analysis of near-field thermophotovoltaic devices considering absorption distribution," *Journal of Quantitative Spectroscopy & Radiative Transfer* **109**, 305–316 (2008).

[39] A. I. Volokitin and B. N. J. Persson, "Resonant photon tunneling enhancement of the radiative heat transfer," *Physical Review B* **63**, 045417 (2004).

[40] A. Kittel, U. F. Wischnath, J. Welker, et al., "Near-field thermal imaging of nanostructured surfaces," *Applied Physics Letters* **93**, 193109 (2008).

[41] M. Laroche, R. Carminati, J-.J. Greffet, et al., "Near-field thermophotovoltaic energy conversion," *Journal of Applied Physics* **100**, 063704 (2006).

[42] S. Jin, M. Lim, S. S. Lee, and B. J. Lee, "Hyperbolic metamaterial-based near-field thermophotovoltaic system for hundred of nanometers vacuum gap," *Optics Express* **24**, A635–A641 (2016).

[43] R. Messina and P. Ben-Abdallah, "Graphene-based photovoltaic cells for near-field thermal energy conversion," *Scientific Reports* **3**, 1383 (2013).

[44] V. B. Svetovoy and G. Palasantzas, "Graphene-on-silicon near-field thermophotovoltaic cell," *Physical Review Applied* **2**, 034006 (2014).

[45] R. S. DiMatteo, M. S. Weinberg, and G. A. Kirkos, Microcavity apparatus and systems for maintaining microcavity over a macroscale area, US Patent 6232546B1 (2001).

[46] P. Greiff, R. DiMatteo, E. Brown, and C. Leitz, Sub-micrometer gap thermophotovoltaic structure (MTPV) and its fabrication method," US Patent 0319749 A1 (2010).

[47] S. Basu, Z. M. Zhang, and C. J. Fu, "Review of near-field thermal radiation and its application to energy conversion," *International Journal of Energy Research* **33**, 1203–1232 (2009).

[48] D. Kirikae, Y. Suzuki, and N. Kasagi, "Selective-emitter-enhanced microthermophotovoltaic power generation system," in *Proceedings of the 23-rd IEEE International Conference on Micro Electro Mechanical Systems*, Jan. 24–28, 2010, Hong Kong, pp. 1195–1198.

[49] R. DiMatteo, P. Greiff, D. Seltzer, et al., "Micron-gap Thermo-PhotoVoltaics (MTPV)," *Proceedings of the Sixth AIP International Conference Thermo-Photo-Voltaic Generation of Electricity*, A. Gopinath, T. J. Coutts, and J. Luther, eds., Sept. 9–10, 2005, pp. 42–52.

[50] Y. X. Yeng, M. Ghebrebrian, P. Bermel, et al., "Enabling high-temperature nanophotonics for energy applications," *Proceedings of the National Academy of Sciences of the United States of America* **109**, 2280–2285 (2012)

[51] M. D. Whale and E. G. Cravalho, "Modeling and performance of microscale thermo-photovoltaic energy conversion devices," *IEEE Transactions on Energy Conversion* **17**, 130–137 (2002).

[52] R. S. Ottens, V. Quetschke, S. Wise, et al., "Near-field radiative heat transfer between macroscopic planar surfaces," *Physical Review Letters* **107**, 014301 (2011).

[53] C. Simovski, S. Maslovski, I. Nefedov, and S. Tretyakov, "Optimization of radiative heat transfer in hyperbolic metamaterials for thermophotovoltaic applications," *Optics Express* **21**, 14988–15013 (2013).

[54] M. S. Mirmoosa and C. Simovski," Micron-gap thermophotovoltaic systems enhanced by nanowires," *Photonics and Nanostructures: Fundamentals and Applications* **13**, 20–30 (2015).

[55] M. S. Mirmoosa, M. Omelyanovich, and C. R. Simovski, "Microgap thermophotovoltaic systems with low emission temperature and high electric output," *Journal of Optics* **19**, 115104 (2016).

[56] M. S. Mirmoosa, S.-A. Biehs, and C. R. Simovski, "Super-Planckian thermophotovoltaics without vacuum gaps," *Physical Review Applied* **8**, 054020 (2017).

Note

1 In fact, it was a ternary structure: W, SiC, and quartz layers. However, quartz is practically lossless, and its absorptivity as well as its emissivity are negligibly small. It does not contribute to thermal emission. The role of the quartz shield is to increase robustness of the emitter to thermal stress.

Index